THE ELEMENTS OF

EUCLID

FOR THE USE OF SCHOOLS AND COLLEGES

COMPRISING THE FIRST SIX BOOKS AND PORTIONS OF THE ELEVENTH AND TWELFTH BOOKS

WITH NOTES, AN APPENDIX, AND EXERCISES

BY

I. TODHUNTER

Elibron Classics

www.elibron.com

Elibron Classics series.

ISBN 1-4021-6900-0 (paperback)
ISBN 1-4021-3237-9 (hardcover)

This Elibron Classics Replica Edition is an unabridged facsimile of the edition published in 1864 by Macmillan and Co., Cambridge and London.

EDUCATIONAL MATHEMATICAL WORKS

BY

I. TODHUNTER, M.A., F.R.S.

1. Euclid for Schools and Colleges. 18mo. Second Edition. 3s. 6d.

2. Algebra for Beginners. With numerous Examples. 18mo. 2s. 6d.

3. Algebra for the use of Colleges and Schools. With numerous Examples. Third Edition. Crown 8vo. 7s. 6d.

4. A Treatise on the Theory of Equations. With a Collection of Examples. Crown 8vo. 7s. 6d.

5. Plane Trigonometry for Colleges and Schools. With numerous Examples. Second Edition. Crown 8vo. 5s.

6. A Treatise on Conic Sections. With numerous Examples. Third Edition. Crown 8vo. 7s. 6d.

EUCLID

FOR THE USE OF SCHOOLS AND COLLEGES.

THE ELEMENTS OF

EUCLID

FOR THE USE OF SCHOOLS AND COLLEGES;

'OMPRISING THE FIRST SIX BOOKS AND PORTIONS OF THE ELEVENT
AND TWELFTH BOOKS;

WITH NOTES, AN APPENDIX, AND EXERCISES

BY

I. TODHUNTER M.A., F.R.S.

NEW EDITION.

MACMILLAN AND CO.

Cambridge and London.

1864

Cambridge:
PRINTED BY C. J. CLAY, M.A.
AT THE UNIVERSITY PRESS.

PREFACE.

In offering to students and teachers a new edition of the Elements of Euclid, it will be proper to give some account of the plan on which it has been arranged, and of the advantages which it hopes to present.

Geometry may be considered to form the real foundation of mathematical instruction. It is true that some acquaintance with Arithmetic and Algebra usually precedes the study of Geometry; but in the former subjects a beginner spends much of his time in gaining a practical facility in the application of rules to examples, while in the latter subject he is wholly occupied in exercising his reasoning faculties.

In England the text-book of Geometry consists of the Elements of Euclid; for nearly every official programme of instruction or examination explicitly includes some portion of this work. Numerous attempts have been made to find an appropriate substitute for the Elements of Euclid; but such attempts, fortunately, have hitherto been made in vain. The advantages attending a common standard of reference in such an important subject, can hardly be overestimated; and it is extremely improbable, if Euclid were once abandoned, that any agreement would exist as to the author who should replace him. It cannot be denied that defects and difficulties occur in the Elements of Euclid, and that these become more obvious as we èxamine the work more closely; but probably during such examination the conviction will grow deeper that these defects and difficulties are due in a great measure to the nature of the subject itself, and to the place which it occupies in a course of education; and it may be readily believed that an equally minute criticism of any other work on Geometry would reveal more and graver blemishes.

Of all the editions of Euclid that of Robert Simson has been the most extensively used in England, and the present edition substantially reproduces Simson's; but his translation has been carefully compared with the original, and some alterations have been made, which it is hoped will be found to be improvements. These alterations, however, are of no great importance; most of them have been introduced with the view of rendering the language more uniform, by constantly using the same words when the same meaning is to be expressed.

As the Elements of Euclid are usually placed in the hands of young students, it is important to exhibit the work in such a form as will assist them in overcoming the difficulties which they experience on their first introduction to processes of continuous argument. No method appears to be so useful as that of breaking up the demonstrations into their constituent parts; this was strongly recommended by Professor De Morgan more than thirty years ago as a suitable exercise for students, and the plan has been adopted more or less closely in some modern editions. An excellent example of this method of exhibiting the Elements of Euclid will be found in an edition in quarto, published at the Hague, in the French language, in 1762. Two persons are named in the title-page as concerned in the work,

Koenig and Blassiere. This edition has served as the model for that which is now offered to the student: some slight modifications have necessarily been made, owing to the difference in the size of the pages.

It will be perceived then, that in the present edition each distinct assertion in the argument begins a new line; and at the ends of the lines are placed the necessary references to the preceding principles on which the assertions depend. Moreover, the longer propositions are distributed into subordinate parts, which are distinguished by breaks at the beginning of the lines.

This edition contains all the propositions which are usually read in the Universities. After the text will be found a selection of notes; these are intended to indicate and explain the principal difficulties which have been noticed in the Elements of Euclid, and to supply the most important inferences which can be drawn from the propositions. The notes relate to Geometry exclusively; they do not introduce developments involving Arithmetic and Algebra, because these latter subjects are always studied in special works, and because Geometry alone presents sufficient matter to occupy the attention of early students. After some hesitation on the point, all remarks relating to Logic have also been excluded. Although the study of Logic appears to be reviving in this country, and may eventually obtain a more assured position than it now holds in a course of liberal education, yet at present few persons take up Logic before Geometry; and it seems therefore premature to devote space to a subject which will be altogether unsuitable to the majority of those who use a work like the present.

After the notes will be found an Appendix, consisting of propositions supplemental to those in the Elements of Euclid; it is hoped that a judicious choice has been made from the abundant materials which exist for such an Appendix. The propositions selected are worthy of notice on various grounds; some for their simplicity, some for their value as geometrical facts, and some as being problems which may naturally suggest themselves, but of which the solutions are not very obvious.

The work finishes with a collection of exercises. Geometrical deductions afford a most valuable discipline for a student of mathematics, especially in the earlier period of his course; the numerous departments of analysis which subsequently demand his attention will leave him but little time then for pure Geometry. It seems however that the habits of mind which the study of pure Geometry tends to form, furnish an advantageous corrective for some of the evils resulting from an exclusive devotion to Analysis, and it is therefore desirable to engage the attention of beginners with geometrical exercises.

Many persons whose duties have rendered them familiar with the examination of large numbers of students in elementary mathematics have noticed with regret the frequent failures in geometrical deductions. Several collections of exercises already exist, but the general complaint is that they are too difficult. Those in the present volume may be divided into two parts; the first part contains 440 exercises, which it is hoped will not be found beyond the power of early students; the second part consists of the remainder, which may be reserved for practice at a later stage. These exercises have been principally selected from College and University examination papers, and have been tested by long experience with pupils. It will be seen that they are distributed into sections according to the propositions in the Elements of Euclid on which they chiefly depend. As far as possible they are arranged in order of difficulty, but it must sometimes happen, as is the case in the Elements of Euclid, that one example prepares the way for a set of others which are much easier than itself. It should be observed that the exercises relate to pure Geometry; all examples which would find a more suitable place in works on Trigonometry or Algebraical Geometry have been carefully rejected.

It only remains to advert to the mechanical execution of the volume, to which great attention has been devoted. The figures will be found to be unusually large and distinct, and they have been repeated when necessary, so that they always occur in immediate connexion with the corresponding text. The type and paper have been chosen so as to render the volume as clear and attractive as possible. The design of the editor and of the publishers has been to produce a practically useful edition of the Elements of Euclid, at a moderate cost; and they trust that the design has been fairly realised.

Any suggestions or corrections relating to the work will be most thankfully received.

I. TODHUNTER.

ST JOHN'S COLLEGE,
October 1862.

CONTENTS.

	PAGE
Introductory Remarks	xv
Book I.	1
Book II.	52
Book III.	71
Book IV.	113
Book V.	134
Book VI.	173
Book XI.	220
Book XII.	244
Notes on Euclid's Elements	250
Appendix	292
Exercises in Euclid	340

INTRODUCTORY REMARKS.

THE subject of Plane Geometry is here presented to the student arranged in six books, and each book is subdivided into propositions. The propositions are of two kinds, *problems* and *theorems*. In a problem something is required to be done; in a theorem some new principle is asserted to be true.

A proposition consists of various parts. We have first the general enunciation of the problem or theorem; as for example, *To describe an equilateral triangle on a given finite straight line*, or *Any two angles of a triangle are together less than two right angles.* After the general enunciation follows the discussion of the proposition. First, the enunciation is repeated and applied to the particular figure which is to be considered; as for example, *Let AB be the given straight line: it is required to describe an equilateral triangle on AB.* The *construction* then usually follows, which states the necessary straight lines and circles which must be drawn in order to constitute the *solution* of the problem, or to furnish assistance in the *demonstration* of the theorem. Lastly, we have the demonstration itself, which shews that the problem has been solved, or that the theorem is true.

Sometimes, however, no construction is required; and sometimes the construction and demonstration are combined.

The demonstration is a process of reasoning in which we draw inferences from results already obtained. These results consist partly of truths established in former propositions, or admitted as obvious in commencing the subject, and partly of truths which follow from the *construction* that has been made, or which are given in the *supposition* of the proposition itself. The word *hypothesis* is used in the same sense as *supposition*.

To assist the student in following the steps of the reasoning, *references* are given to the results already obtained which are required in the demonstration. Thus I. 5 indicates that we appeal to the result established in the fifth proposition of the First Book; *Constr.* is sometimes used as an abbreviation of *Construction*, and *Hyp.* as an abbreviation of *Hypothesis*.

It is usual to place the letters Q.E.F. at the end of the discussion of a problem, and the letters Q.E.D. at the end of the discussion of a theorem. Q.E.F. is an abbreviation for *quod erat faciendum*, that is, *which was to be done;* and Q.E.D. is an abbreviation for *quod erat demonstrandum*, that is, *which was to be proved.*

EUCLID'S ELEMENTS.

BOOK I.

DEFINITIONS.

1. A POINT is that which has no parts, or which has no magnitude.

2. A line is length without breadth.

3. The extremities of a line are points.

4. A straight line is that which lies evenly between its extreme points.

5. A superficies is that which has only length and breadth.

6. The extremities of a superficies are lines.

7. A plane superficies is that in which any two points being taken, the straight line between them lies wholly in that superficies.

8. A plane angle is the inclination of two lines to one another in a plane, which meet together, but are not in the same direction.

9. A plane rectilineal angle is the inclination of two straight lines to one another, which meet together, but are not in the same straight line.

Note. When several angles are at one point B, any one of them is expressed by three letters, of which the letter which is at the vertex of the angle, that is, at the point at which the straight lines that contain the angle meet one another, is put between the other two letters, and one of these two letters is somewhere on one of those straight lines, and the other letter on the other straight line. Thus, the angle which is contained by the

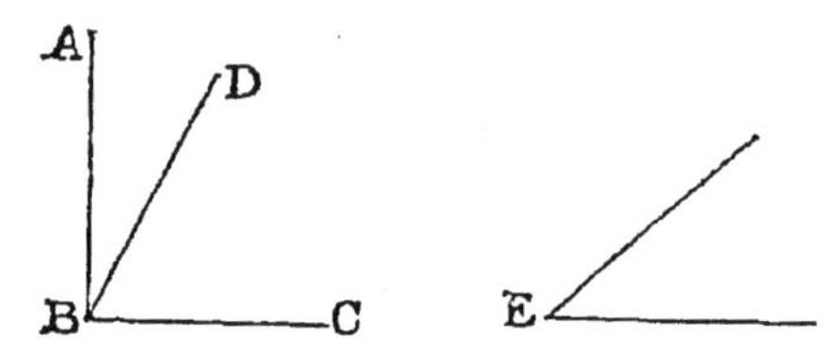

straight lines AB, CB is named the angle ABC, or CBA; the angle which is contained by the straight lines AB, DB is named the angle ABD, or DBA; and the angle which is contained by the straight lines DB, CB is named the angle DBC, or CBD; but if there be only one angle at a point, it may be expressed by a letter placed at that point; as the angle at E.

10. When a straight line standing on another straight line, makes the adjacent angles equal to one another, each of the angles is called a right angle; and the straight line which stands on the other is called a perpendicular to it.

11. An obtuse angle is that which is greater than a right angle.

12. An acute angle is that which is less than a right angle.

13. A term or boundary is the extremity of any thing.

14. A figure is that which is enclosed by one or more boundaries.

15. A circle is a plane figure contained by one line, which is called the circumference, and is such, that all straight lines drawn from a certain point within the figure to the circumference are equal to one another:

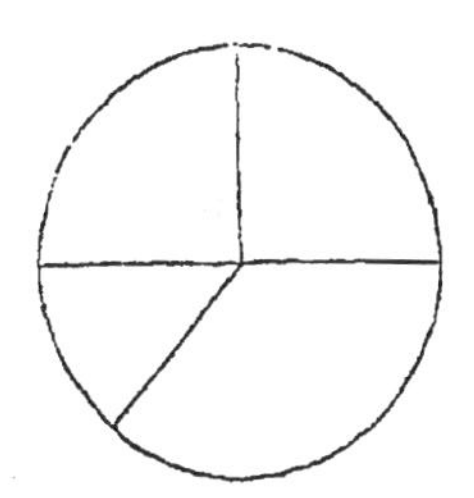

16. And this point is called the centre of the circle.

17. A diameter of a circle is a straight line drawn through the centre, and terminated both ways by the circumference.

[A radius of a circle is a straight line drawn from the centre to the circumference.]

18. A semicircle is the figure contained by a diameter and the part of the circumference cut off by the diameter.

19. A segment of a circle is the figure contained by a straight line and the circumference which it cuts off.

20. Rectilineal figures are those which are contained by straight lines:

21. Trilateral figures, or triangles, by three straight lines:

22. Quadrilateral figures by four straight lines:

23. Multilateral figures, or polygons, by more than four straight lines.

24. Of three-sided figures,

An equilateral triangle is that which has three equal sides:

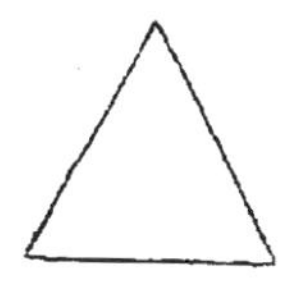

25. An isosceles triangle is that which has two sides equal:

26. A scalene triangle is that which has three unequal sides:

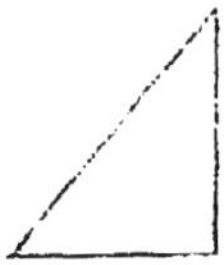

27. A right-angled triangle is that which has a right angle:

[The side opposite to the right angle in a right-angled triangle is frequently called the hypotenuse.]

28. An obtuse-angled triangle is that which has an obtuse angle:

29. An acute-angled triangle is that which has three acute angles.

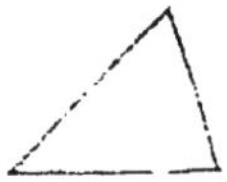

Of four-sided figures,

30. A square is that which has all its sides equal, and all its angles right angles:

31. An oblong is that which has all its angles right angles, but not all its sides equal:

32. A rhombus is that which has all its sides equal, but its angles are not right angles:

33. A rhomboid is that which has its opposite sides equal to one another, but all its sides are not equal, nor its angles right angles:

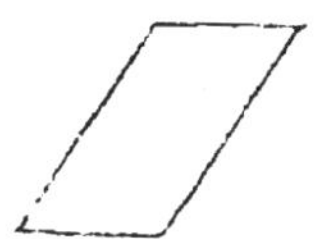

34. All other four-sided figures besides these are called trapeziums.

35. Parallel straight lines are such as are in the same plane, and which being produced ever so far both ways do not meet.

[*Note.* The terms *oblong* and *rhomboid* are not often used. Practically the following definitions are used. Any four-sided figure is called a *quadrilateral.* A line joining two opposite angles of a quadrilateral is called a *diagonal.* A quadrilateral which has its opposite sides parallel is called a *parallelogram.* The words *square* and *rhombus* are used in the sense defined by Euclid; and the word *rectangle* is used instead of the word *oblong.*

Some writers propose to restrict the word *trapezium* to a quadrilateral which has two of its sides parallel; and it would certainly be convenient if this restriction were universally adopted.]

POSTULATES.

Let it be granted,

1. That a straight line may be drawn from any one point to any other point:

2. That a terminated straight line may be produced to any length in a straight line:

3. And that a circle may be described from any centre, at any distance from that centre.

AXIOMS.

1. Things which are equal to the same thing are equal to one another.

2. If equals be added to equals the wholes are equal.

3. If equals be taken from equals the remainders are equal.

4. If equals be added to unequals the wholes are unequal.

5. If equals be taken from unequals the remainders are unequal.

6. Things which are double of the same thing are equal to one another.

7. Things which are halves of the same thing are equal to one another.

8. Magnitudes which coincide with one another, that is, which exactly fill the same space, are equal to one another.

9. The whole is greater than its part.

10. Two straight lines cannot enclose a space.

11. All right angles are equal to one another.

12. If a straight line meet two straight lines, so as to make the two interior angles on the same side of it taken together less than two right angles, these straight lines, being continually produced, shall at length meet on that side on which are the angles which are less than two right angles.

PROPOSITION 1. *PROBLEM.*

To describe an equilateral triangle on a given finite straight line.

Let AB be the given straight line: it is required to describe an equilateral triangle on AB.

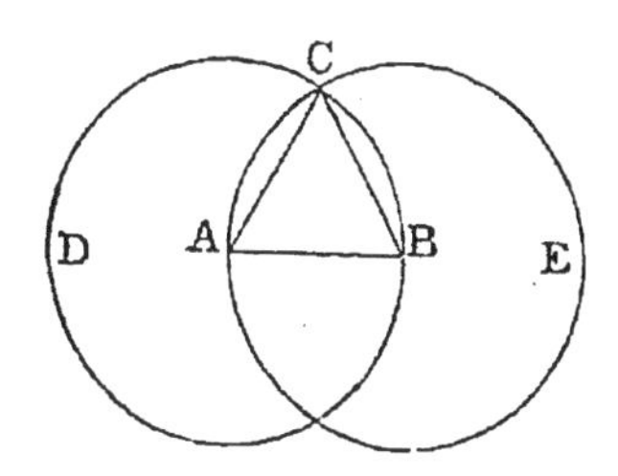

From the centre A, at the distance AB, describe the circle BCD. [*Postulate* 3.

From the centre B, at the distance BA, describe the circle ACE. [*Postulate* 3.

From the point C, at which the circles cut one another, draw the straight lines CA and CB to the points A and B. [*Post.* 1.

ABC shall be an equilateral triangle.

Because the point A is the centre of the circle BCD, AC is equal to AB. [*Definition* 15.

And because the point B is the centre of the circle ACE, BC is equal to BA. [*Definition* 15.

But it has been shewn that CA is equal to AB;

therefore CA and CB are each of them equal to AB.

But things which are equal to the same thing are equal to one another. [*Axiom* 1.

Therefore CA is equal to CB.

Therefore CA, AB, BC are equal to one another.

. Wherefore *the triangle ABC is equilateral,* [*Def.* 24.
and it is described on the given straight line AB. Q.E.F.

PROPOSITION 2. *PROBLEM.*

From a given point to draw a straight line equal to a given straight line.

Let A be the given point, and BC the given straight line: it is required to draw from the point A a straight line equal to BC.

From the point A to B draw the straight line AB; [*Post.* 1.
and on it describe the equilateral triangle DAB, [I. 1.
and produce the straight lines DA, DB to E and F. [*Post.* 2.
From the centre B, at the distance BC, describe the circle CGH, meeting DF at G. [*Post.* 3.
From the centre D, at the distance DG, describe the circle GKL, meeting DE at L. [*Post.* 3.
AL shall be equal to BC.

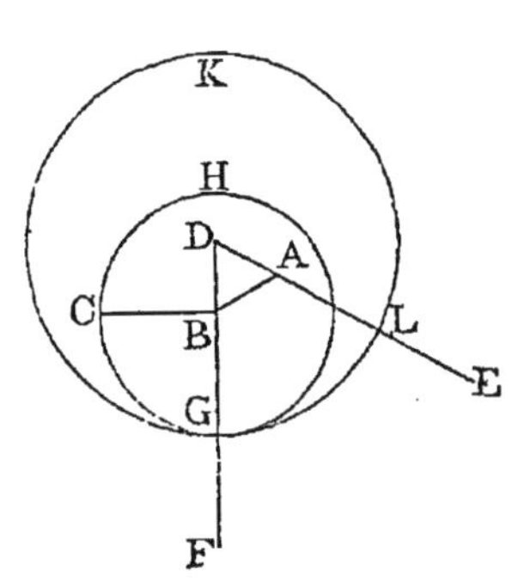

Because the point B is the centre of the circle CGH, BC is equal to BG. [*Definition* 15.
And because the point D is the centre of the circle GKL, DL is equal to DG; [*Definition* 15.
and DA, DB parts of them are equal; [*Definition* 24.
therefore the remainder AL is equal to the remainder BG. [*Axiom* 3.
But it has been shewn that BC is equal to BG;
therefore AL and BC are each of them equal to BG.
But things which are equal to the same thing are equal to one another. [*Axiom* 1.
Therefore AL is equal to BC.

Wherefore *from the given point A a straight line AL has been drawn equal to the given straight line BC.* Q.E.F.

PROPOSITION 3. *PROBLEM.*

From the greater of two given straight lines to cut off a part equal to the less.

Let AB and C be the two given straight lines, of which

AB is the greater: it is required to cut off from AB, the greater, a part equal to C the less.

From the point A draw the straight line AD equal to C; [I. 2.

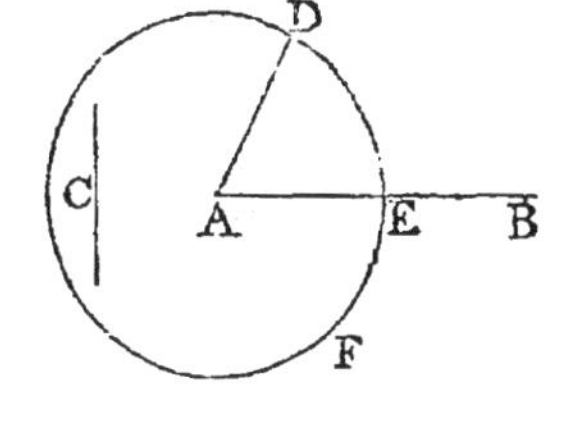

and from the centre A, at the distance AD, describe the circle DEF meeting AB at E. [*Postulate* 3.

AE shall be equal to C.

Because the point A is the centre of the circle DEF, AE is equal to AD. [*Definition* 15.

But C is equal to AD. [*Construction.*

Therefore AE and C are each of them equal to AD.

Therefore AE is equal to C. [*Axiom* 1.

Wherefore *from AB the greater of two given straight lines a part AE has been cut off equal to C the less.* Q.E.F.

PROPOSITION 4. *THEOREM.*

If two triangles have two sides of the one equal to two sides of the other, each to each, and have also the angles contained by those sides equal to one another, they shall also have their bases or third sides equal; and the two triangles shall be equal, and their other angles shall be equal, each to each, namely those to which the equal sides are opposite.

Let ABC, DEF be two triangles which have the two sides AB, AC equal to the two sides DE, DF, each to each, namely, AB to DE, and AC to DF, and the angle BAC equal to the angle EDF: the base BC shall be equal to the base EF, and the triangle ABC to the triangle DEF, and the other angles shall be equal, each to each, to which the equal sides are opposite, namely, the angle ABC to the angle DEF, and the angle ACB to the angle DFE.

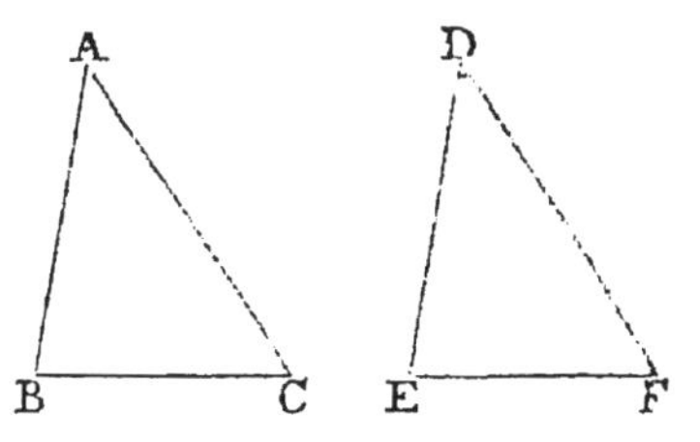

For if the triangle ABC be applied to the triangle DEF, so that the point A may be on the point D, and the straight line AB on the straight line DE, the point B will coincide with the point E, because AB is equal to DE. [*Hyp.*

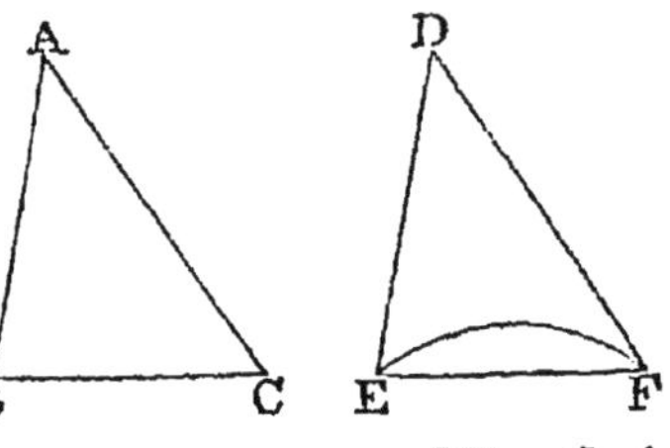

And, AB coinciding with DE, AC will fall on DF, because the angle BAC is equal to the angle EDF. [*Hypothesis.*

Therefore also the point C will coincide with the point F, because AC is equal to DF. [*Hypothesis.*

But the point B was shewn to coincide with the point E, therefore the base BC will coincide with the base EF;

because, B coinciding with E and C with F, if the base BC does not coincide with the base EF, two straight lines will enclose a space; which is impossible. [*Axiom* 10.

Therefore the base BC coincides with the base EF, and is equal to it. [*Axiom* 8.

Therefore the whole traingle ABC coincides with the whole triangle DEF, and is equal to it. [*Axiom* 8.

And the other angles of the one coincide with the other angles of the other, and are equal to them, namely, the angle ABC to the angle DEF, and the angle ACB to the angle DFE.

Wherefore, *if two triangles* &c. Q.E.D.

PROPOSITION 5. *THEOREM.*

The angles at the base of an isosceles triangle are equal to one another; and if the equal sides be produced the angles on the other side of the base shall be equal to one another.

Let ABC be an isosceles triangle, having the side AB equal to the side AC, and let the straight lines AB, AC be produced to D and E: the angle ABC shall be equal to the angle ACB, and the angle CBD to the angle BCE.

In BD take any point F,

and from AE the greater cut off AG equal to AF the less, [I. 3.

and join FC, GB.

Because AF is equal to AG, [*Constr.*
and AB to AC, [*Hypothesis.*
the two sides FA, AC are equal to the two sides GA, AB, each to each; and they contain the angle FAG common to the two triangles AFC, AGB;
therefore the base FC is equal to the base GB, and the triangle AFC to the triangle AGB, and the remaining angles of the one to the remaining angles of the other, each to each, to which the equal sides are opposite,

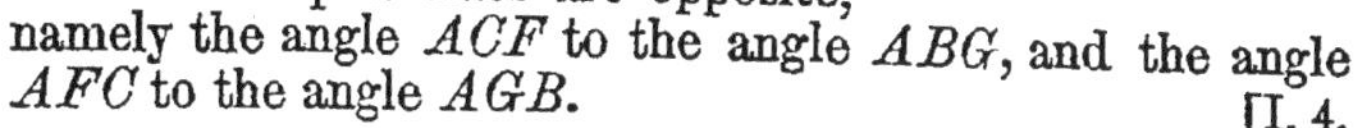

namely the angle ACF to the angle ABG, and the angle AFC to the angle AGB. [I. 4.

And because the whole AF is equal to the whole AG, of which the parts AB, AC are equal, [*Hypothesis.*
the remainder BF is equal to the remainder CG. [*Axiom* 3.
And FC was shewn to be equal to GB;
therefore the two sides BF, FC are equal to the two sides CG, GB, each to each;
and the angle BFC was shewn to be equal to the angle CGB;
therefore the triangles BFC, CGB are equal, and their other angles are equal, each to each, to which the equal sides are opposite, namely the angle FBC to the angle GCB, and the angle BCF to the angle CBG. [I. 4.

And since it has been shewn that the whole angle ABG is equal to the whole angle ACF,
and that the parts of these, the angles CBG, BCF are also equal;
therefore the remaining angle ABC is equal to the remaining angle ACB, which are the angles at the base of the triangle ABC. [*Axiom* 3.

And it has also been shewn that the angle FBC is equal to the angle GCB, which are the angles on the other side of the base.

Wherefore, *the angles* &c. Q.E.D.

Corollary. Hence every equilateral triangle is also equiangular.

PROPOSITION 6. *THEOREM.*

If two angles of a triangle be equal to one another, the sides also which subtend, or are opposite to, the equal angles, shall be equal to one another.

Let ABC be a triangle, having the angle ABC equal to the angle ACB: the side AC shall be equal to the side AB.

For if AC be not equal to AB, one of them must be greater than the other.

Let AB be the greater, and from it cut off DB equal to AC the less, [I. 3.
and join DC.

Then, because in the triangles DBC, ACB,
DB is equal to AC, [*Construction.*
and BC is common to both,
the two sides DB, BC are equal to the two sides AC, CB, each to each;
and the angle DBC is equal to the angle ACB; [*Hypothesis.*
therefore the base DC is equal to the base AB, and the triangle DBC is equal to the triangle ACB, [I. 4.
the less to the greater; which is absurd. [*Axiom* 9.
Therefore AB is not unequal to AC, that is, it is equal to it.

Wherefore, *if two angles* &c. Q.E.D.

Corollary. Hence every equiangular triangle is also equilateral.

PROPOSITION 7. *THEOREM.*

On the same base, and on the same side of it, there cannot be two triangles having their sides which are terminated at one extremity of the base equal to one another, and likewise those which are terminated at the other extremity.

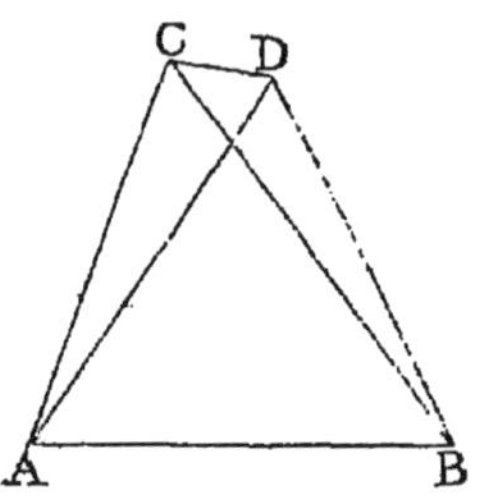

If it be possible, on the same base AB, and on the same side of it, let there be two triangles ACB, ADB, having their sides CA, DA, which are terminated at the extremity A of the base, equal to one another, and likewise their sides *CB*, *DB*, which are terminated at *B*.

Join *CD*. In the case in which the vertex of each triangle is without the other triangle;

because *AC* is equal to *AD*, [*Hypothesis.*

the angle *ACD* is equal to the angle *ADC*. [I. 5.

But the angle *ACD* is greater than the angle *BCD*, [*Ax.* 9.

therefore the angle *ADC* is also greater than the angle *BCD*;

much more then is the angle *BDC* greater than the angle *BCD*.

Again, because *BC* is equal to *BD*, [*Hypothesis.*

the angle *BDC* is equal to the angle *BCD*. [I. 5.

But it has been shewn to be greater; which is impossible.

But if one of the vertices as *D*, be within the other triangle *ACB*, produce *AC*, *AD* to *E*, *F*.

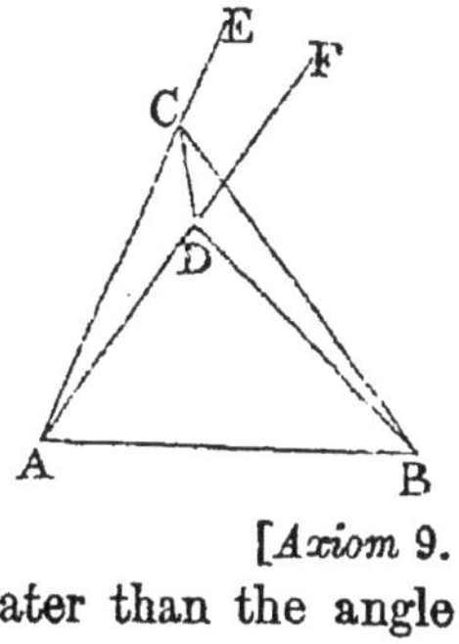

Then because *AC* is equal to *AD*, in the triangle *ACD*, [*Hyp.*

the angles *ECD*, *FDC*, on the other side of the base *CD*, are equal to one another. [I. 5.

But the angle *ECD* is greater than the angle *BCD*, [*Axiom* 9.

therefore the angle *FDC* is also greater than the angle *BCD*;

much more then is the angle *BDC* greater than the angle *BCD*.

Again, because *BC* is equal to *BD*, [*Hypothesis.*

the angle *BDC* is equal to the angle *BCD*. [I. 5.

But it has been shewn to be greater; which is impossible.

The case in which the vertex of one triangle is on a side of the other needs no demonstration.

Wherefore, *on the same base* &c. Q.E.D.

PROPOSITION 8. *THEOREM.*

If two triangles have two sides of the one equal to two sides of the other, each to each, and have likewise their bases equal, the angle which is contained by the two sides of the one shall be equal to the angle which is contained by the two sides, equal to them, of the other.

Let *ABC*, *DEF* be two triangles, having the two sides *AB*, *AC* equal to the two sides *DE*, *DF*, each to each, namely *AB* to *DE*, and *AC* to *DF*, and also the base *BC* equal to the base *EF*: the angle *BAC* shall be equal to the angle *EDF*.

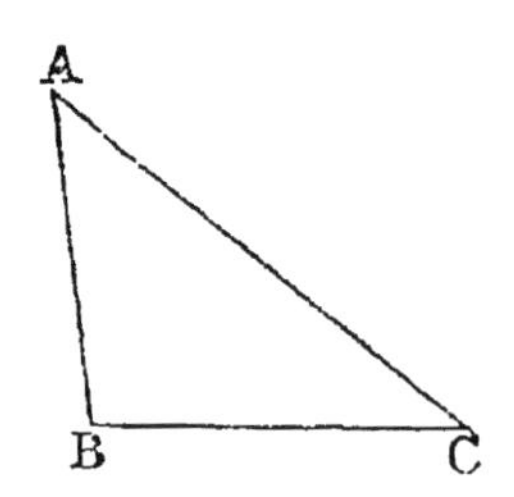

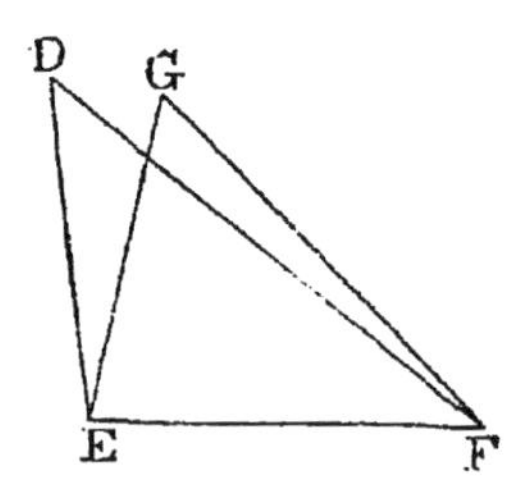

For if the triangle *ABC* be applied to the triangle *DEF*, so that the point *B* may be on the point *E*, and the straight line *BC* on the straight line *EF*, the point *C* will also coincide with the point *F*, because *BC* is equal to *EF*. [*Hyp.*
Therefore, *BC* coinciding with *EF*, *BA* and *AC* will coincide with *ED* and *DF*.

For if the base *BC* coincides with the base *EF*, but the sides *BA*, *CA* do not coincide with the sides *ED*, *FD*, but have a different situation as *EG*, *FG*; then on the same base and on the same side of it there will be two triangles having their sides which are terminated at one extremity of the base equal to one another, and likewise their sides which are terminated at the other extremity.
But this is impossible. [I. 7.
Therefore since the base *BC* coincides with the base *EF*, the sides *BA*, *AC* must coincide with the sides *ED*, *DF*. Therefore also the angle *BAC* coincides with the angle *EDF*, and is equal to it. [*Axiom* 8.

Wherefore, *if two triangles* &c. Q.E.D.

PROPOSITION 9. *PROBLEM.*

To bisect a given rectilineal angle, that is to divide it into two equal angles.

Let BAC be the given rectilineal angle: it is required to bisect it.

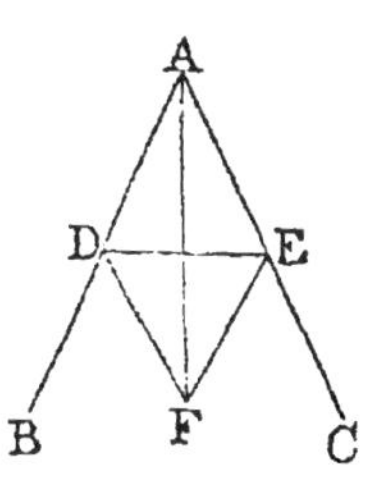

Take any point D in AB, and from AC cut off AE equal to AD; [I. 3.

join DE, and on DE, on the side remote from A, describe the equilateral triangle DEF. [I. 1.

Join AF. The straight line AF shall bisect the angle BAC.

Because AD is equal to AE, [*Construction.*
and AF is common to the two triangles DAF, EAF,
the two sides DA, AF are equal to the two sides EA, AF, each to each;

and the base DF is equal to the base EF; [*Definition* 24.
therefore the angle DAF is equal to the angle EAF. [I. 8.

Wherefore *the given rectilineal angle BAC is bisected by the straight line AF.* Q.E.F.

PROPOSITION 10. *PROBLEM.*

To bisect a given finite straight line, that is to divide it into two equal parts.

Let AB be the given straight line: it is required to divide it into two equal parts.

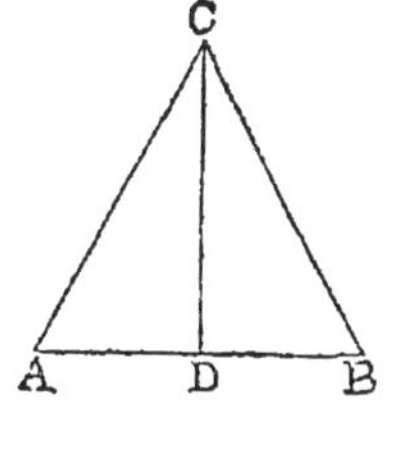

Describe on it an equilateral triangle ABC, [I. 1.

and bisect the angle ACB by the straight line CD, meeting AB at D. [I. 9.

AB shall be cut into two equal parts at the point D.

Because AC is equal to CB, [*Definition* 24.
and CD is common to the two triangles ACD, BCD,
the two sides AC, CD are equal to the two sides BC, CD, each to each;

and the angle ACD is equal to the angle BCD; [*Constr.*
therefore the base AD is equal to the base DB. [I. 4.

Wherefore *the given straight line AB is divided into two equal parts at the point D.* Q.E.F.

PROPOSITION 11. *PROBLEM.*

To draw a straight line at right angles to a given straight line, from a given point in the same.

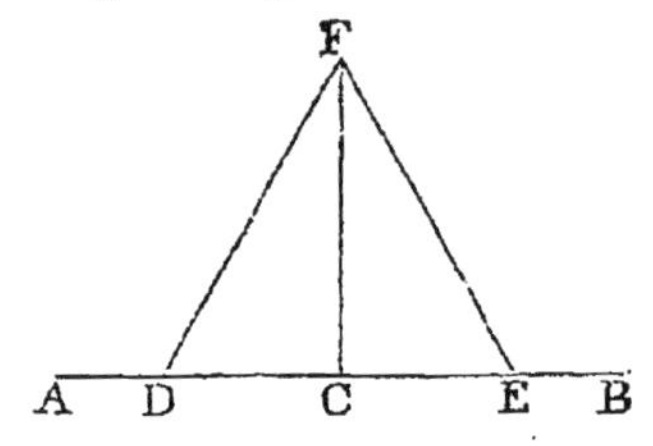

Let AB be the given straight line, and C the given point in it: it is required to draw from the point C a straight line at right angles to AB.

Take any point D in AC, and make CE equal to CD. [I. 3.
On DE describe the equilateral triangle DFE, [I. 1.
and join CF.
The straight line CF drawn from the given point C shall be at right angles to the given straight line AB.

Because DC is equal to CE, [*Construction.*
and CF is common to the two triangles DCF, ECF;
the two sides DC, CF are equal to the two sides EC, CF, each to each;
and the base DF is equal to the base EF; [*Definition* 24.
therefore the angle DCF is equal to the angle ECF; [I. 8.
and they are adjacent angles.
But when a straight line, standing on another straight line, makes the adjacent angles equal to one another, each of the angles is called a right angle; [*Definition* 10.
therefore each of the angles DCF, ECF is a right angle.

Wherefore *from the given point C in the given straight line AB, CF has been drawn at right angles to AB.* Q.E.F.

Corollary. By the help of this problem it may be shewn that two straight lines cannot have a common segment.

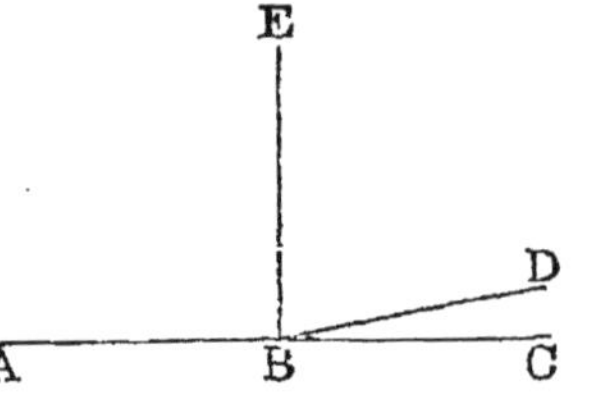

If it be possible, let the two straight lines ABC, ABD have the segment AB common to both of them.

From the point B draw BE at right angles to AB.

Then, because ABC is a straight line, [*Hypothesis.*
the angle CBE is equal to the angle EBA. [*Definition* 10.

Also, because ABD is a straight line, [*Hypothesis.*
the angle DBE is equal to the angle EBA.
Therefore the angle DBE is equal to the angle CBE, [*Ax.* 1.
the less to the greater; which is impossible. [*Axiom* 9.

Wherefore *two straight lines cannot have a common segment.*

PROPOSITION 12. *PROBLEM.*

To draw a straight line perpendicular to a given straight line of an unlimited length, from a given point without it.

Let AB be the given straight line, which may be produced to any length both ways, and let C be the given point without it: it is required to draw from the point C a straight line perpendicular to AB.

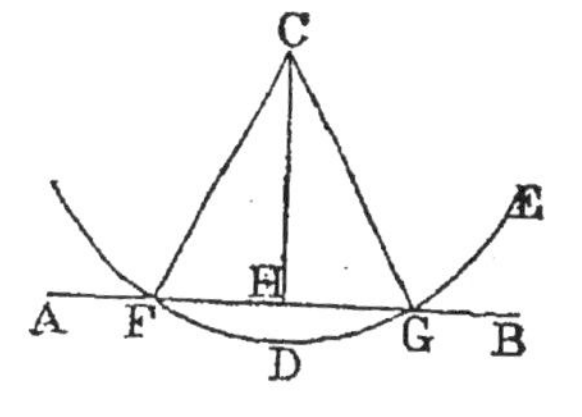

Take any point D on the other side of AB, and from the centre C, at the distance CD, describe the circle EGF, meeting AB at F and G. [*Postulate* 3.

Bisect FG at H, [I. 10.
and join CH.

The straight line CH drawn from the given point C shall be perpendicular to the given straight line AB.

Join CF, CG.

Because FH is equal to HG, [*Construction.*
and HC is common to the two triangles FHC, GHC;
the two sides FH, HC are equal to the two sides GH, HC, each to each;
and the base CF is equal to the base CG; [*Definition* 15.
therefore the angle CHF is equal to the angle CHG; [I. 8.
and they are adjacent angles.
But when a straight line, standing on another straight line, makes the adjacent angles equal to one another, each of the angles is called a right angle, and the straight line which stands on the other is called a perpendicular to it. [*Def.* 10.

Wherefore *a perpendicular CH has been drawn to the given straight line AB from the given point C without it.* Q.E.F.

PROPOSITION 13. *THEOREM.*

The angles which one straight line makes with another straight line on one side of it, either are two right angles, or are together equal to two right angles.

Let the straight line *AB* make with the straight line *CD*, on one side of it, the angles *CBA*, *ABD*: these either are two right angles, or are together equal to two right angles.

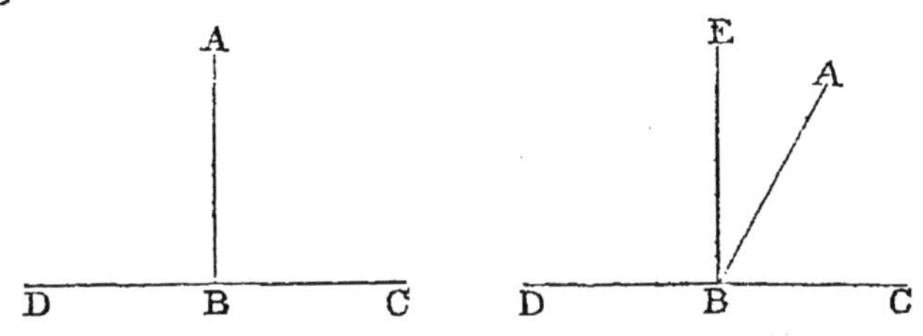

For if the angle *CBA* is equal to the angle *ABD*, each of them is a right angle. [*Definition* 10.

But if not, from the point *B* draw *BE* at right angles to *CD*; [I. 11.

therefore the angles *CBE*, *EBD* are two right angles. [*Def.* 10.

Now the angle *CBE* is equal to the two angles *CBA*, *ABE*; to each of these equals add the angle *EBD*;

therefore the angles *CBE*, *EBD* are equal to the three angles *CBA*, *ABE*, *EBD*. [*Axiom* 2.

Again, the angle *DBA* is equal to the two angles *DBE*, *EBA*;

to each of these equals add the angle *ABC*;

therefore the angles *DBA*, *ABC* are equal to the three angles *DBE*, *EBA*, *ABC*. [*Axiom* 2.

But the angles *CBE*, *EBD* have been shewn to be equal to the same three angles.

Therefore the angles *CBE*, *EBD* are equal to the angles *DBA*, *ABC*. [*Axiom* 1.

But *CBE*, *EBD* are two right angles;

therefore *DBA*, *ABC* are together equal to two right angles.

Wherefore, *the angles* &c. Q.E.D.

PROPOSITION 14. *THEOREM.*

If, at a point in a straight line, two other straight lines, on the opposite sides of it, make the adjacent angles together equal to two right angles, these two straight lines shall be in one and the same straight line.

At the point B in the straight line AB, let the two straight lines BC, BD, on the opposite sides of AB, make the adjacent angles ABC, ABD together equal to two right angles: BD shall be in the same straight line with CB.

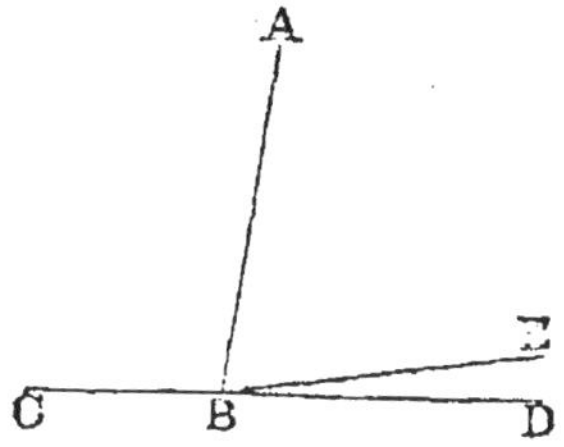

For if BD be not in the same straight line with CB, let BE be in the same straight line with it.

Then because the straight line AB makes with the straight line CBE, on one side of it, the angles ABC, ABE, these angles are together equal to two right angles. [I. 13.

But the angles ABC, ABD are also together equal to two right angles. [*Hypothesis.*

Therefore the angles ABC, ABE are equal to the angles ABC, ABD.

From each of these equals take away the common angle ABC, and the remaining angle ABE is equal to the remaining angle ABD, [*Axiom* 3.

the less to the greater; which is impossible.

Therefore BE is not in the same straight line with CB.

And in the same manner it may be shewn that no other can be in the same straight line with it but BD;

therefore BD is in the same straight line with CB.

Wherefore, *if at a point* &c. Q.E.D.

PROPOSITION 15. *THEOREM.*

If two straight lines cut one another, the vertical, or opposite, angles shall be equal.

Let the two straight lines AB, CD cut one another at the point E; the angle AEC shall be equal to the angle DEB, and the angle CEB to the angle AED.

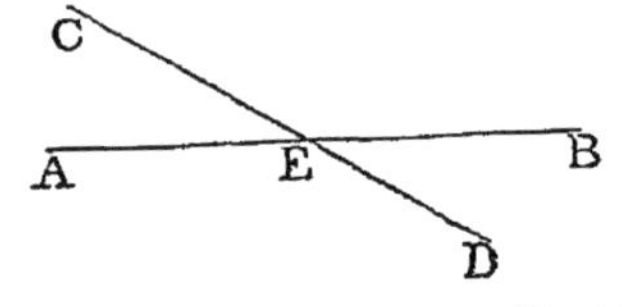

Because the straight line AE makes with the straight line CD the angles CEA, AED, these angles are together equal to two right angles. [I. 13.

Again, because the straight line DE makes with the straight line AB the angles AED, DEB, these also are together equal to two right angles. [I. 13.

But the angles CEA, AED have been shewn to be together equal to two right angles.

Therefore the angles CEA, AED are equal to the angles AED, DEB.

From each of these equals take away the common angle AED, and the remaining angle CEA is equal to the remaining angle DEB. [*Axiom* 3.

In the same manner it may be shewn that the angle CEB is equal to the angle AED.

Wherefore, *if two straight lines* &c. Q.E.D.

Corollary 1. From this it is manifest that, if two straight lines cut one another, the angles which they make at the point where they cut, are together equal to four right angles.

Corollary 2. And consequently, that all the angles made by any number of straight lines meeting at one point, are together equal to four right angles.

PROPOSITION 16. *THEOREM.*

If one side of a triangle be produced, the exterior angle shall be greater than either of the interior opposite angles.

Let ABC be a triangle, and let one side BC be produced to D: the exterior angle ACD shall be greater than either of the interior opposite angles CBA, BAC.

Bisect AC at E, [I. 10.

join BE and produce it to F, making EF equal to EB, [I. 3.

and join FC.

Because AE is equal to EC, and BE to EF; [*Constr.*

the two sides AE, EB are equal to the two sides CE, EF, each to each;

and the angle *AEB* is equal to the angle *CEF*, because they are opposite vertical angles; [I. 15.

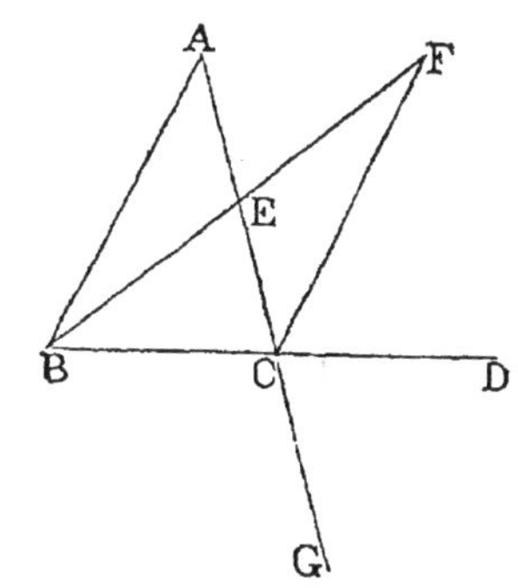

therefore the triangle *AEB* is equal to the triangle *CEF*, and the remaining angles to the remaining angles, each to each, to which the equal sides are opposite; [I. 4.

therefore the angle *BAE* is equal to the angle *ECF*.

But the angle *ECD* is greater than the angle *ECF*. [*Axiom* 9.

Therefore the angle *ACD* is greater than the angle *BAE*.

In the same manner if *BC* be bisected, and the side *AC* be produced to *G*, it may be shewn that the angle *BCG*, that is the angle *ACD*, is greater than the angle *ABC*. [I. 15.

Wherefore, *if one side* &c. Q.E.D.

PROPOSITION 17. *THEOREM.*

Any two angles of a triangle are together less than two right angles.

Let *ABC* be a triangle: any two of its angles are together less than two right angles.

Produce *BC* to *D*.

Then because *ACD* is the exterior angle of the triangle *ABC*, it is greater than the interior opposite angle *ABC*. [I. 16.

To each of these add the angle *ACB*

Therefore the angles *ACD*, *ACB* are greater than the angles *ABC*, *ACB*.

But the angles *ACD*, *ACB* are together equal to two right angles. [I. 13.

Therefore the angles *ABC*, *ACB* are together less than two right angles.

In the same manner it may be shewn that the angles *BAC*, *ACB*, as also the angles *CAB*, *ABC*, are together less than two right angles.

Wherefore, *any two angles* &c. Q.E.D.

PROPOSITION 18. *THEOREM.*

The greater side of every triangle has the greater angle opposite to it.

Let ABC be a triangle, of which the side AC is greater than the side AB: the angle ABC is also greater than the angle ACB.

Because AC is greater than AB, make AD equal to AB, [I. 3. and join BD.

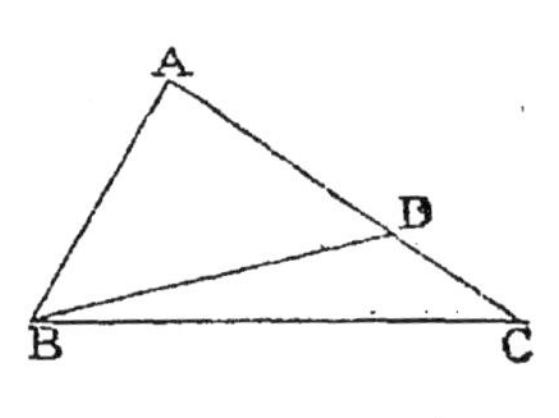

Then, because ADB is the exterior angle of the triangle BDC, it is greater than the interior opposite angle DCB. [I. 16.

But the angle ADB is equal to the angle ABD, [I. 5.
because the side AD is equal to the side AB. [*Constr.*
Therefore the angle ABD is also greater than the angle ACB.
Much more then is the angle ABC greater than the angle ACB. [*Axiom* 9.

Wherefore, *the greater side* &c. Q.E.D.

PROPOSITION 19. *THEOREM.*

The greater angle of every triangle is subtended by the greater side, or has the greater side opposite to it.

Let ABC be a triangle, of which the angle ABC is greater than the angle ACB: the side AC is also greater than the side AB.

For if not, AC must be either equal to AB or less than AB.
But AC is not equal to AB,
for then the angle ABC would be equal to the angle ACB; [I. 5.
but it is not; [*Hypothesis.*
therefore AC is not equal to AB.

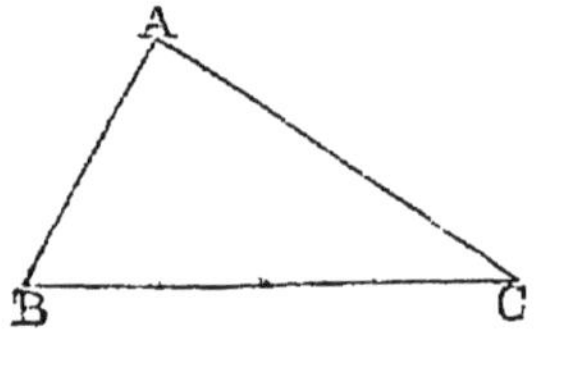

Neither is AC less than AB,
for then the angle ABC would be less than the angle ACB; [I. 18.
but it is not; [*Hypothesis.*

therefore AC is not less than AB.
And it has been shewn that AC is not equal to AB.
Therefore AC is greater than AB.

Wherefore, *the greater angle* &c. Q.E.D.

PROPOSITION 20. *THEOREM.*

Any two sides of a triangle are together greater than the third side.

Let ABC be a triangle: any two sides of it are together greater than the third side; namely, BA, AC greater than BC; and AB, BC greater than AC; and BC, CA greater than AB.

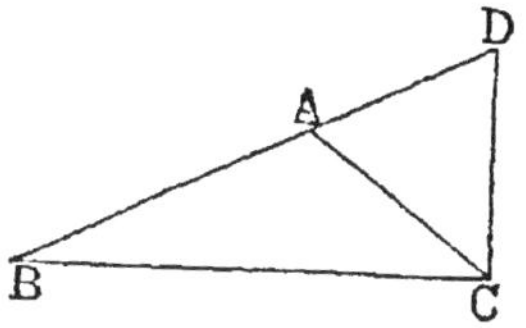

Produce BA to D, making AD equal to AC, [I. 3.
and join DC.

Then, because AD is equal to AC, [*Construction.*
the angle ADC is equal to the angle ACD. [I. 5.
But the angle BCD is greater than the angle ACD. [*Ax.* 9.
Therefore the angle BCD is greater than the angle BDC.
And because the angle BCD of the triangle BCD is greater than its angle BDC, and that the greater angle is subtended by the greater side; [I. 19.
therefore the side BD is greater than the side BC.
But BD is equal to BA and AC.
Therefore BA, AC are greater than BC.

In the same manner it may be shewn that AB, BC are greater than AC, and BC, CA greater than AB.

Wherefore, *any two sides* &c. Q.E.D.

PROPOSITION 21. *THEOREM.*

If from the ends of the side of a triangle there be drawn two straight lines to a point within the triangle, these shall be less than the other two sides of the triangle, but shall contain a greater angle.

Let ABC be a triangle, and from the points B, C, the ends of the side BC, let the two straight lines BD, CD be drawn to the point D within the triangle: BD, DC shall be less than the other two sides BA, AC of the triangle, but shall contain an angle BDC greater than the angle BAC.

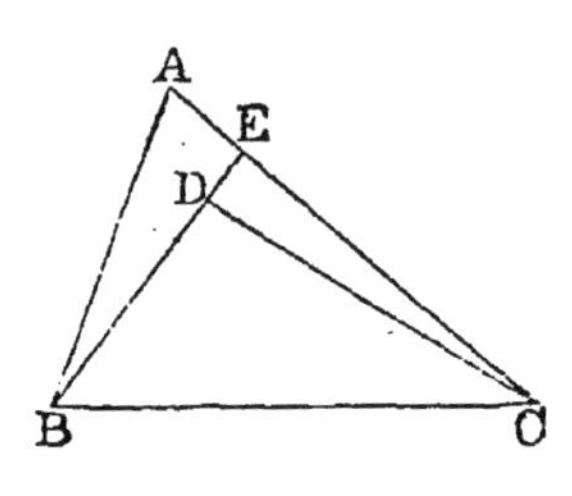

Produce BD to meet AC at E.

Because two sides of a triangle are greater than the third side, the two sides BA, AE of the triangle ABE are greater than the side BE. [I. 20.

To each of these add EC.

Therefore BA, AC are greater than BE, EC.

Again; the two sides CE, ED of the triangle CED are greater than the third side CD. [I. 40.

To each of these add DB.

Therefore CE, EB are greater than CD, DB.

But it has been shewn that BA, AC are greater than BE, EC;

much more then are BA, AC greater than BD, DC.

Again, because the exterior angle of any triangle is greater than the interior opposite angle, the exterior angle BDC of the triangle CDE is greater than the angle CED. [I. 16.

For the same reason, the exterior angle CEB of the triangle ABE is greater than the angle BAE.

But it has been shewn that the angle BDC is greater than the angle CEB;

much more then is the angle BDC greater than the angle BAC.

Wherefore, *if from the ends* &c. Q.E.D.

PROPOSITION 22. *PROBLEM.*

To make a triangle of which the sides shall be equal to three given straight lines, but any two whatever of these must be greater than the third.

Let *A*, *B*, *C* be the three given straight lines, of which any two whatever are greater than the third; namely, *A* and *B* greater than *C*; *A* and *C* greater than *B*; and *B* and *C* greater than *A*: it is required to make a triangle of which the sides shall be equal to *A*, *B*, *C*, each to each.

Take a straight line *DE* terminated at the point *D*, but unlimited towards *E*, and make *DF* equal to *A*, *FG* equal to *B*, and *GH* equal to *C*. [I. 3.

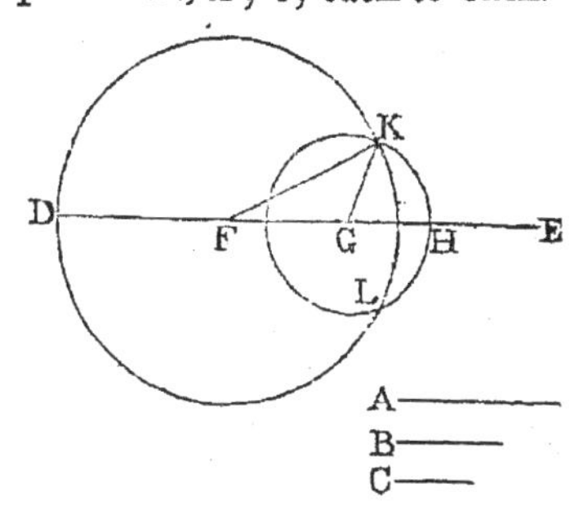

From the centre *F*, at the distance *FD*, describe the circle *DKL*. [*Post.* 3.

From the centre *G*, at the distance *GH*, describe the circle *HLK*, cutting the former circle at *K*.

Join *KF*, *KG*. The triangle *KFG* shall have its sides equal to the three straight lines *A*, *B*, *C*.

Because the point *F* is the centre of the circle *DKL*, *FD* is equal to *FK*. [*Definition* 15.

But *FD* is equal to *A*. [*Construction.*

Therefore *FK* is equal to *A*. [*Axiom* 1.

Again, because the point *G* is the centre of the circle *HLK*, *GH* is equal to *GK*. [*Definition* 15.

But *GH* is equal to *C*. [*Construction.*

Therefore *GK* is equal to *C*. [*Axiom* 1.

And *FG* is equal to *B*. [*Construction.*

Therefore the three straight lines *KF*, *FG*, *GK* are equal to the three *A*, *B*, *C*.

Wherefore *the triangle KFG has its three sides KF, FG, GK equal to the three given straight lines A, B, C.* Q.E.F.

PROPOSITION 23. *PROBLEM.*

At a given point in a given straight line, to make a rectilineal angle equal to a given rectilineal angle.

Let *AB* be the given straight line, and *A* the given point in it, and *DCE* the given rectilineal angle: it is required to make at the given point *A*, in the given straight line *AB*, an angle equal to the given rectilineal angle *DCE*.

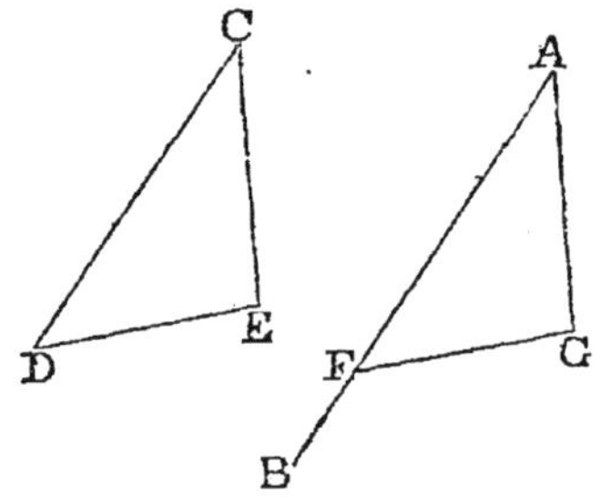

In *CD*, *CE* take any points *D*, *E*, and join *DE*. Make the triangle *AFG* the sides of which shall be equal to the three straight lines *CD*, *DE*, *EC*; so that *AF* shall be equal to *CD*, *AG* to *CE*, and *FG* to *DE*. [I. 22.

The angle *FAG* shall be equal to the angle *DCE*.

Because *FA*, *AG* are equal to *DC*, *CE*, each to each, and the base *FG* equal to the base *DE*; [*Construction.*

therefore the angle *FAG* is equal to the angle *DCE*. [I. 8.

Wherefore *at the given point A in the given straight line AB, the angle FAG has been made equal to the given rectilineal angle DCE.* Q.E.F.

PROPOSITION 24. *THEOREM.*

If two triangles have two sides of the one equal to two sides of the other, each to each, but the angle contained by the two sides of one of them greater than the angle contained by the two sides equal to them, of the other, the base of that which has the greater angle shall be greater than the base of the other.

Let *ABC*, *DEF* be two triangles, which have the two sides *AB*, *AC*, equal to the two sides *DE*, *DF*, each to each, namely, *AB* to *DE*, and *AC* to *DF*, but the angle *BAC* greater than the angle *EDF*: the base *BC* shall be greater than the base EF.

Of the two sides DE, DF, let DE be the side which is not greater than the other. At the point D in the straight line DE, make the angle EDG equal to the angle BAC, [I. 23.

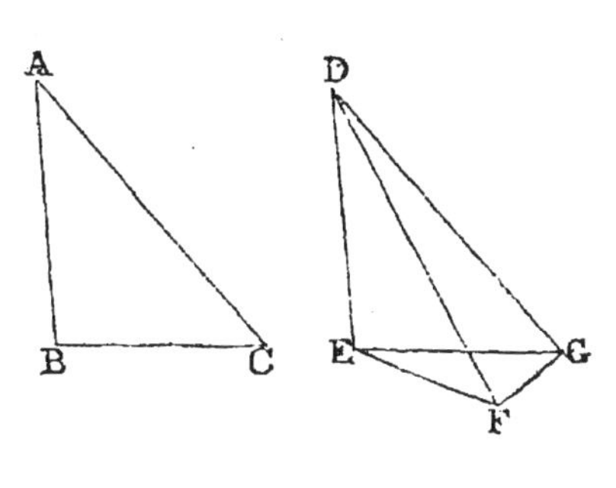

and make DG equal to AC or DF, [I. 3.
and join EG, GF.

Because AB is equal to DE, [*Hypothesis.*
and AC to DG; [*Construction.*
the two sides BA, AC are equal to the two sides ED, DG, each to each;
and the angle BAC is equal to the angle EDG; [*Constr.*
therefore the base BC is equal to the base EG. [I. 4.

And because DG is equal to DF, [*Construction.*
the angle DGF is equal to the angle DFG. [I. 5.
But the angle DGF is greater than the angle EGF. [*Ax.* 9.
Therefore the angle DFG is greater than the angle EGF.
Much more then is the angle EFG greater than the angle EGF. [*Axiom* 9.

And because the angle EFG of the triangle EFG is greater than its angle EGF, and that the greater angle is subtended by the greater side, [I. 19.
therefore the side EG is greater than the side EF.
But EG was shewn to be equal to BC;
therefore BC is greater than EF.

Wherefore, *if two triangles* &c. Q.E.D.

PROPOSITION 25. *THEOREM.*

If two triangles have two sides of the one equal to two sides of the other, each to each, but the base of the one greater than the base of the other, the angle contained by the sides of that which has the greater base, shall be greater than the angle contained by the sides equal to them, of the other.

Let *ABC*, *DEF* be two triangles, which have the two sides *AB*, *AC* equal to the two sides *DE*, *DF*, each to each, namely, *AB* to *DE*, and *AC* to *DF*, but the base *BC* greater than the base *EF*: the angle *BAC* shall be greater than the angle *EDF*.

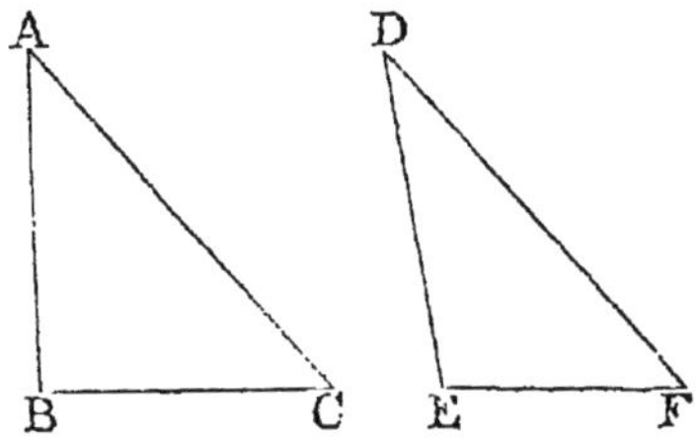

For if not, the angle *BAC* must be either equal to the angle *EDF* or less than the angle *EDF*.

But the angle *BAC* is not equal to the angle *EDF*, for then the base *BC* would be equal to the base *EF*; [I. 4.
but it is not; [*Hypothesis.*
therefore the angle *BAC* is not equal to the angle *EDF*.
Neither is the angle *BAC* less than the angle *EDF*,
for then the base *BC* would be less than the base *EF*; [I. 24.
but it is not; [*Hypothesis.*
therefore the angle *BAC* is not less than the angle *EDF*.
And it has been shewn that the angle *BAC* is not equal to the angle *EDF*.
Therefore the angle *BAC* is greater than the angle *EDF*.

Wherefore, *if two triangles* &c. Q.E.D.

PROPOSITION 26. *THEOREM.*

If two triangles have two angles of the one equal to two angles of the other, each to each, and one side equal to one side, namely, either the sides adjacent to the equal angles, or sides which are opposite to equal angles in each, then shall the other sides be equal, each to each, and also the third angle of the one equal to the third angle of the other.

Let *ABC*, *DEF* be two triangles, which have the angles *ABC*, *BCA* equal to the angles *DEF*, *EFD*, each to each, namely, ABC to DEF, and BCA to EFD; and let them have also one side equal to one side; and first let those sides be equal which are adjacent to the equal angles in the two triangles, namely, BC to EF: the other sides shall be equal, each to each, namely, AB to DE, and AC to DF, and the third angle BAC equal to the third angle EDF.

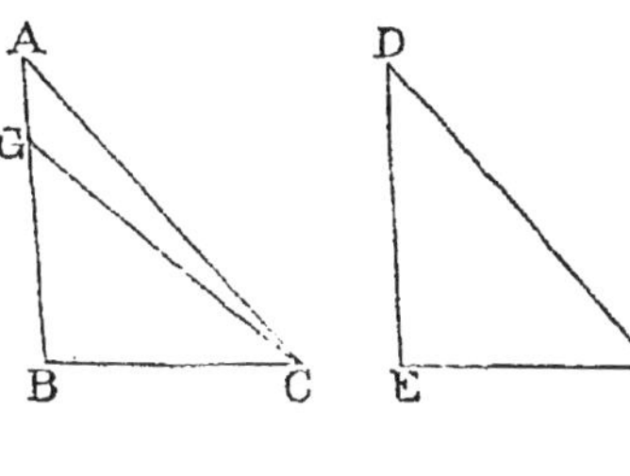

For if AB be not equal to DE, one of them must be greater than the other. Let AB be the greater, and make BG equal to DE, [I. 3.

and join GC.

Then because GB is equal to DE, [*Construction.*

and BC to EF; [*Hypothesis.*

the two sides GB, BC are equal to the two sides DE, EF, each to each;

and the angle GBC is equal to the angle DEF; [*Hypothesis.*

therefore the triangle GBC is equal to the triangle DEF, and the other angles to the other angles, each to each, to which the equal sides are opposite; [I. 4.

therefore the angle GCB is equal to the angle DFE.

But the angle DFE is equal to the angle ACB. [*Hypothesis.*

Therefore the angle GCB is equal to the angle ACB, [*Ax.* 1.

the less to the greater; which is impossible.

Therefore AB is not unequal to DE,

that is, it is equal to it;

and BC is equal to EF; [*Hypothesis.*

therefore the two sides AB, BC are equal to the two sides DE, EF, each to each;

and the angle ABC is equal to the angle DEF; [*Hypothesis.*

therefore the base AC is equal to the base DF, and the third angle BAC to the third angle EDF. [I. 4.

Next, let sides which are opposite to equal angles in each triangle be equal to one another, namely, *AB* to *DE*: likewise in this case the other sides shall be equal, each to each, namely, *BC* to *EF*, and *AC* to *DF*, and also the third angle *BAC* equal to the third angle *EDF*.

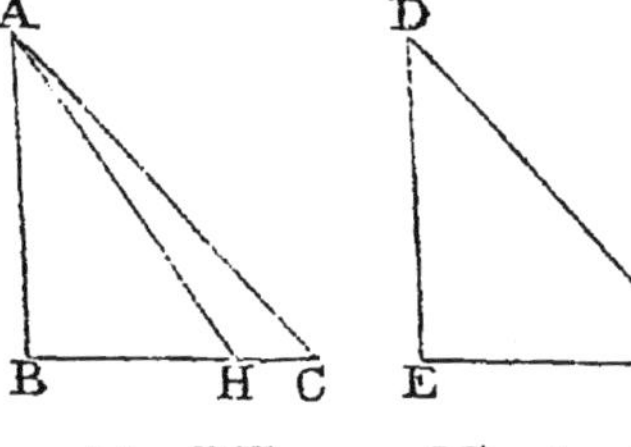

For if *BC* be not equal to *EF*, one of them must be greater than the other. Let *BC* be the greater, and make *BH* equal to *EF*, [I. 3.
and join *AH*.

Then because *BH* is equal to *EF*, [*Construction.*
and *AB* to *DE*; [*Hypothesis.*
the two sides *AB*, *BH* are equal to the two sides *DE*, *EF*, each to each;
and the angle *ABH* is equal to the angle *DEF*; [*Hypothesis.*
therefore the triangle *ABH* is equal to the triangle *DEF*, and the other angles to the other angles, each to each, to which the equal sides are opposite; [I. 4.
therefore the angle *BHA* is equal to the angle *EFD*.
But the angle *EFD* is equal to the angle *BCA*. [*Hypothesis.*
Therefore the angle *BHA* is equal to the angle *BCA*; [*Ax.* 1.
that is, the exterior angle *BHA* of the triangle *AHC* is equal to its interior opposite angle *BCA*;
which is impossible. [I. 16.
Therefore *BC* is not unequal to *EF*,
that is, it is equal to it;
and *AB* is equal to *DE*; [*Hypothesis.*
therefore the two sides *AB*, *BC* are equal to the two sides *DE*, *EF*, each to each;
and the angle *ABC* is equal to the angle *DEF*; [*Hypothesis.*
therefore the base *AC* is equal to the base *DF*, and the third angle *BAC* to the third angle *EDF*. [I. 4.

Wherefore, *if two triangles* &c. Q.E.D.

PROPOSITION 27. *THEOREM.*

If a straight line falling on two other straight lines, make the alternate angles equal to one another, the two straight lines shall be parallel to one another.

Let the straight line *EF*, which falls on the two straight lines *AB*, *CD*, make the alternate angles *AEF*, *EFD* equal to one another: *AB* shall be parallel to *CD*.

For if not, *AB* and *CD*, being produced, will meet either towards *B*, *D* or towards *A*, *C*. Let them be produced and meet towards *B*, *D* at the point *G*.

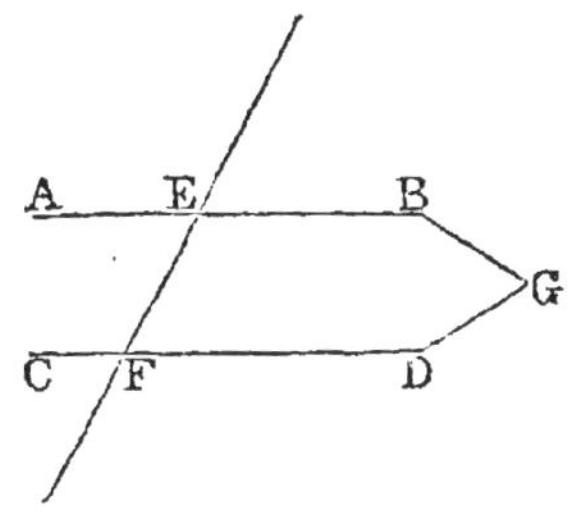

Therefore *GEF* is a triangle, and its exterior angle *AEF* is greater than the interior opposite angle *EFG*; [I. 16.
But the angle *AEF* is also equal to the angle *EFG*; [*Hyp.*
which is impossible.
Therefore *AB* and *CD* being produced, do not meet towards *B*, *D*.

In the same manner, it may be shewn that they do not meet towards *A*, *C*.
But those straight lines which being produced ever so far both ways do not meet, are parallel. [*Definition* 35.
Therefore *AB* is parallel to *CD*.

Wherefore, *if a straight line* &c. Q.E.D.

PROPOSITION 28. *THEOREM.*

If a straight line falling on two other straight lines, make the exterior angle equal to the interior and opposite angle on the same side of the line, or make the interior angles on the same side together equal to two right angles, the two straight lines shall be parallel to one another.

Let the straight line *EF*, which falls on the two straight lines *AB*, *CD*, make the exterior angle *EGB* equal to the interior and opposite angle *GHD* on the same side, or make the interior angles on the same side *BGH*, *GHD* together equal to two right angles: *AB* shall be parallel to *CD*.

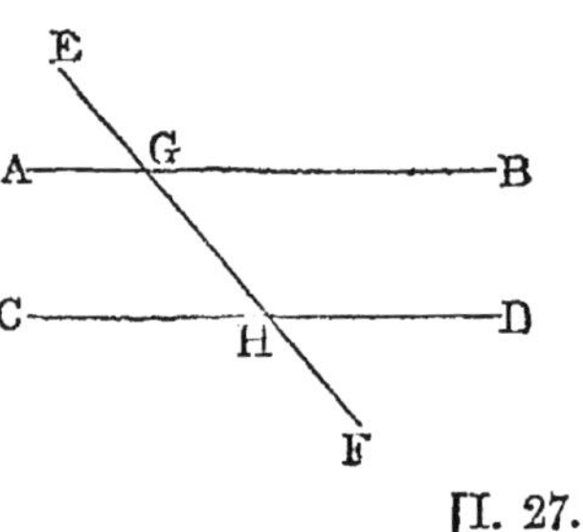

Because the angle *EGB* is equal to the angle *GHD*, [*Hyp.*
and the angle *EGB* is also equal to the angle *AGH*, [I. 15.
therefore the angle *AGH* is equal to the angle *GHD*; [*Ax.* 1.
and they are alternate angles;
therefore *AB* is parallel to *CD*. [I. 27.

Again; because the angles *BGH*, *GHD* are together equal to two right angles, [*Hypothesis.*
and the angles *AGH*, *BGH* are also together equal to two right angles, [I. 13.
therefore the angles *AGH*, *BGH* are equal to the angles *BGH*, *GHD*.
Take away the common angle *BGH*; therefore the remaining angle *AGH* is equal to the remaining angle *GHD*; [*Axiom* 3.
and they are alternate angles;
therefore *AB* is parallel to *CD*. [I. 27.

Wherefore, *if a straight line* &c. Q.E.D.

PROPOSITION 29. *THEOREM.*

If a straight line fall on two parallel straight lines, it makes the alternate angles equal to one another, and the exterior angle equal to the interior and opposite angle on the same side; and also the two interior angles on the same side together equal to two right angles.

Let the straight line *EF* fall on the two parallel straight lines *AB*, *CD*: the alternate angles *AGH*, *GHD* shall be equal to one another, and the exterior angle *EGB* shall be equal to the interior and opposite angle on the same side, *GHD*, and the two interior angles on the same side, *BGH*, *GHD*, shall be together equal to two right angles.

For if the angle *AGH* be not equal to the angle *GHD*, one of them must be greater than the other; let the angle *AGH* be the greater.

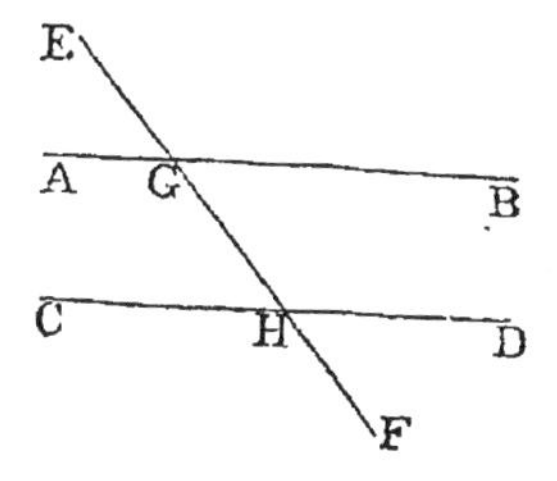

Then the angle *AGH* is greater than the angle *GHD*;

to each of them add the angle *BGH*;

therefore the angles *AGH*, *BGH* are greater than the angles *BGH*, *GHD*.

But the angles *AGH*, *BGH* are together equal to two right angles; [I. 13.

therefore the angles *BGH*, *GHD* are together less than two right angles.

But if a straight line meet two straight lines, so as to make the two interior angles on the same side of it, taken together, less than two right angles, these straight lines being continually produced, shall at length meet on that side on which are the angles which are less than two right angles. [*Axiom* 12.

Therefore the straight lines *AB*, *CD*, if continually produced, will meet.

But they never meet, since they are parallel by hypothesis.

Therefore the angle *AGH* is not unequal to the angle *GHD*; that is, it is equal to it.

But the angle *AGH* is equal to the angle *EGB*. [I. 15.

Therefore the angle *EGB* is equal to the angle *GHD*. [*Ax.* 1.

Add to each of these the angle *BGH*.

Therefore the angles *EGB*, *BGH* are equal to the angles *BGH*, *GHD*. [*Axiom* 2.

But the angles *EGB*, *BGH* are together equal to two right angles. [I. 13.

Therefore the angles *BGH*, *GHD* are together equal to two right angles. [*Axiom* 1.

Wherefore, *if a straight line* &c. Q.E.D.

PROPOSITION 30. *THEOREM.*

Straight lines which are parallel to the same straight line are parallel to each other.

Let *AB*, *CD* be each of them parallel to *EF*: *AB* shall be parallel to *CD*.

Let the straight line *GHK* cut *AB*, *EF*, *CD*.

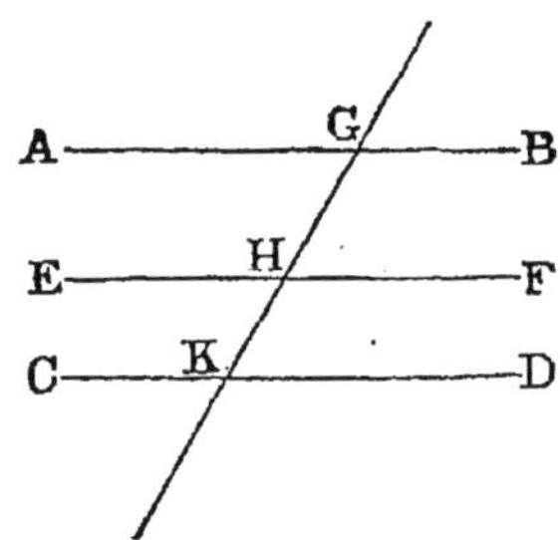

Then, because *GHK* cuts the parallel straight lines *AB*, *EF*, the angle *AGH* is equal to the angle *GHF*. [I. 29.

Again, because *GK* cuts the parallel straight lines *EF*, *CD*, the angle *GHF* is equal to the angle *GKD*. [I. 29.

And it was shewn that the angle *AGK* is equal to the angle *GHF*.

Therefore the angle *AGK* is equal to the angle *GKD*; [*Ax.* 1.

and they are alternate angles;

therefore *AB* is parallel to *CD*. [I. 27.

Wherefore, *straight lines* &c. Q.E.D.

PROPOSITION 31. *PROBLEM.*

To draw a straight line through a given point parallel to a given straight line.

Let *A* be the given point, and *BC* the given straight line: it is required to draw a straight line through the point *A* parallel to the straight line *BC*.

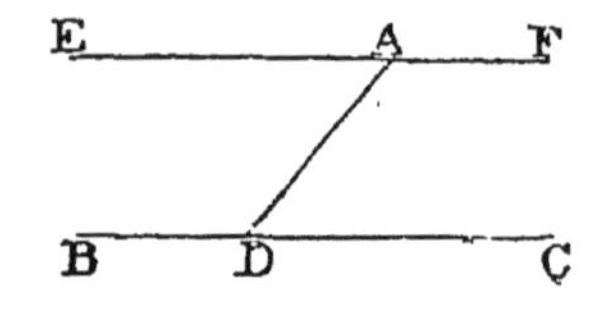

In *BC* take any point *D*, and join *AD*; at the point *A* in the straight line *AD*, make the angle *DAE* equal to the angle *ADC*; [I. 23.

and produce the straight line *EA* to *F*.

EF shall be parallel to *BC*.

Because the straight line *AD*, which meets the two straight lines *BC*, *EF*, makes the alternate angles *EAD*, *ADC* equal to one another, [*Construction.*

EF is parallel to *BC*. [I. 27.

Wherefore *the straight line EAF is drawn through the given point A, parallel to the given straight line BC.* Q.E.F.

PROPOSITION 32. *THEOREM.*

If a side of any triangle be produced, the exterior angle is equal to the two interior and opposite angles; and the three interior angles of every triangle are together equal to two right angles.

Let *ABC* be a triangle, and let one of its sides *BC* be produced to *D*: the exterior angle *ACD* shall be equal to the two interior and opposite angles *CAB*, *ABC*; and the three interior angles of the triangle, namely, *ABC*, *BCA*, *CAB* shall be equal to two right angles.

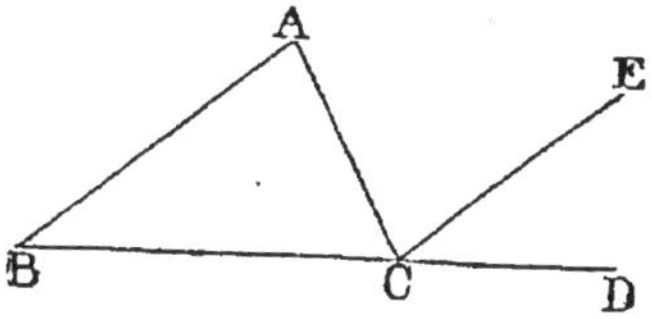

Through the point *C* draw *CE* parallel to *AB*. [I. 31.

A
E
B
C
D

Then, because *AB* is parallel to *CE*, and *AC* falls on them, the alternate angles *BAC*, *ACE* are equal. [I. 29.

Again, because *AB* is parallel to *CE*, and *BD* falls on them, the exterior angle *ECD* is equal to the interior and opposite angle *ABC*. [I. 29.

But the angle *ACE* was shewn to be equal to the angle *BAC*;

therefore the whole exterior angle *ACD* is equal to the two interior and opposite angles *CAB*, *ABC*. [*Axiom* 2.

To each of these equals add the angle *ACB*;

therefore the angles *ACD*, *ACB* are equal to the three angles *CBA*, *BAC*, *ACB*. [*Axiom* 2.

But the angles *ACD*, *ACB* are together equal to two right angles; [I. 13.

therefore also the angles *CBA*, *BAC*, *ACB* are together equal to two right angles. [*Axiom* 1.

Wherefore, *if a side of any triangle* &c. Q.E.D.

COROLLARY 1. *All the interior angles of any rectilineal figure, together with four right angles, are equal to twice as many right angles as the figure has sides.*

For any rectilineal figure $ABCDE$ can be divided into as many triangles as the figure has sides, by drawing straight lines from a point F within the figure to each of its angles.

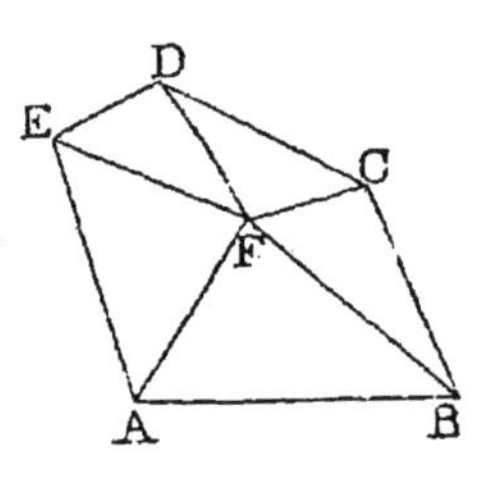

And by the preceding proposition, all the angles of these triangles are equal to twice as many right angles as there are triangles, that is, as the figure has sides.

And the same angles are equal to the interior angles of the figure, together with the angles at the point F, which is the common vertex of the triangles, that is, together with four right angles. [I. 15. *Corollary* 2.

Therefore all the interior angles of the figure, together with four right angles, are equal to twice as many right angles as the figure has sides.

COROLLARY 2. *All the exterior angles of any rectilineal figure are together equal to four right angles.*

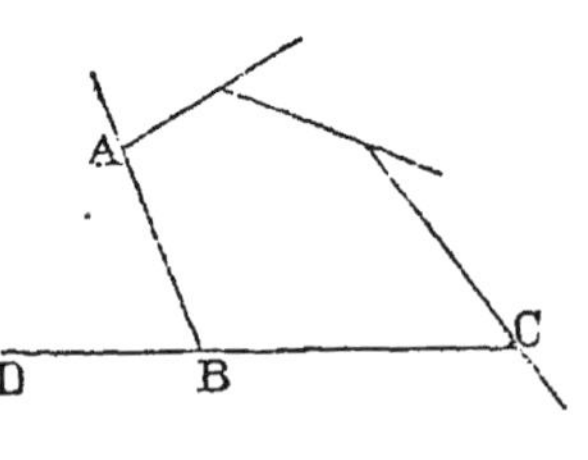

Because every interior angle ABC, with its adjacent exterior angle ABD, is equal to two right angles; [I. 13.

therefore all the interior angles of the figure, together with all its exterior angles, are equal to twice as many right angles as the figure has sides.

But, by the foregoing Corollary all the interior angles of the figure, together with four right angles, are equal to twice as many right angles as the figure has sides.

Therefore all the interior angles of the figure, together with all its exterior angles, are equal to all the interior angles of the figure, together with four right angles.

Therefore all the exterior angles are equal to four right angles.

PROPOSITION 33. *THEOREM.*

The straight lines which join the extremities of two equal and parallel straight lines towards the same parts, are also themselves equal and parallel.

Let AB and CD be equal and parallel straight lines, and let them be joined towards the same parts by the straight lines AC and BD: AC and BD shall be equal and parallel.

Join BC.

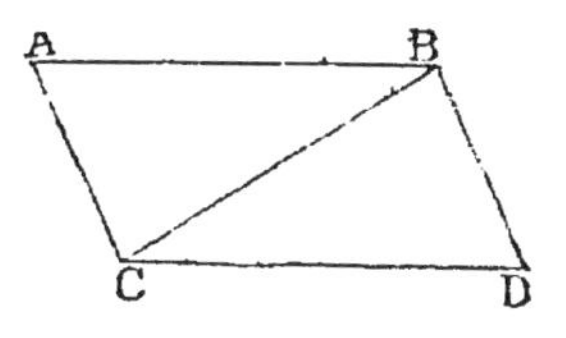

Then because AB is parallel to CD, [*Hypothesis.*
and BC meets them,
the alternate angles ABC, BCD are equal. [I. 29.
And because AB is equal to CD, [*Hypothesis.*
and BC is common to the two triangles ABC, DCB;
the two sides AB, BC are equal to the two sides DC, CB, each to each;
and the angle ABC was shewn to be equal to the angle BCD;
therefore the base AC is equal to the base BD, and the triangle ABC to the triangle BCD, and the other angles to the other angles, each to each, to which the equal sides are opposite; [I. 4.
therefore the angle ACB is equal to the angle CBD.

And because the straight line BC meets the two straight lines AC, BD, and makes the alternate angles ACB, CBD equal to one another, AC is parallel to BD. [I. 27.
And it was shewn to be equal to it.

Wherefore, *the straight lines* &c. Q.E.D.

PROPOSITION 34. *THEOREM.*

The opposite sides and angles of a parallelogram are equal to one another, and the diameter bisects the parallelogram, that is, divides it into two equal parts.

Note. A parallelogram is a four-sided figure of which the opposite sides are parallel; and a diameter is the straight line joining two of its opposite angles.

Let *ACDB* be a parallelogram, of which *BC* is a diameter; the opposite sides and angles of the figure shall be equal to one another, and the diameter *BC* shall bisect it.

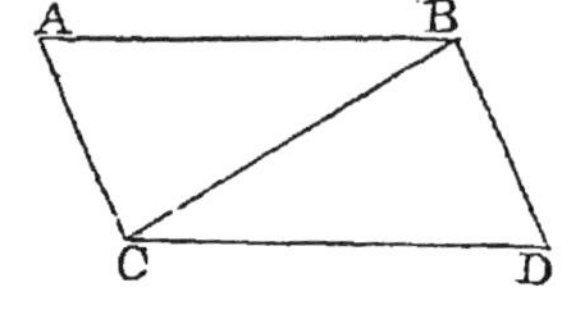

Because *AB* is parallel to *CD*, and *BC* meets them, the alternate angles *ABC*, *BCD* are equal to one another. [I. 29.

And because *AC* is parallel to *BD*, and *BC* meets them, the alternate angles *ACB*, *CBD* are equal to one another. [I. 29.

Therefore the two triangles *ABC*, *BCD* have two angles *ABC*, *BCA* in the one, equal to two angles *DCB*, *CBD* in the other, each to each, and one side *BC* is common to the two triangles, which is adjacent to their equal angles;

therefore their other sides are equal, each to each, and the third angle of the one to the third angle of the other, namely, the side *AB* equal to the side *CD*, and the side *AC* equal to the side *BD*, and the angle *BAC* equal to the angle *CDB*. [I. 26.

And because the angle *ABC* is equal to the angle *BCD*, and the angle *CBD* to the angle *ACB*,

the whole angle *ABD* is equal to the whole angle *ACD*. [*Ax.* 2.

And the angle *BAC* has been shewn to be equal to the angle *CDB*.

Therefore the opposite sides and angles of a parallelogram are equal to one another.

Also the diameter bisects the parallelogram.

For *AB* being equal to *CD*, and *BC* common,

the two sides *AB*, *BC* are equal to the two sides *DC*, *CB* each to each;

and the angle *ABC* has been shewn to be equal to the angle *BCD*;

therefore the triangle *ABC* is equal to the triangle *BCD*, [I. 4.

and the diameter *BC* divides the parallelogram *ACDB* into two equal parts.

Wherefore, *the opposite sides* &c. Q.E.D.

PROPOSITION 35. *THEOREM.*

Parallelograms on the same base, and between the same parallels, are equal to one another.

Let the parallelograms *ABCD, EBCF* be on the same base *BC*, and between the same parallels *AF, BC*: the parallelogram *ABCD* shall be equal to the parallelogram *EBCF*.

If the sides *AD, DF* of the parallelograms *ABCD, DBCF*, opposite to the base *BC*, be terminated at the same point *D*, it is plain that each of the parallelograms is double of the triangle *BDC*; [I. 34.

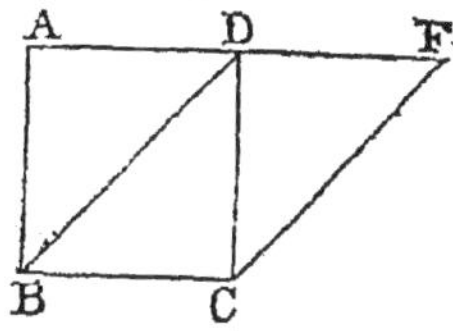

and they are therefore equal to one another. [*Axiom* 6.

But if the sides *AD, EF*, opposite to the base *BC* of the parallelograms *ABCD, EBCF* be not terminated at the same point, then, because *ABCD* is a parallelogram *AD* is equal to *BC*; [I. 34.

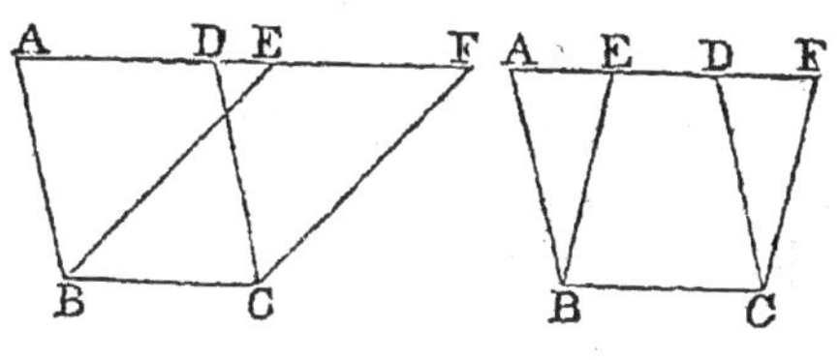

for the same reason *EF* is equal to *BC*;

therefore *AD* is equal to *EF*; [*Axiom* 1.

therefore the whole, or the remainder, *AE* is equal to the whole, or the remainder, *DF*. [*Axioms* 2, 3.

And *AB* is equal to *DC*; [I. 34.

therefore the two sides *EA, AB* are equal to the two sides *FD, DC* each to each;

and the exterior angle *FDC* is equal to the interior and opposite angle *EAB*; [I. 29.

therefore the triangle *EAB* is equal to the triangle *FDC*. [I. 4.

Take the triangle *FDC* from the trapezium *ABCF*, and from the same trapezium take the triangle *EAB*, and the remainders are equal; [*Axiom* 3.

that is, the parallelogram *ABCD* is equal to the parallelogram *EBCF*.

Wherefore, *parallelograms on the same base &c.* Q.E.D.

PROPOSITION 36. *THEOREM.*

Parallelograms on equal bases, and between the same parallels, are equal to one another.

Let *ABCD*, *EFGH* be parallelograms on equal bases *BC*, *FG*, and between the same parallels *AH*, *BG*: the parallelogram *ABCD* shall be equal to the parallelogram *EFGH*.

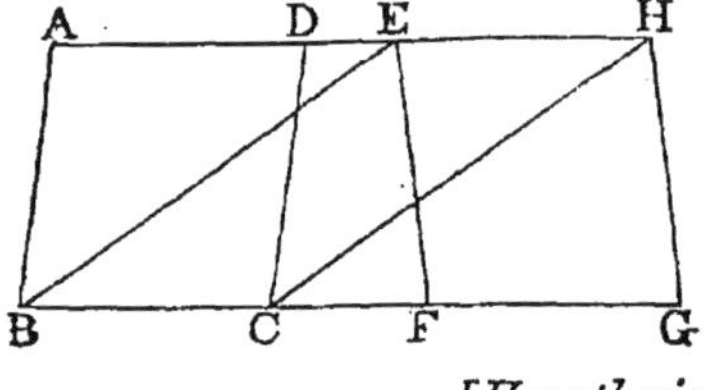

Join *BE*, *CH*.

Then, because *BC* is equal to *FG*, [*Hyp.*
and *FG* to *EH*, [I. 34.
BC is equal to *EH*; [*Axiom* 1.
and they are parallels, [*Hypothesis.*
and joined towards the same parts by the straight lines *BE*, *CH*.
But straight lines which join the extremities of equal and parallel straight lines towards the same parts are themselves equal and parallel. [I. 33.
Therefore *BE*, *CH* are both equal and parallel.
Therefore *EBCH* is a parallelogram. [*Definition.*
And it is equal to *ABCD*, because they are on the same base *BC*, and between the same parallels *BC*, *AH*. [I. 35.

For the same reason the parallelogram *EFGH* is equal to the same *EBCH*.
Therefore the parallelogram *ABCD* is equal to the parallelogram *EFGH*. [*Axiom* 1.

Wherefore, *parallelograms* &c. Q.E.D.

PROPOSITION 37. *THEOREM.*

Triangles on the same base, and between the same parallels, are equal.

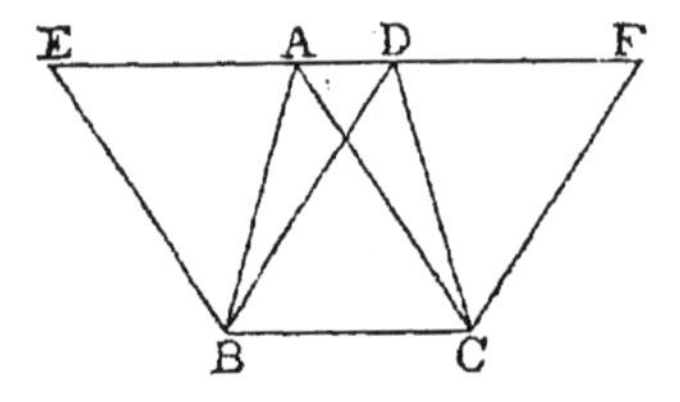

Let the triangles *ABC*, *DBC* be on the same base *BC*, and between the same parallels *AD*, *BC*: the triangle *ABC* shall be equal to the triangle *DBC*.

Produce *AD* both ways to the points *E*, *F*; [*Post.* 2.

through B draw BE parallel to CA, and through C draw CF parallel to BD. [I. 31.

Then each of the figures $EBCA$, $DBCF$ is a parallelogram; [*Definition.*

and $EBCA$ is equal to $DBCF$, because they are on the same base BC, and between the same parallels BC, EF. [I. 35.

And the triangle ABC is half of the parallelogram $EBCA$, because the diameter AB bisects the parallelogram; [I. 34.

and the triangle DBC is half of the parallelogram $DBCF$, because the diameter DC bisects the parallelogram. [I. 34.

But the halves of equal things are equal. [*Axiom* 7.

Therefore the triangle ABC is equal to the triangle DBC.

Wherefore, *triangles* &c. Q.E.D.

PROPOSITION 38. *THEOREM.*

Triangles on equal bases, and between the same parallels, are equal to one another.

Let the triangles ABC, DEF be on equal bases BC, EF, and between the same parallels BF, AD: the triangle ABC shall be equal to the triangle DEF.

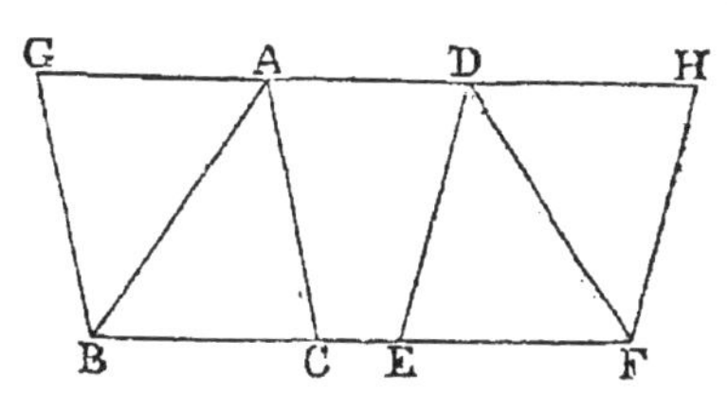

Produce AD both ways to the points G, H;

through B draw BG parallel to CA, and through F draw FH parallel to ED. [I. 31.

Then each of the figures $GBCA$, $DEFH$ is a parallelogram. [*Definition.*

And they are equal to one another because they are on equal bases BC, EF, and between the same parallels BF, GH. [I. 36.

And the triangle ABC is half of the parallelogram $GBCA$, because the diameter AB bisects the parallelogram; [I. 34.

and the triangle DEF is half of the parallelogram $DEFH$, because the diameter DF bisects the parallelogram.

But the halves of equal things are equal. [*Axiom* 7.

Therefore the triangle ABC is equal to the triangle DEF.

Wherefore, *triangles* &c. Q.E.D.

PROPOSITION 39. *THEOREM.*

Equal triangles on the same base, and on the same side of it, are between the same parallels.

Let the equal triangles *ABC*, *DBC* be on the same base *BC*, and on the same side of it: they shall be between the same parallels.

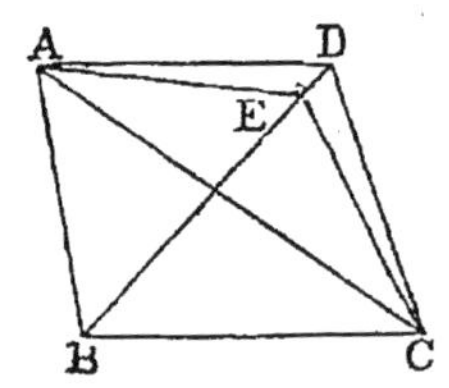

Join *AD*.

AD shall be parallel to *BC*.

For if it is not, through *A* draw *AE* parallel to *BC*, meeting *BD* at *E*. [I. 31.

and join *EC*.

Then the triangle *ABC* is equal to the triangle *EBC*, because they are on the same base *BC*, and between the same parallels *BC*, *AE*. [I. 37.

But the triangle *ABC* is equal to the triangle *DBC*. [*Hyp.*

Therefore also the triangle *DBC* is equal to the triangle *EBC*, [*Axiom* 1.

the greater to the less; which is impossible.

Therefore *AE* is not parallel to *BC*.

In the same manner it can be shewn, that no other straight line through *A* but *AD* is parallel to *BC*; therefore *AD* is parallel to *BC*.

Wherefore, *equal triangles* &c. Q.E.D.

PROPOSITION 40. *THEOREM.*

Equal triangles, on equal bases, in the same straight line, and on the same side of it, are between the same parallels.

Let the equal triangles *ABC*, *DEF* be on equal bases *BC*, *EF*, in the same straight line *BF*, and on the same side of it: they shall be between the same parallels.

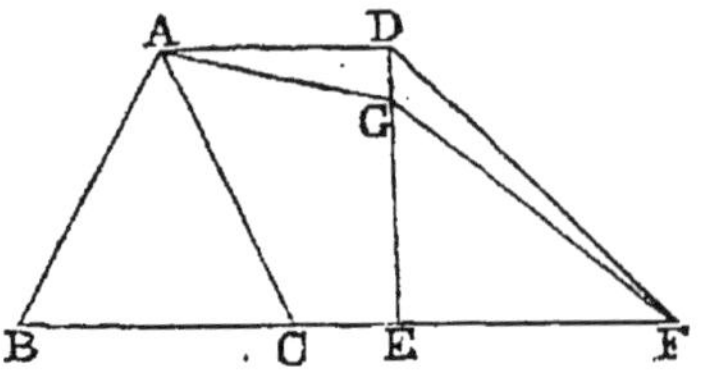

Join *AD*.

AD shall be parallel to *BF*.

For if it is not, through *A* draw *AG* parallel to *BF*, meeting *ED* at *G* [I. 31.

and join *GF*.

Then the triangle ABC is equal to the triangle GEF, because they are on equal bases BC, EF, and between the same parallels. [I. 38.

But the triangle ABC is equal to the triangle DEF. [*Hyp.*

Therefore also the triangle DEF is equal to the triangle GEF, [*Axiom* 1.

the greater to the less; which is impossible.

Therefore AG is not parallel to BF.

In the same manner it can be shewn that no other straight line through A but AD is parallel to BF; therefore AD is parallel to BF.

Wherefore, *equal triangles* &c. Q.E.D.

PROPOSITION 41. *THEOREM.*

If a parallelogram and a triangle be on the same base and between the same parallels, the parallelogram shall be double of the triangle.

Let the parallelogram $ABCD$ and the triangle EBC be on the same base BC, and between the same parallels BC, AE: the parallelogram $ABCD$ shall be double of the triangle EBC.

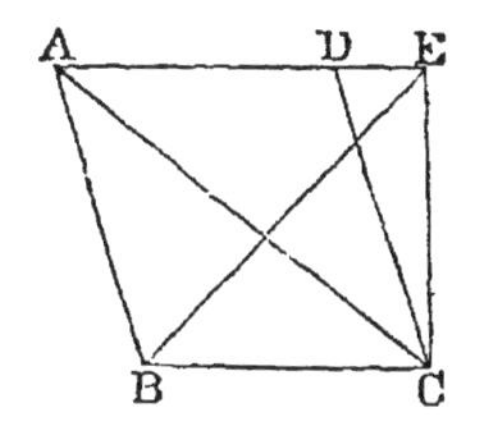

Join AC.

Then the triangle ABC is equal to the triangle EBC, because they are on the same base BC, and between the same parallels BC, AE. [I. 37.

But the parallelogram $ABCD$ is double of the triangle ABC,

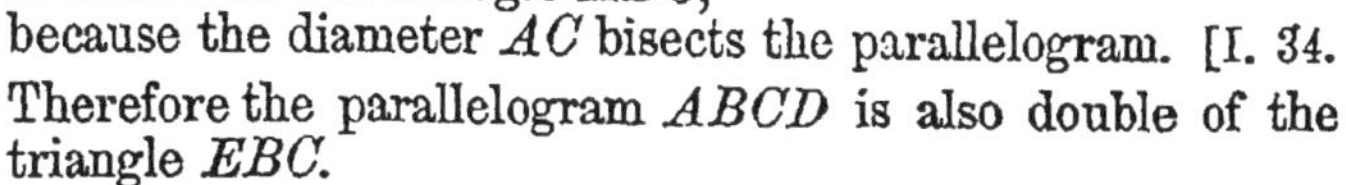

because the diameter AC bisects the parallelogram. [I. 34.

Therefore the parallelogram $ABCD$ is also double of the triangle EBC.

Wherefore, *if a parallelogram* &c. Q.E.D.

PROPOSITION 42. *PROBLEM.*

To describe a parallelogram that shall be equal to a given triangle, and have one of its angles equal to a given rectilineal angle.

Let ABC be the given triangle, and D the given rectilineal angle: it is required to describe a parallelogram that shall be equal to the given triangle ABC, and have one of its angles equal to D.

Bisect BC at E: [I. 10.
join AE, and at the point E, in the straight line EC, make the angle CEF equal to D; [I. 23.
through A draw AFG parallel to EC, and through C draw CG parallel to EF. [I. 31.

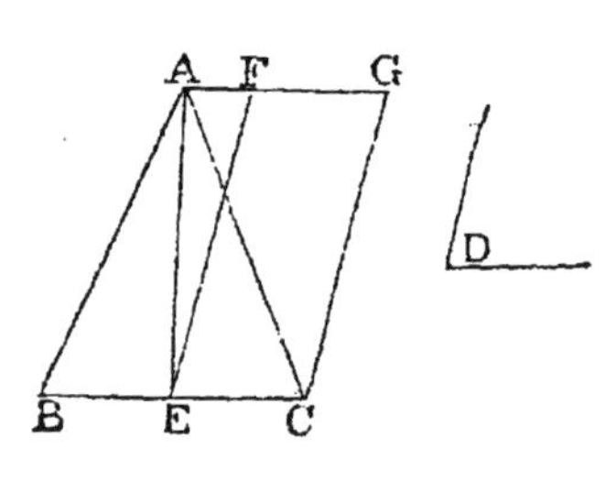

Therefore $FECG$ is a parallelogram. [*Definition.*

And, because BE is equal to EC, [*Construction.*
the triangle ABE is equal to the triangle AEC, because they are on equal bases BE, EC, and between the same parallels BC, AG. [I. 38.

Therefore the triangle ABC is double of the triangle AEC.

But the parallelogram $FECG$ is also double of the triangle AEC, because they are on the same base EC, and between the same parallels EC, AG. [I. 41.

Therefore the parallelogram $FECG$ is equal to the triangle ABC; [*Axiom* 6.

and it has one of its angles CEF equal to the given angle D. [*Construction.*

Wherefore *a parallelogram FECG has been described equal to the given triangle ABC, and having one of its angles CEF equal to the given angle D.* Q.E.F.

PROPOSITION 43. *THEOREM.*

The complements of the parallelograms which are about the diameter of any parallelogram, are equal to one another.

Let $ABCD$ be a parallelogram, of which the diameter is AC; and EH, GF parallelograms about AC, that is, through which AC passes; and BK, KD the other parallelograms which make up the whole figure $ABCD$, and which are therefore called the complements: the complement BK shall be equal to the complement KD.

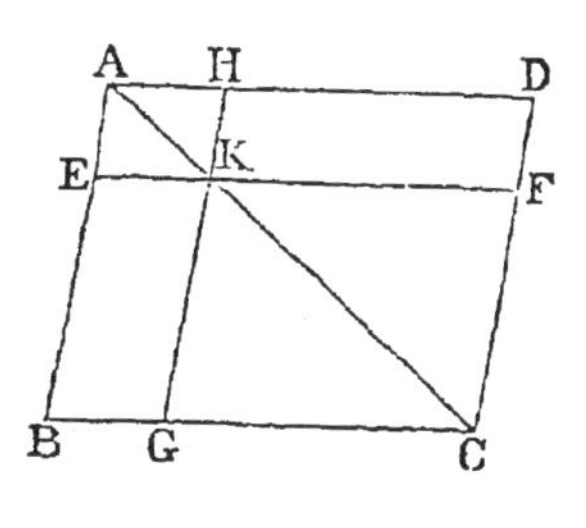

Because $ABCD$ is a parallelogram, and AC its diameter, the triangle ABC is equal to the triangle ADC. [I. 34.

Again, because $AEKH$ is a parallelogram, and AK its diameter, the triangle AEK is equal to the triangle AHK. [I. 34.

For the same reason the triangle KGC is equal to the triangle KFC.

Therefore, because the triangle AEK is equal to the triangle AHK, and the triangle KGC to the triangle KFC; the triangle AEK together with the triangle KGC is equal to the triangle AHK together with the triangle KFC. [*Ax.* 2.

But the whole triangle ABC was shewn to be equal to the whole triangle ADC.

Therefore the remainder, the complement BK, is equal to the remainder, the complement KD. [*Axiom* 3.

Wherefore, *the complements* &c. Q.E.D.

PROPOSITION 44. *PROBLEM.*

To a given straight line to apply a parallelogram, which shall be equal to a given triangle, and have one of its angles equal to a given rectilineal angle.

Let AB be the given straight line, and C the given triangle, and D the given rectilineal angle: it is required to apply to the straight line AB a parallelogram equal to the triangle C, and having an angle equal to D.

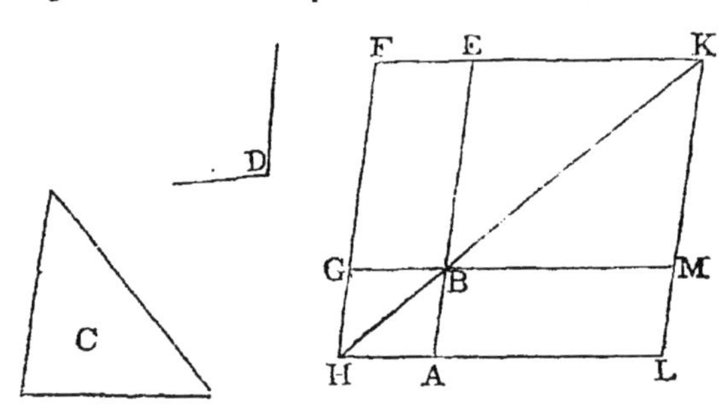

Make the parallelogram $BEFG$ equal to the triangle C, and having the angle EBG equal to the angle D, so that BE may be in the same straight line with AB; [I. 42.
produce FG to H;
through A draw AH parallel to BG or EF, [I. 31.
and join HB.

Then, because the straight line HF falls on the parallels AH, EF, the angles AHF, HFE are together equal to two right angles. [I. 29.
Therefore the angles BHF, HFE are together less than two right angles.
But straight lines which with another straight line make the interior angles on the same side together less than two right angles will meet on that side, if produced far enough. [*Ax.* 12.
Therefore HB and FE will meet if produced;
let them meet at K.
Through K draw KL parallel to EA or FH; [I. 31.
and produce HA, GB to the points L, M.

Then $HLKF$ is a parallelogram, of which the diameter is HK; and AG, ME are parallelograms about HK; and LB, BF are the complements.
Therefore LB is equal to BF. [I. 43.
But BF is equal to the triangle C. [*Construction.*
Therefore LB is equal to the triangle C. [*Axiom* 1.

And because the angle GBE is equal to the angle ABM, [I. 15.
and likewise to the angle D; [*Construction.*
the angle ABM is equal to the angle D. [*Axiom* 1.

Wherefore *to the given straight line AB the parallelogram LB is applied, equal to the triangle C, and having the angle ABM equal to the angle D.* Q.E.F.

PROPOSITION 45. *PROBLEM.*

To describe a parallelogram equal to a given rectilineal figure, and having an angle equal to a given rectilineal angle.

Let $ABCD$ be the given rectilineal figure, and E the given rectilineal angle: it is required to describe a parallelogram equal to $ABCD$, and having an angle equal to E.

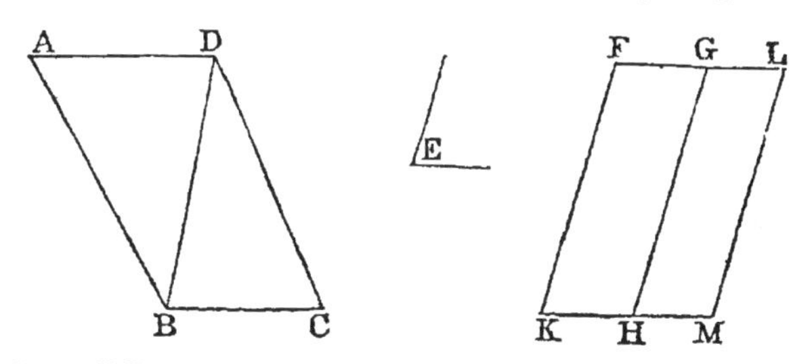

Join DB, and describe the parallelogram FH equal to the triangle ADB, and having the angle FKH equal to the angle E; [I. 42.
and to the straight line GH apply the parallelogram GM equal to the triangle DBC, and having the angle GHM equal to the angle E. [I. 44.
The figure $FKML$ shall be the parallelogram required.

Because the angle E is equal to each of the angles FKH, GHM, [*Construction.*
the angle FKH is equal to the angle GHM. [*Axiom* 1.
Add to each of these equals the angle KHG;
therefore the angles FKH, KHG are equal to the angles KHG, GHM. [*Axiom* 2.
But FKH, KHG are together equal to two right angles; [I. 29.
therefore KHG, GHM are together equal to two right angles.

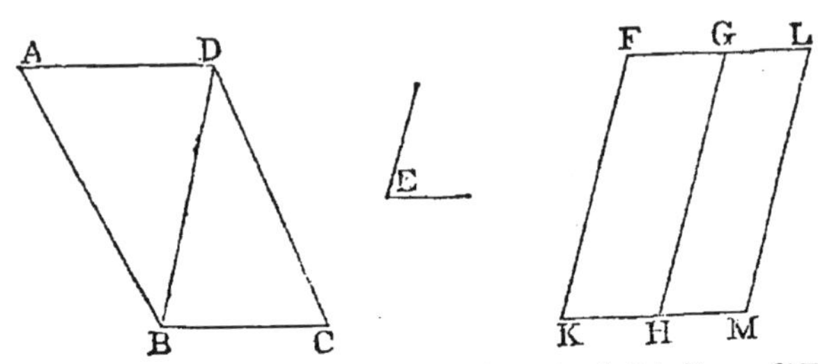

And because at the point H in the straight line GH, the two straight lines KH, HM, on the opposite sides of it, make the adjacent angles together equal to two right angles, KH is in the same straight line with HM. [I. 14.

And because the straight line HG meets the parallels KM, FG, the alternate angles MHG, HGF are equal. [I. 29.

Add to each of these equals the angle HGL;

therefore the angles MHG, HGL, are equal to the angles HGF, HGL. [*Axiom* 2.

But MHG, HGL are together equal to two right angles; [I. 29.

therefore HGF, HGL are together equal to two right angles.

Therefore FG is in the same straight line with GL. [I. 14.

And because KF is parallel to HG, and HG to ML, [*Constr.*

KF is parallel to ML; [I. 30.

and KM, FL are parallels; [*Construction.*

therefore $KFLM$ is a parallelogram. [*Definition.*

And because the triangle ABD is equal to the parallelogram HF, [*Construction.*

and the triangle DBC to the parallelogram GM; [*Constr.*

the whole rectilineal figure $ABCD$ is equal to the whole parallelogram $KFLM$. [*Axiom* 2.

Wherefore, *the parallelogram KFLM has been described equal to the given rectilineal figure ABCD, and having the angle FKM equal to the given angle E.* Q.E.F.

COROLLARY. From this it is manifest, how to a given straight line, to apply a parallelogram, which shall have an angle equal to a given rectilineal angle, and shall be equal to a given rectilineal figure; namely, by applying to the given straight line a parallelogram equal to the first triangle ABD, and having an angle equal to the given angle; and so on. [I. 44.

PROPOSITION 46. *PROBLEM.*

To describe a square on a given straight line.

Let AB be the given straight line: it is required to describe a square on AB.

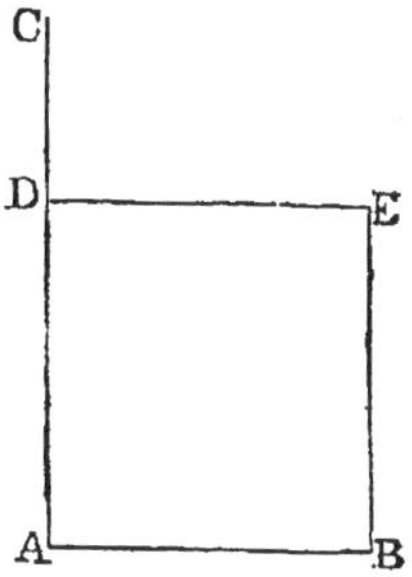

From the point A draw AC at right angles to AB; [I. 11.
and make AD equal to AB; [I. 3.
through D draw DE parallel to AB; and through B draw BE parallel to AD. [I. 31.
$ADEB$ shall be a square.

For $ADEB$ is by construction a parallelogram;
therefore AB is equal to DE, and AD to BE. [I. 34.
But AB is equal to AD. [*Construction.*
Therefore the four straight lines BA, AD, DE, EB are equal to one another, and the parallelogram $ADEB$ is equilateral. [*Axiom* 1.

Likewise all its angles are right angles.
For since the straight line AD meets the parallels AB, DE, the angles BAD, ADE are together equal to two right angles; [I. 29.
but BAD is a right angle; [*Construction.*
therefore also ADE is a right angle. [*Axiom* 3.
But the opposite angles of parallelograms are equal. [I. 34.
Therefore each of the opposite angles ABE, BED is a right angle. [*Axiom* 1.
Therefore the figure $ADEB$ is rectangular;
and it has been shewn to be equilateral.
Therefore *it is a square.* [*Definition* 30.
And it is described on the given straight line AB. Q.E.F.

Corollary. From the demonstration it is manifest that every parallelogram which has one right angle has all its angles right angles.

PROPOSITION 47. *THEOREM.*

In any right-angled triangle, the square which is described on the side subtending the right angle is equal to the squares described on the sides which contain the right angle.

Let ABC be a right-angled triangle, having the right angle BAC: the square described on the side BC shall be equal to the squares described on the sides BA, AC.

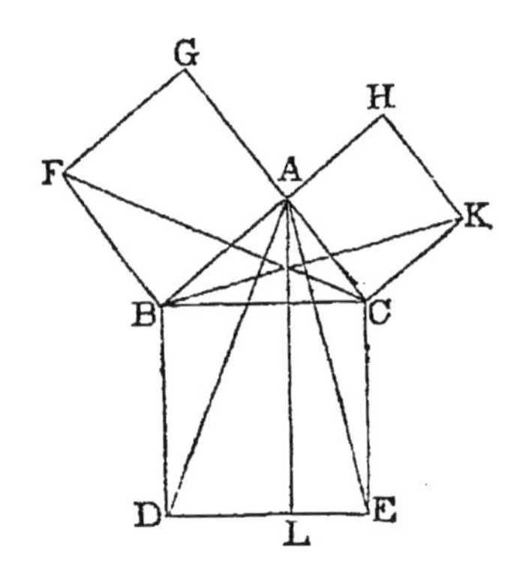

On BC describe the square $BDEC$, and on BA, AC describe the squares GB, HC; [I. 46.
through A draw AL parallel to BD or CE; [I. 31.
and join AD, FC.

Then, because the angle BAC is a right angle, [*Hypothesis.*
and that the angle BAG is also a right angle, [*Definition* 30.
the two straight lines AC, AG, on the opposite sides of AB, make with it at the point A the adjacent angles equal to two right angles;
therefore CA is in the same straight line with AG. [I. 14.
For the same reason, AB and AH are in the same straight line.

Now the angle DBC is equal to the angle FBA, for each of them is a right angle. [*Axiom* 11.
Add to each the angle ABC.
Therefore the whole angle DBA is equal to the whole angle FBC. [*Axiom* 2.
And because the two sides AB, BD are equal to the two sides FB, BC, each to each; [*Definition* 30.
and the angle DBA is equal to the angle FBC;
therefore the triangle ABD is equal to the triangle FBC. [I. 4.

Now the parallelogram BL is double of the triangle ABD, because they are on the same base BD, and between the same parallels BD, AL. [I. 41.

And the square GB is double of the triangle FBC, because they are on the same base FB, and between the same parallels FB, GC. [I. 41.

But the doubles of equals are equal to one another. [*Ax.* 6.

Therefore the parallelogram BL is equal to the square GB.

In the same manner, by joining AE, BK, it can be shewn, that the parallelogram CL is equal to the square CH.

Therefore the whole square $BDEC$ is equal to the two squares GB, HC. [*Axiom* 2.

And the square $BDEC$ is described on BC, and the squares GB, HC on BA, AC.

Therefore the square described on the side BC is equal to the squares described on the sides BA, AC.

Wherefore, *in any right-angled triangle* &c. Q.E.D.

PROPOSITION 48. *THEOREM.*

If the square described on one of the sides of a triangle be equal to the squares described on the other two sides of it, the angle contained by these two sides is a right angle.

Let the square described on BC, one of the sides of the triangle ABC, be equal to the squares described on the other sides BA, AC: the angle BAC shall be a right angle.

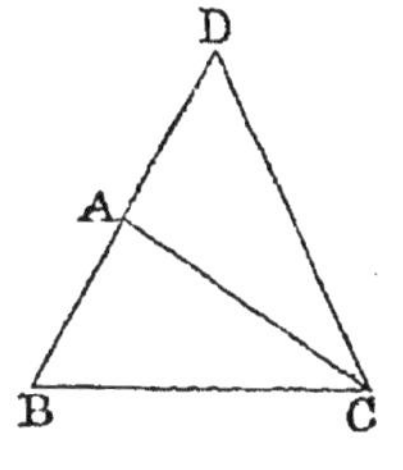

From the point A draw AD at right angles to AC; [I. 11.

and make AD equal to BA; [I. 3.

and join DC.

Then because DA is equal to BA, the square on DA is equal to the square on BA.

To each of these add the square on AC.

Therefore the squares on DA, AC are equal to the squares on BA, AC. [*Axiom* 2.

But because the angle DAC is a right angle, [*Construction.*
the square on DC is equal to the squares on DA, AC. [I. 47.
And, by hypothesis, the square on BC is equal to the squares on BA, AC.
Therefore the square on DC is equal to the square on BC. [*Ax.* 1.
Therefore also the side DC is equal to the side BC.

And because the side DA is equal to the side AB; [*Constr.*
and the side AC is common to the two triangles DAC, BAC;
the two sides DA, AC are equal to the two sides BA, AC, each to each;
and the base DC has been shewn to be equal to the base BC;

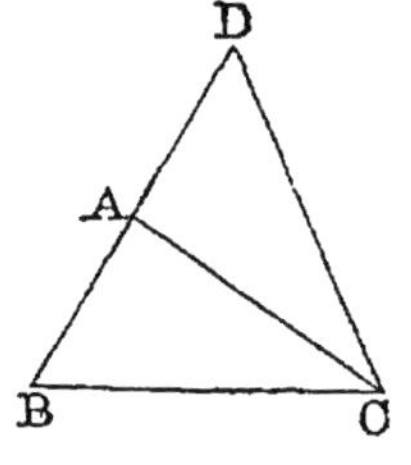

therefore the angle DAC is equal to the angle BAC. [I. 8.
But DAC is a right angle; [*Construction.*
therefore also BAC is a right angle. [*Axiom* 1.

Wherefore, *if the square* &c. Q.E.D.

BOOK II.

DEFINITIONS.

1. EVERY right-angled parallelogram, or rectangle, is said to be contained by any two of the straight lines which contain one of the right angles.

2. In every parallelogram, any of the parallelograms about a diameter, together with the two complements, is called a Gnomon.

Thus the parallelogram *HG*, together with the complements *AF*, *FC*, is the gnomon, which is more briefly expressed by the letters *AGK*, or *EHC*, which are at the opposite angles of the parallelograms which make the gnomon.

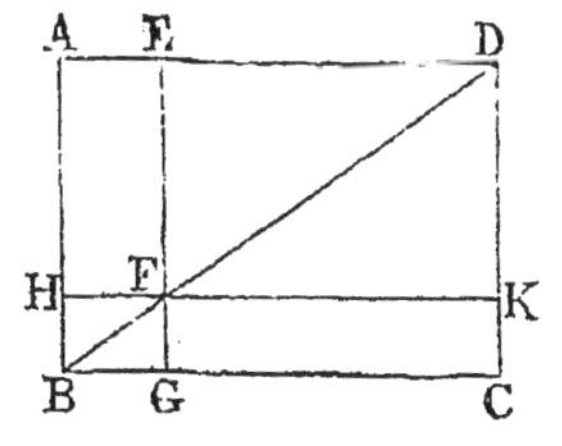

PROPOSITION 1. *THEOREM.*

If there be two straight lines, one of which is divided into any number of parts, the rectangle contained by the two straight lines is equal to the rectangles contained by the undivided line, and the several parts of the divided line.

Let *A* and *BC* be two straight lines; and let *BC* be divided into any number of parts at the points *D*, *E*: the rectangle contained by the straight lines *A*, *BC*, shall be equal to the rectangle contained by *A*, *BD*, together with that contained by *A*, *DE*, and that contained by *A*, *EC*.

From the point *B* draw *BF* at right angles to *BC*; [I. 11.
and make *BG* equal to *A*; [I. 3.
through *G* draw *GH* parallel to *BC*; and through *D*, *E*, *C* draw *DK*, *EL*, *CH*, parallel to *BG*. [I. 31.

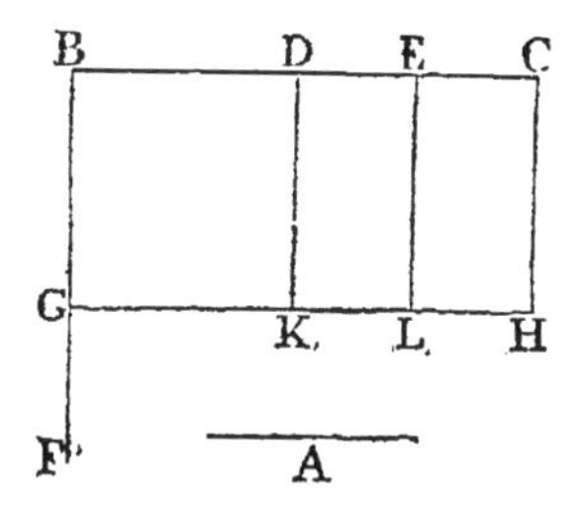

Then the rectangle *BH* is equal to the rectangles *BK*, *DL*, *EH*.

But *BH* is contained by *A*, *BC*, for it is contained by *GB*, *BC*, and *GB* is equal to *A*. [*Construction.*

And *BK* is contained by *A*, *BD*, for it is contained by *GB*, *BD*, and *GB* is equal to *A*;

and *DL* is contained by *A*, *DE*, because *DK* is equal to *BG*, which is equal to *A*; [I. 34.

and in like manner *EH* is contained by *A*, *EC*.

Therefore the rectangle contained by *A*, *BC* is equal to the rectangles contained by *A*, *BD*, and by *A*, *DE*, and by *A*, *EC*.

Wherefore, *if there be two straight lines* &c. Q.E.D.

PROPOSITION 2. *THEOREM.*

If a straight line be divided into any two parts, the rectangles contained by the whole and each of the parts, are together equal to the square on the whole line.

Let the straight line AB be divided into any two parts at the point C: the rectangle contained by AB, BC, together with the rectangle AB, AC, shall be equal to the square on AB.

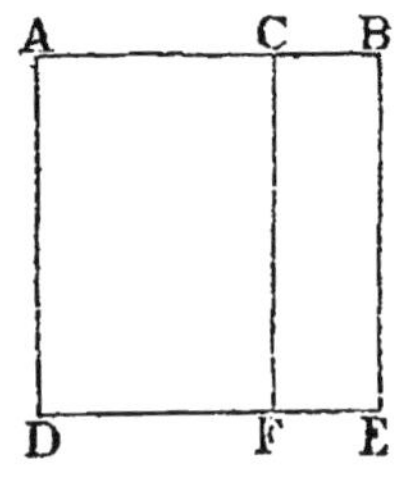

[*Note.* To avoid repeating the word *contained* too frequently, the rectangle contained by two straight lines AB, AC is sometimes simply called the rectangle AB, AC.]

On AB describe the square $ADEB$; [I. 46.

and through C draw CF parallel to AD or BE. [I. 31.

Then AE is equal to the rectangles AF, CE.

But AE is the square on AB.

And AF is the rectangle contained by BA, AC, for it is contained by DA, AC, of which DA is equal to BA;

and CE is contained by AB, BC, for BE is equal to AB.

Therefore the rectangle AB, AC, together with the rectangle AB, BC, is equal to the square on AB.

Wherefore, *if a straight line* &c. Q.E.D.

PROPOSITION 3. *THEOREM.*

If a straight line be divided into any two parts, the rectangle contained by the whole and one of the parts, is equal to the rectangle contained by the two parts, together with the square on the aforesaid part.

Let the straight line AB be divided into any two parts at the point C: the rectangle AB, BC shall be equal to the rectangle AC, CB, together with the square on BC.

On BC describe the square $CDEB$; [I. 46.

produce ED to F, and through A draw AF parallel to CD or BE. [I. 31.

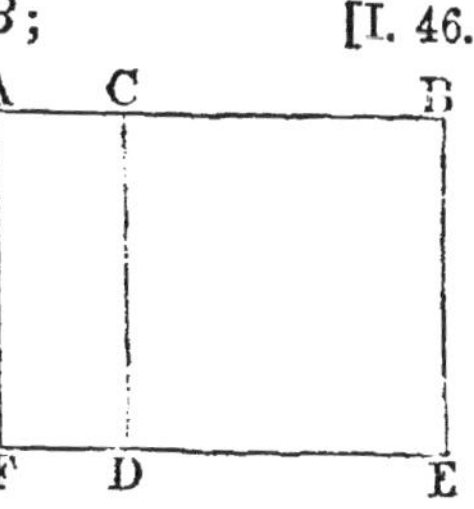

Then the rectangle AE is equal to the rectangles AD, CE.

But AE is the rectangle contained by AB, BC, for it is contained by AB, BE, of which BE is equal to BC;

and AD is contained by AC, CB, for CD is equal to CB; and CE is the square on BC.

Therefore the rectangle AB, BC is equal to the rectangle AC, CB, together with the square on BC.

Wherefore, *if a straight line* &c. Q.E.D.

PROPOSITION 4. *THEOREM.*

If a straight line be divided into any two parts, the square on the whole line is equal to the squares on the two parts, together with twice the rectangle contained by the two parts.

Let the straight line AB be divided into any two parts at the point C: the square on AB shall be equal to the squares on AC, CB, together with twice the rectangle contained by AC, CB.

On AB describe the square $ADEB$; [I. 46.

join BD; through C draw CGF parallel to AD or BE, and through G draw HK parallel to AB or DE. [I. 31.

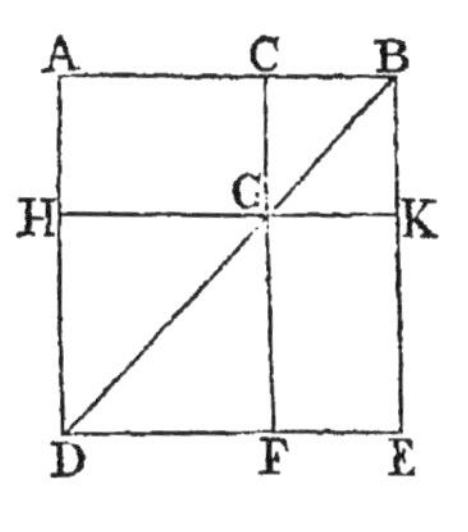

Then, because CF is parallel to AD, and BD falls on them, the exterior angle CGB is equal to the interior and opposite angle ADB; [I. 29.

but the angle ADB is equal to the angle ABD, [I. 5.

because BA is equal to AD, being sides of a square;

therefore the angle CGB is equal to the angle CBG; [*Ax.* 1.

and therefore the side CG is equal to the side CB. [I. 6.

But CB is also equal to GK, and CG to BK; [I. 34.

therefore the figure $CGKB$ is equilateral.

It is likewise rectangular. For since *CG* is parallel to *BK*, and *CB* meets them, the angles *KBC*, *GCB* are together equal to two right angles. [I. 29.

But *KBC* is a right angle. [I. *Definition* 30.

Therefore *GCB* is a right angle. [*Axiom* 3.

And therefore also the angles *CGK*, *GKB* opposite to these are right angles. [I. 34. and *Axiom* 1.

Therefore *CGKB* is rectangular; and it has been shewn to be equilateral; therefore it is a square, and it is on the side *CB*.

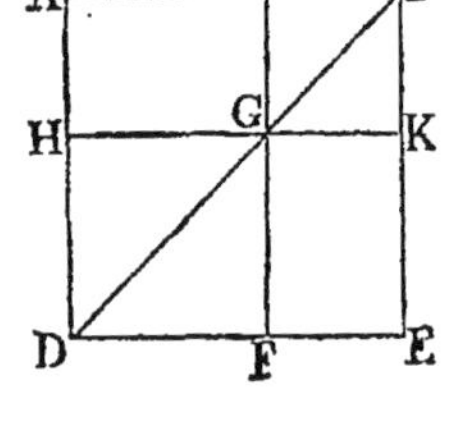

For the same reason *HF* is also a square, and it is on the side *HG*, which is equal to *AC*. [I. 34.

Therefore *HF*, *CK* are the squares on *AC*, *CB*.

And because the complement *AG* is equal to the complement *GE*; [I. 43.

and that *AG* is the rectangle contained by *AC*, *CB*, for *CG* is equal to *CB*;

therefore *GE* is also equal to the rectangle *AC*, *CB*. [*Ax.* 1.

Therefore *AG*, *GE* are equal to twice the rectangle *AC*, *CB*.

And *HF*, *CK* are the squares on *AC*, *CB*.

Therefore the four figures *HF*, *CK*, *AG*, *GE* are equal to the squares on *AC*, *CB*, together with twice the rectangle *AC*, *CB*.

But *HF*, *CK*, *AG*, *GE* make up the whole figure *ADEB*, which is the square on *AB*.

Therefore the square on *AB* is equal to the squares on *AC*, *CB*, together with twice the rectangle *AC*, *CB*.

Wherefore, *if a straight line* &c. Q.E.D.

COROLLARY. From the demonstration it is manifest, that parallellograms about the diameter of a square are likewise squares.

PROPOSITION 5. *THEOREM.*

If a straight line be divided into two equal parts and also into two unequal parts, the rectangle contained by the unequal parts, together with the square on the line between the points of section, is equal to the square on half the line.

Let the straight line AB be divided into two equal parts at the point C, and into two unequal parts at the point D: the rectangle AD, DB, together with the square on CD, shall be equal to the square on CB.

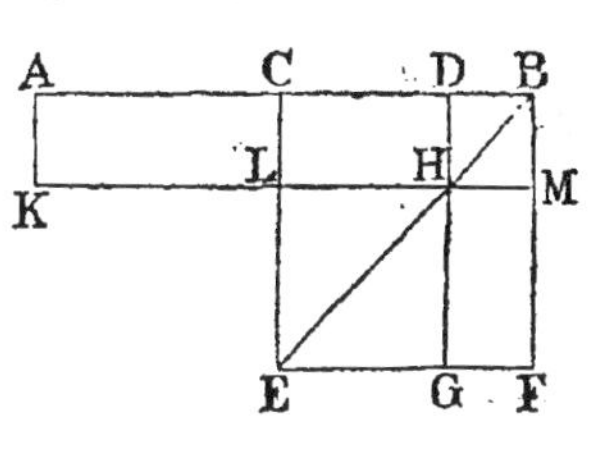

On CB describe the square $CEFB$; [I. 46.
join BE; through D draw DHG parallel to CE or BF; through H draw KLM parallel to CB or EF; and through A draw AK parallel to CL or BM. [I. 31.

Then the complement CH is equal to the complement HF; [I. 43.

to each of these add DM; therefore the whole CM is equal to the whole DF. [*Axiom* 2.

But CM is equal to AL, [I. 36.

because AC is equal to CB. [*Hypothesis.*

Therefore also AL is equal to DF. [*Axiom* 1.

To each of these add CH; therefore the whole AH is equal to DF and CH. [*Axiom* 2.

But AH is the rectangle contained by AD, DB, for DH is equal to DB; [II. 4, *Corollary.*

and DF together with CH is the gnomon CMG;

therefore the gnomon CMG is equal to the rectangle AD, DB.

To each of these add LG, which is equal to the square on CD. [II. 4, *Corollary*, and I. 34.

Therefore the gnomon CMG, together with LG, is equal to the rectangle AD, DB, together with the square on CD. [*Ax.* 2.

But the gnomon CMG and LG make up the whole figure $CEFB$, which is the square on CB.

Therefore the rectangle AD, DB, together with the square on CD, is equal to the square on CB.

Wherefore, *if a straight line* &c. Q.E.D.

From this proposition it is manifest that the difference of the squares on two unequal straight lines AC, CD, is equal to the rectangle contained by their sum and difference.

PROPOSITION 6. *THEOREM.*

If a straight line be bisected, and produced to any point, the rectangle contained by the whole line thus produced, and the part of it produced, together with the square on half the line bisected, is equal to the square on the straight line which is made up of the half and the part produced.

Let the straight line *AB* be bisected at the point *C*, and produced to the point *D*: the rectangle *AD*, *DB*, together with the square on *CB*, shall be equal to the square on *CD*.

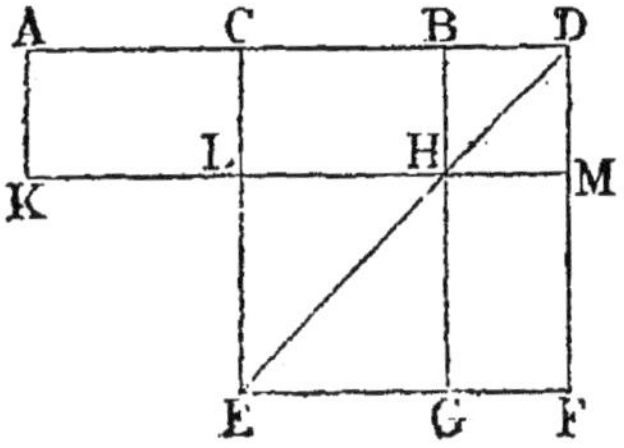

On *CD* describe the square *CEFD*; [I. 46.
join *DE*; through *B* draw *BHG* parallel to *CE* or *DF*; through *H* draw *KLM* parallel to *AD* or *EF*; and through *A* draw *AK* parallel to *CL* or *DM*. [I. 31.

Then, because *AC* is equal to *CB*, [*Hypothesis.*
the rectangle *AL* is equal to the rectangle *CH*; [I. 36.
but *CH* is equal to *HF*; [I. 43.
therefore also *AL* is equal to *HF*. [*Axiom* 1.
To each of these add *CM*;
therefore the whole *AM* is equal to the gnomon *CMG*. [*Ax.* 2.
But *AM* is the rectangle contained by *AD*, *DB*, for *DM* is equal to *DB*. [II. 4, *Corollary.*
Therefore the rectangle *AD*, *DB* is equal to the gnomon *CMG*. [*Axiom* 1.
To each of these add *LG*, which is equal to the square on *CB*. [II. 4, *Corollary*, and I. 34.
Therefore the rectangle *AD*, *DB*, together with the square on *CB*, is equal to the gnomon *CMG* and the figure *LG*.
But the gnomon *CMG* and *LG* make up the whole figure *CEFD*, which is the square on *CD*.
Therefore the rectangle *AD*, *DB*, together with the square on *CB*, is equal to the square on *CD*.

Wherefore, *if a straight line* &c. Q.E.D.

PROPOSITION 7. *THEOREM.*

If a straight line be divided into any two parts, the squares on the whole line, and on one of the parts, are equal to twice the rectangle contained by the whole and that part, together with the square on the other part.

Let the straight line AB be divided into any two parts at the point C: the squares on AB, BC shall be equal to twice the rectangle AB, BC, together with the square on AC.

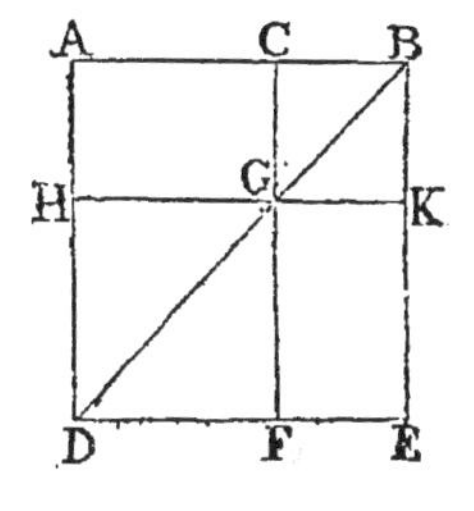

On AB describe the square $ADEB$, and construct the figure as in the preceding propositions.

Then AG is equal to GE; [I. 43.
to each of these add CK;
therefore the whole AK is equal to the whole CE;
therefore AK, CE are double of AK.
But AK, CE are the gnomon AKF, together with the square CK;
therefore the gnomon AKF, together with the square CK, is double of AK.
But twice the rectangle AB, BC is double of AK, for BK is equal to BC. [II. 4, *Corollary.*
Therefore the gnomon AKF, together with the square CK, is equal to twice the rectangle AB, BC.
To each of these equals add HF, which is equal to the square on AC. [II. 4, *Corollary*, and I. 34.
Therefore the gnomon AKF, together with the squares CK, HF, is equal to twice the rectangle AB, BC, together with the square on AC.
But the gnomon AKF together with the squares CK, HF, make up the whole figure $ADEB$ and CK, which are the squares on AB and BC.
Therefore the squares on AB, BC, are equal to twice the rectangle AB, BC, together with the square on AC.

Wherefore, *if a straight line* &c. Q.E.D.

PROPOSITION 8. *THEOREM.*

If a straight line be divided into any two parts, four times the rectangle contained by the whole line and one of the parts, together with the square on the other part, is equal to the square on the straight line which is made up of the whole and that part.

Let the straight line *AB* be divided into any two parts at the point *C*: four times the rectangle *AB*, *BC*, together with the square on *AC*, shall be equal to the square on the straight line made up of *AB* and *BC* together.

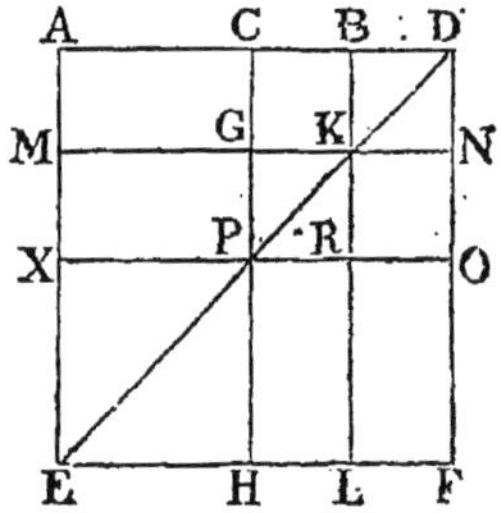

A C B D
M G K N
X P R O
E H L F

Produce *AB* to *D*, so that *BD* may be equal to *CB*; [*Post.* 2. and I. 3.
on *AD* describe the square *AEFD*;
and construct two figures such as in the preceding propositions.

Then, because *CB* is equal to *BD*, [*Construction.*
and that *CB* is equal to *GK*, and *BD* to *KN*, [I. 34.
therefore *GK* is equal to *KN*. [*Axiom* 1.
For the same reason *PR* is equal to *RO*.
And because *CB* is equal to *BD*, and *GK* to *KN*, the rectangle *CK* is equal to the rectangle *BN*, and the rectangle *GR* to the rectangle *RN*. [I. 36.
But *CK* is equal to *RN*, because they are the complements of the parallelogram *CO*; [I. 43.
therefore also *BN* is equal to *GR*. [*Axiom* 1.
Therefore the four rectangles *BN*, *CK*, *GR*, *RN* are equal to one another, and so the four are quadruple of one of them *CK*.

Again, because *CB* is equal to *BD*, [*Construction.*
and that *BD* is equal to *BK*, [II. 4, *Corollary.*
that is to *CG*, [I. 34.
and that *CB* is equal to *GK*, [I. 34.

that is to GP; [II. 4, *Corollary.*
therefore CG is equal to GP. [*Axiom* 1.

And because CG is equal to GP, and PR to RO, the rectangle AG is equal to the rectangle MP, and the rectangle PL to the rectangle RF. [I. 36.

But MP is equal to PL, because they are the complements of the parallelogram ML; [I. 43.

therefore also AG is equal to RF. [*Axiom* 1.

Therefore the four rectangles AG, MP, PL, RF are equal to one another, and so the four are quadruple of one of them AG.

And it was shewn that the four CK, BN, GR and RN are quadruple of CK; therefore the eight rectangles which make up the gnomon AOH are quadruple of AK.

And because AK is the rectangle contained by AB, BC, for BK is equal to BC;

therefore four times the rectangle AB, BC is quadruple of AK.

But the gnomon AOH was shewn to be quadruple of AK.

Therefore four times the rectangle AB, BC is equal to the gnomon AOH. [*Axiom* 1.

To each of these add XH, which is equal to the square on AC. [II. 4, *Corollary*, and I. 34.

Therefore four times the rectangle AB, BC, together with the square on AC, is equal to the gnomon AOH and the square XH.

But the gnomon AOH and the square XH make up the figure $AEFD$, which is the square on AD.

Therefore four times the rectangle AB, BC, together with the square on AC, is equal to the square on AD, that is to the square on the line made of AB and BC together.

Wherefore, *if a straight line* &c. Q.E.D.

PROPOSITION 9. *THEOREM.*

If a straight line be divided into two equal, and also into two unequal parts, the squares on the two unequal parts are together double of the square on half the line and of the square on the line between the points of section.

Let the straight line AB be divided into two equal parts at the point C, and into two unequal parts at the point D: the squares on AD, DB shall be together double of the squares on AC, CD.

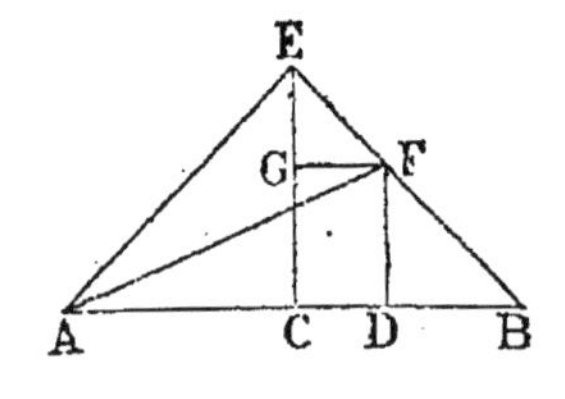

From the point C draw CE at right angles to AB, [I. 11.
and make it equal to AC or CB, [I. 3.
and join EA, EB; through D draw DF parallel to CE, and through F draw FG parallel to BA; [I. 31.
and join AF.

Then, because AC is equal to CE, [*Construction.*
the angle EAC is equal to the angle AEC. [I. 5.
And because the angle ACE is a right angle, [*Construction.*
the two other angles AEC, EAC are together equal to one right angle; [I. 32.
and they are equal to one another;
therefore each of them is half a right angle.
For the same reason each of the angles CEB, EBC is half a right angle.
Therefore the whole angle AEB is a right angle.

And because the angle GEF is half a right angle, and the angle EGF a right angle, for it is equal to the interior and opposite angle ECB; [I. 29.
therefore the remaining angle EFG is half a right angle.
Therefore the angle GEF is equal to the angle EFG, and the side EG is equal to the side GF. [I. 6.
Again, because the angle at B is half a right angle, and the

angle *FDB* a right angle, for it is equal to the interior and opposite angle *ECB*; [I. 29.
therefore the remaining angle *BFD* is half a right angle.
Therefore the angle at *B* is equal to the angle *BFD*, and the side *DF* is equal to the side *DB*. [I. 6.

And because *AC* is equal to *CE*, [*Construction.*
the square on *AC* is equal to the square on *CE*;
therefore the squares on *AC*, *CE* are double of the square on *AC*.
But the square on *AE* is equal to the squares on *AC*, *CE*, because the angle *ACE* is a right angle; [I. 47.
therefore the square on *AE* is double of the square on *AC*.
Again, because *EG* is equal to *GF*, [*Construction.*
the square on *EG* is equal to the square on *GF*;
therefore the squares on *EG*, *GF* are double of the square on *GF*.
But the square on *EF* is equal to the squares on *EG*, *GF*, because the angle *EGF* is a right angle; [I. 47.
therefore the square on *EF* is double of the square on *GF*.
And *GF* is equal to *CD*; [I. 34.
therefore the square on *EF* is double of the square on *CD*.
But it has been shewn that the square on *AE* is also double of the square on *AC*.
Therefore the squares on *AE*, *EF* are double of the squares on *AC*, *CD*.

But the square on *AF* is equal to the squares on *AE*, *EF*, because the angle *AEF* is a right angle. [I. 47.
Therefore the square on *AF* is double of the squares on *AC*, *CD*.
But the squares on *AD*, *DF* are equal to the square on *AF*, because the angle *ADF* is a right angle. [I. 47.
Therefore the squares on *AD*, *DF* are double of the squares on *AC*, *CD*.
And *DF* is equal to *DB*;
therefore the squares on *AD*, *DB* are double of the squares on *AC*, *CD*.

Wherefore, *if a straight line* &c. Q.E.D.

PROPOSITION 10. *THEOREM.*

If a straight line be bisected, and produced to any point, the square on the whole line thus produced, and the square on the part of it produced, are together double of the square on half the line bisected and of the square on the line made up of the half and the part produced.

Let the straight line *AB* be bisected at *C*, and produced to *D*: the squares on *AD*, *DB* shall be together double of the squares on *AC*, *CD*.

From the point *C* draw *CE* at right angles to *AB*, [I. 11.
and make it equal to *AC* or *CB*; [I. 3.
and join *AE*, *EB*; through *E* draw *EF* parallel to *AB*, and through *D* draw *DF* parallel to *CE*. [I. 31.

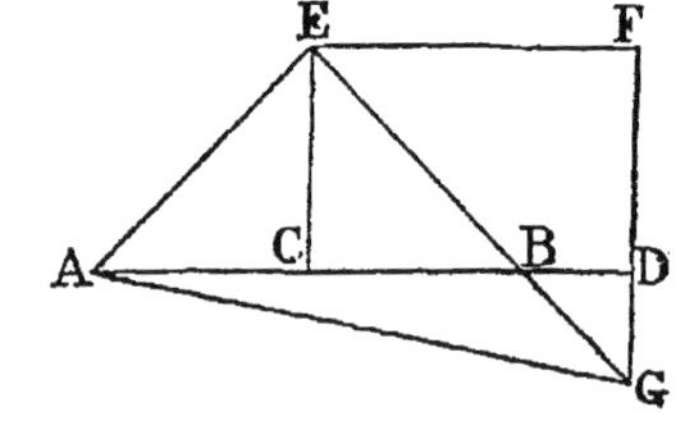

Then because the straight line *EF* meets the parallels *EC*, *FD*, the angles *CEF*, *EFD* are together equal to two right angles; [I. 29.
and therefore the angles *BEF*, *EFD* are together less than two right angles.
Therefore the straight lines *EB*, *FD* will meet, if produced, towards *B*, *D*. [*Axiom* 12.
Let them meet at *G*, and join *AG*.

Then because *AC* is equal to *CE*, [*Construction.*
the angle *CEA* is equal to the angle *EAC*; [I. 5.
and the angle *ACE* is a right angle; [*Construction.*
therefore each of the angles *CEA*, *EAC* is half a right angle. [I. 32.
For the same reason each of the angles *CEB*, *EBC* is half a right angle.
Therefore the angle *AEB* is a right angle.

And because the angle *EBC* is half a right angle, the angle *DBG* is also half a right angle, for they are vertically opposite; [I. 15.
but the angle *BDG* is a right angle, because it is equal to the alternate angle *DCE*; [I. 29.
therefore the remaining angle *DGB* is half a right angle, [I. 32.

and is therefore equal to the angle DBG;
therefore also the side BD is equal to the side DG. [I. 6.
Again, because the angle EGF is half a right angle, and the angle at F a right angle, for it is equal to the opposite angle ECD; [I. 34.
therefore the remaining angle FEG is half a right angle, [I. 32.
and is therefore equal to the angle EGF;
therefore also the side GF is equal to the side FE. [I. 6.

And because EC is equal to CA, the square on EC is equal to the square on CA;
therefore the squares on EC, CA are double of the square on CA.
But the square on AE is equal to the squares on EC, CA. [I. 47.
Therefore the square on AE is double of the square on AC.
Again, because GF is equal to FE, the square on GF is equal to the square on FE;
therefore the squares on GF, FE are double of the square on FE.
But the square on EG is equal to the squares on GF, FE. [I. 47.
Therefore the square on EG is double of the square on FE.
And FE is equal to CD; [I. 34.
therefore the square on EG is double of the square on CD.
But it has been shewn that the square on AE is double of the square on AC.
Therefore the squares on AE, EG are double of the squares on AC, CD.

But the square on AG is equal to the squares on AE, EG. [I. 47.
Therefore the square on AG is double of the squares on AC, CD.
But the squares on AD, DG are equal to the square on AG. [I. 47.
Therefore the squares on AD, DG are double of the squares on AC, CD.
And DG is equal to DB;
therefore the squares on AD, DB are double of the squares on AC, CD.

Wherefore, *if a straight line &c.* Q.E.D.

PROPOSITION 11. *PROBLEM.*

To divide a given straight line into two parts, so that the rectangle contained by the whole and one of the parts may be equal to the square on the other part.

Let *AB* be the given straight line: it is required to divide it into two parts, so that the rectangle contained by the whole and one of the parts may be equal to the square on the other part.

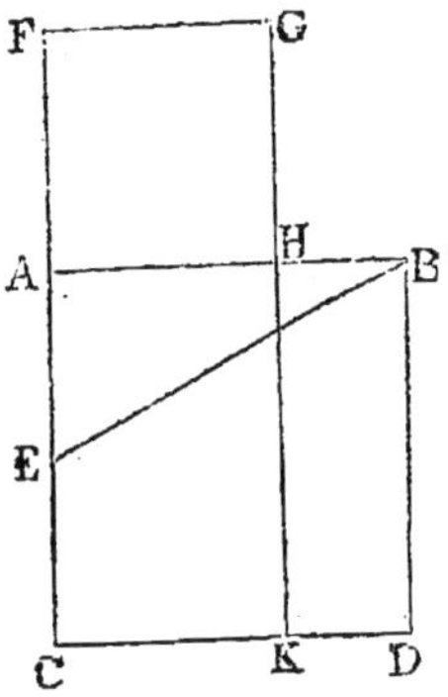

On *AB* describe the square *ABDC*; [I. 46.

bisect *AC* at *E*; [I. 10.

join *BE*; produce *CA* to *F*, and make *EF* equal to *EB*; [I. 3.

and on *AF* describe the square *AFGH*. [I. 46.

AB shall be divided at *H* so that the rectangle *AB*, *BH* is equal to the square on *AH*.

F G H A B E C K D

Produce *GH* to *K*.

Then, because the straight line *AC* is bisected at *E*, and produced to *F*, the rectangle *CF*, *FA*, together with the square on *AE*, is equal to the square on *EF*. [II. 6.

But *EF* is equal to *EB*. [*Construction.*

Therefore the rectangle *CF*, *FA*, together with the square on *AE*, is equal to the square on *EB*.

But the square on *EB* is equal to the squares on *AE*, *AB*, because the angle *EAB* is a right angle. [I. 47.

Therefore the rectangle *CF*, *FA*, together with the square on *AE*, is equal to the squares on *AE*, *AB*.

Take away the square on *AE*, which is common to both; therefore the remainder, the rectangle *CF*, *FA*, is equal to the square on *AB*. [*Axiom* 3.

But the figure *FK* is the rectangle contained by *CF*, *FA*, for *FG* is equal to *FA*;

and *AD* is the square on *AB*;

therefore *FK* is equal to *AD*.

Take away the common part *AK*, and the remainder *FH* is equal to the remainder *HD*. [*Axiom* 3.

But *HD* is the rectangle contained by *AB*, *BH*, for *AB* is equal to *BD*;
and *FH* is the square on *AH*;
therefore the rectangle *AB*,*BH* is equal to the square on *AH*.

Wherefore *the straight line AB is divided at H, so that the rectangle AB, BH is equal to the square on AH.* Q.E.F.

PROPOSITION 12. *THEOREM.*

In obtuse-angled triangles, if a perpendicular be drawn from either of the acute angles to the opposite side produced, the square on the side subtending the obtuse angle is greater than the squares on the sides containing the obtuse angle, by twice the rectangle contained by the side on which, when produced, the perpendicular falls, and the straight line intercepted without the triangle, between the perpendicular and the obtuse angle.

Let *ABC* be an obtuse-angled triangle, having the obtuse angle *ACB*, and from the point *A* let *AD* be drawn perpendicular to *BC* produced: the square on *AB* shall be greater than the squares on *AC*, *CB*, by twice the rectangle *BC*, *CD*.

A

B C D

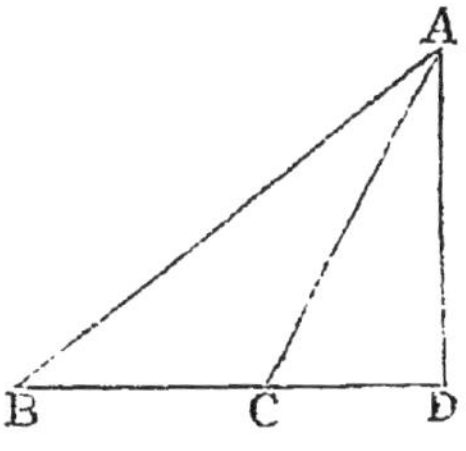

Because the straight line *BD* is divided into two parts at the point *C*, the square on *BD* is equal to the squares on *BC*, *CD*, and twice the rectangle *BC*, *CD*. [II. 4.

To each of these equals add the square on *DA*.

Therefore the squares on *BD*, *DA* are equal to the squares on *BC*, *CD*, *DA*, and twice the rectangle *BC*, *CD*. [*Axiom* 2.

But the square on *BA* is equal to the squares on *BD*, *DA*, because the angle at *D* is a right angle; [I. 47.

and the square on *CA* is equal to the squares on *CD*, *DA*. [I. 47.

Therefore the square on *BA* is equal to the squares on *BC*, *CA*, and twice the rectangle *BC*, *CD*;

that is, the square on *BA* is greater than the squares on *BC*, *CA* by twice the rectangle *BC*, *CD*.

Wherefore, *in obtuse-angled triangles* &c. Q.E.D.

PROPOSITION 13. *THEOREM.*

In every triangle, the square on the side subtending an acute angle, is less than the squares on the sides containing that angle, by twice the rectangle contained by either of these sides, and the straight line intercepted between the perpendicular let fall on it from the opposite angle, and the acute angle.

Let ABC be any triangle, and the angle at B an acute angle; and on BC one of the sides containing it, let fall the perpendicular AD from the opposite angle: the square on AC, opposite to the angle B, shall be less than the squares on CB, BA, by twice the rectangle CB, BD.

First, let AD fall within the triangle ABC.

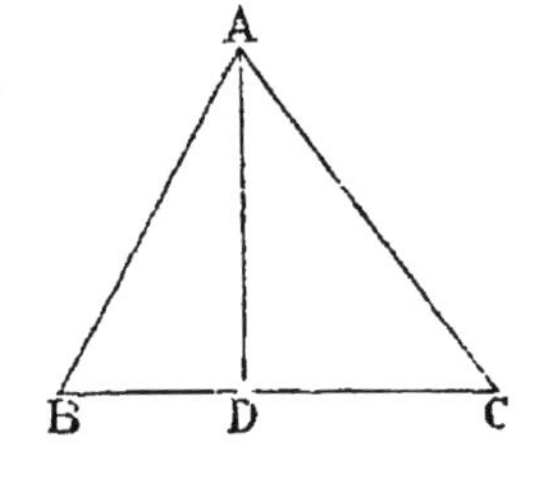

Then, because the straight line CB is divided into two parts at the point D, the squares on CB, BD are equal to twice the rectangle contained by CB, BD and the square on CD. [II. 7.

To each of these equals add the square on DA.

Therefore the squares on CB, BD, DA are equal to twice the rectangle CB, BD and the squares on CD, DA. [*Ax.* 2.

But the square on AB is equal to the squares on BD, DA, because the angle BDA is a right angle; [I. 47.

and the square on AC is equal to the squares on CD, DA. [I. 47.

Therefore the squares on CB, BA are equal to the square on AC and twice the rectangle CB, BD;

that is, the square on AC alone is less than the squares on CB, BA by twice the rectangle CB, BD.

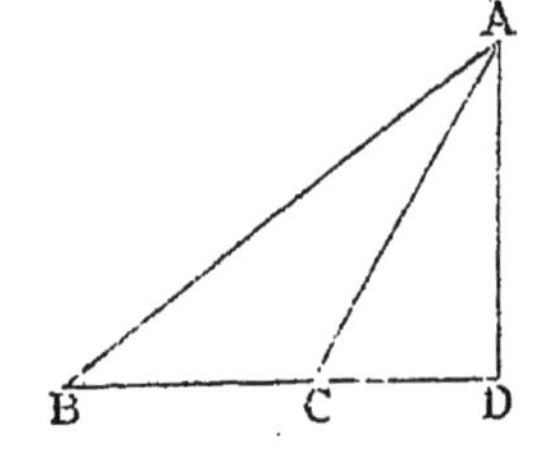

Secondly, let AD fall without the triangle ABC.

Then because the angle at D is a right angle, [*Construction.*

the angle ACB is greater than a right angle; [I. 16.

and therefore the square on AB is equal to the squares on AC, CB, and twice the rectangle BC, CD. [II. 12.

To each of these equals add the square on BC.

Therefore the squares on AB, BC are equal to the square on AC, and twice the square on BC, and twice the rectangle BC, CD. [*Axiom* 2.

But because BD is divided into two parts at C, the rectangle DB, BC is equal to the rectangle BC, CD and the square on BC; [II. 3.

and the doubles of these are equal,

that is, twice the rectangle DB, BC is equal to twice the rectangle BC, CD and twice the square on BC.

Therefore the squares on AB, BC are equal to the square on AC, and twice the rectangle DB, BC;

that is, the square on AC alone is less than the squares on AB, BC by twice the rectangle DB, BC.

Lastly, let the side AC be perpendicular to BC.

Then BC is the straight line between the perpendicular and the acute angle at B;

and it is manifest, that the squares on AB, BC are equal to the square on AC, and twice the square on BC. [I. 47 and *Ax.* 2.

Wherefore, *in every triangle* &c. Q.E.D.

PROPOSITION 14. *PROBLEM.*

To describe a square that shall be equal to a given rectilineal figure.

Let A be the given rectilineal figure: it is required to describe a square that shall be equal to A.

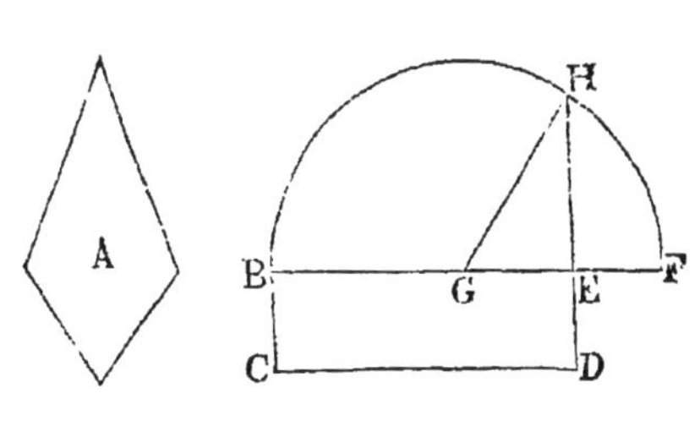

Describe the rectangular parallelogram $BCDE$ equal to the rectilineal figure A. [I. 45.

Then if the sides of it, BE, ED are equal to one another, it is a square, and what was required is now done.

But if they are not equal, produce one of them *BE* to *F*, make *EF* equal to *ED*, [I. 3. and bisect *BF* at *G*; [I. 10. from the centre *G*, at the distance *GB*, or *GF*, describe the semicircle *BHF*, and produce *DE* to *H*.

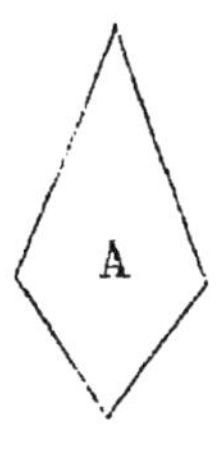

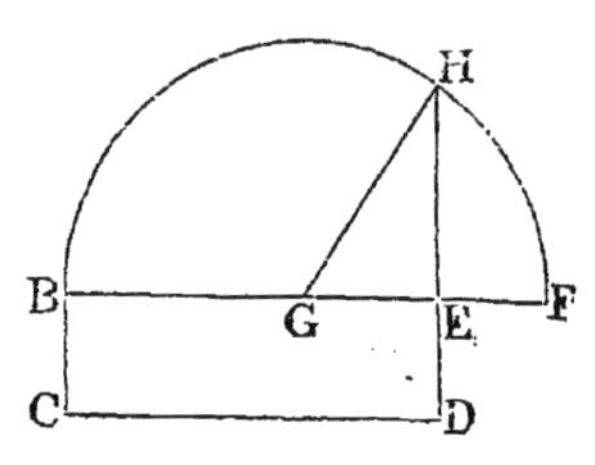

The square described on *EH* shall be equal to the given rectilineal figure *A*.

Join *GH*. Then, because the straight line *BF* is divided into two equal parts at the point *G*, and into two unequal parts at the point *E*, the rectangle *BE*, *EF*, together with the square on *GE*, is equal to the square on *GF*. [II. 5.

But *GF* is equal to *GH*.

Therefore the rectangle *BE*, *EF*, together with the square on *GE*, is equal to the square on *GH*.

But the square on *GH* is equal to the squares on *GE*, *EH*; [I. 47.

therefore the rectangle *BE*, *EF*, together with the square on *GE*, is equal to the squares on *GE*, *EH*.

Take away the square on *GE*, which is common to both;

therefore the rectangle *BE*, *EF* is equal to the square on *EH*. [*Axiom* 3.

But the rectangle contained by *BE*, *EF* is the parallelogram *BD*,

because *EF* is equal to *ED*. [*Construction.*

Therefore *BD* is equal to the square on *EH*.

But *BD* is equal to the rectilineal figure *A*. [*Construction.*

Therefore the square on *EH* is equal to the rectilineal figure *A*.

Wherefore *a square has been made equal to the given rectilineal figure A, namely, the square described on EH.* Q.E.F.

BOOK III.

DEFINITIONS.

1. EQUAL circles are those of which the diameters are equal, or from the centres of which the straight lines to the circumferences are equal.

This is not a definition, but a theorem, the truth of which is evident; for, if the circles be applied to one another, so that their centres coincide, the circles must likewise coincide, since the straight lines from the centres are equal.

2. A straight line is said to touch a circle, when it meets the circle, and being produced does not cut it.

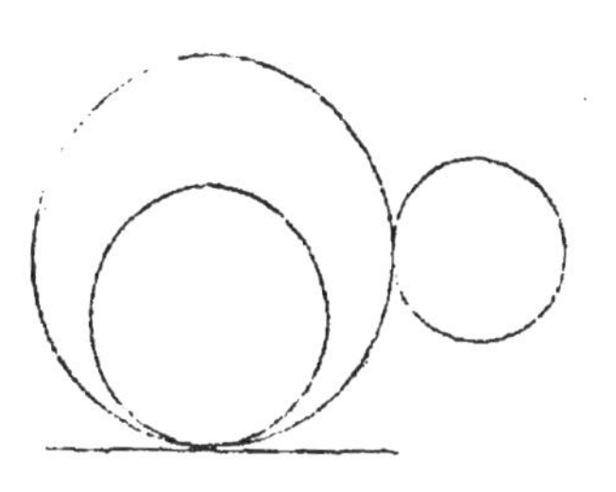

3. Circles are said to touch one another, which meet but do not cut one another.

4. Straight lines are said to be equally distant from the centre of a circle, when the perpendiculars drawn to them from the centre are equal.

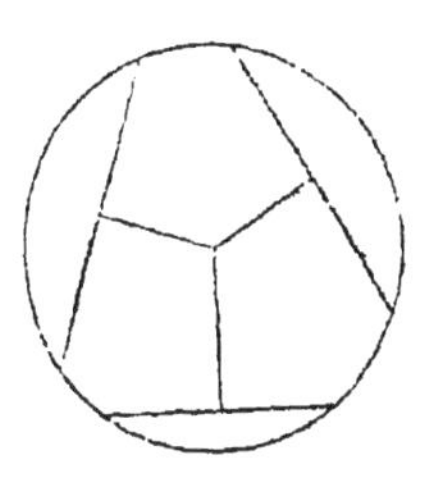

5. And the straight line on which the greater perpendicular falls, is said to be farther from the centre.

6. A segment of a circle is the figure contained by a straight line and the circumference it cuts off.

7. The angle of a segment is that which is contained by the straight line and the circumference.

8. An angle in a segment is the angle contained by two straight lines drawn from any point in the circumference of the segment to the extremities of the straight line which is the base of the segment.

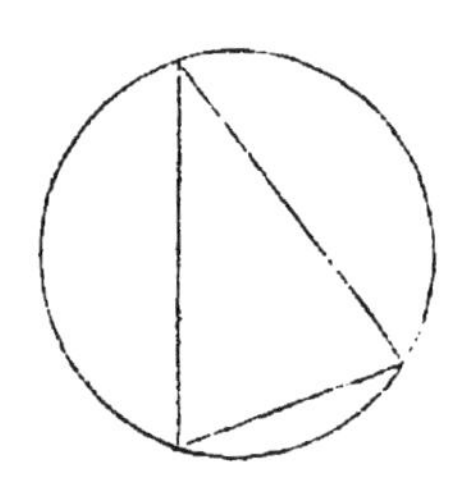

9. And an angle is said to insist or stand on the circumference intercepted between the straight lines which contain the angle.

10. A sector of a circle is the figure contained by two straight lines drawn from the centre, and the circumference between them.

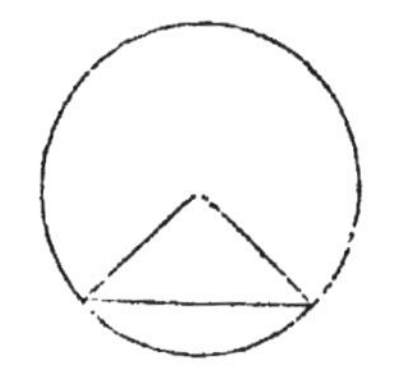

11. Similar segments of circles are those in which the angles are equal, or which contain equal angles.

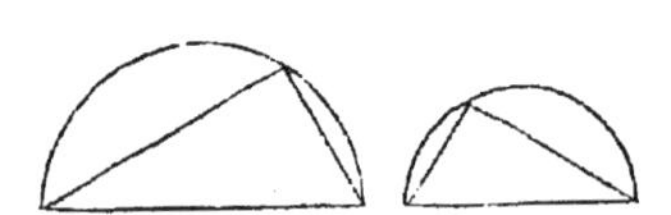

[*Note.* In the following propositions, whenever the expression "straight lines from the centre," or "drawn from the centre," occurs, it is to be understood that the lines are drawn to the circumference.

Any portion of the circumference is called an *arc.*]

PROPOSITION 1. *PROBLEM.*

To find the centre of a given circle.

Let ABC be the given circle: it is required to find its centre.

Draw within it any straight line AB, and bisect AB at D; [I. 10.
from the point D draw DC at right angles to AB; [I. 11.
produce CD to meet the circumference at E, and bisect CE at F. [I. 10.
The point F shall be the centre of the circle ABC.

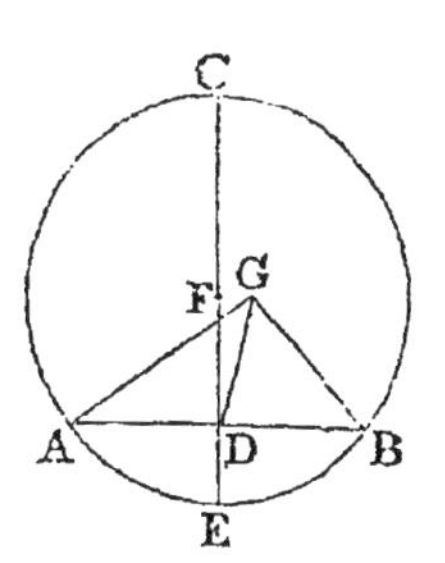

For if F be not the centre, if possible, let G be the centre; and join GA, GD, GB.
Then, because DA is equal to DB, [*Construction.*
and DG is common to the two triangles ADG, BDG;
the two sides AD, DG are equal to the two sides BD, DG, each to each;
and the base GA is equal to the base GB, because they are drawn from the centre G; [I. *Definition* 15.
therefore the angle ADG is equal to the angle BDG. [I. 8.
But when a straight line, standing on another straight line, makes the adjacent angles equal to one another, each of the angles is called a right angle; [I. *Definition* 10.
therefore the angle BDG is a right angle.
But the angle BDF is also a right angle. [*Construction.*
Therefore the angle BDG is equal to the angle BDF, [*Ax.* 11.
the less to the greater; which is impossible.
Therefore G is not the centre of the circle ABC.

In the same manner it may be shewn that no other point out of the line CE is the centre;
and since CE is bisected at F, any other point in CE divides it into unequal parts, and cannot be the centre.
Therefore no point but F is the centre;
that is, F is the centre of the circle ABC:
which was to be found.

Corollary. From this it is manifest, that if in a circle a straight line bisect another at right angles, the centre of the circle is in the straight line which bisects the other.

PROPOSITION 2. *THEOREM.*

If any two points be taken in the circumference of a circle, the straight line which joins them shall fall within the circle.

Let ABC be a circle, and A and B any two points in the circumference: the straight line drawn from A to B shall fall within the circle.

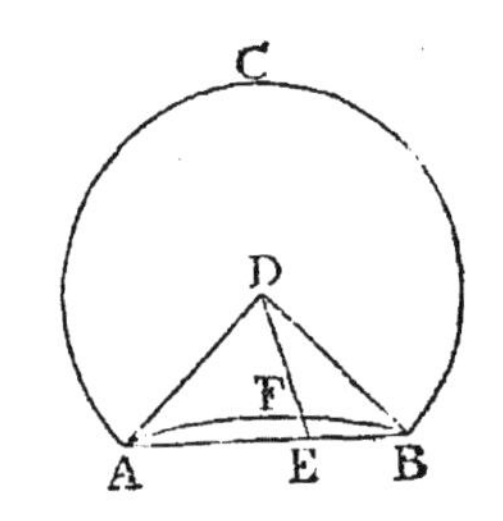

For if it do not, let it fall, if possible, without, as AEB. Find D the centre of the circle ABC; [III. 1.
and join DA, DB; in the arc AB take any point F, join DF, and produce it to meet the straight line AB at E.

Then, because DA is equal to DB, [I. *Definition* 15.
the angle DAB is equal to the angle DBA. [I. 5.
And because AE, a side of the triangle DAE, is produced to B, the exterior angle DEB is greater than the interior opposite angle DAE. [I. 16.
But the angle DAE was shewn to be equal to the angle DBE;
therefore the angle DEB is greater than the angle DBE.
But the greater angle is subtended by the greater side; [I. 19.
therefore DB is greater than DE.
But DB is equal to DF; [I. *Definition* 15.
therefore DF is greater than DE, the less than the greater; which is impossible.
Therefore the straight line drawn from A to B does not fall without the circle.

In the same manner it may be shewn that it does not fall on the circumference.
Therefore it falls within it.

Wherefore, *if any two points* &c. Q.E.D.

PROPOSITION 3. *THEOREM.*

*If a straight line drawn through the centre of a circle, bisect a straight line in it which does not pass through the *centre, it shall cut it at right angles; and if it cut it at right angles it shall bisect it.*

Let *ABC* be a circle; and let *CD*, a straight line drawn through the centre, bisect any straight line *AB*, which does not pass through the centre, at the point *F*: *CD* shall cut *AB* at right angles.

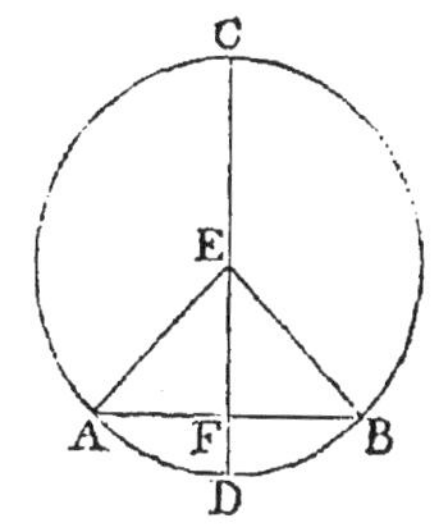

Take *E* the centre of the circle; and join *EA*, *EB*. [III. 1.

Then, because *AF* is equal to *FB*, [*Hypothesis.*

and *FE* is common to the two triangles *AFE*, *BFE*;

the two sides *AF*, *FE* are equal to the two sides *BF*, *FE*, each to each;

and the base *EA* is equal to the base *EB*; [I. *Def.* 15.

therefore the angle *AFE* is equal to the angle *BFE*. [I. 8.

But when a straight line, standing on another straight line, makes the adjacent angles equal to one another, each of the angles is called a right angle; [I. *Definition* 10.

therefore each of the angles *AFE*, *BFE* is a right angle.

Therefore the straight line *CD*, drawn through the centre, bisecting another *AB* which does not pass through the centre, also cuts it at right angles.

But let *CD* cut *AB* at right angles: *CD* shall also bisect *AB*; that is, *AF* shall be equal to *FB*.

The same construction being made, because *EA*, *EB*, drawn from the centre, are equal to one another, [I. *Def.* 15.

the angle *EAF* is equal to the angle *EBF*. [I. 5.

And the right angle *AFE* is equal to the right angle *BFE*.

Therefore in the two triangles *EAF*, *EBF*, there are two angles in the one equal to two angles in the other, each to each;

and the side *EF*, which is opposite to one of the equal angles in each, is common to both;

therefore their other sides are equal; [I. 26.

therefore *AF* is equal to *FB*.

Wherefore, *if a straight line* &c. Q.E.D.

PROPOSITION 4. *THEOREM.*

If in a circle two straight lines cut one another, which do not pass through the centre, they do not bisect one another.

Let $ABCD$ be a circle, and AC, BD two straight lines in it, which cut one another at the point E, and do not both pass through the centre: AC, BD shall not bisect one another.

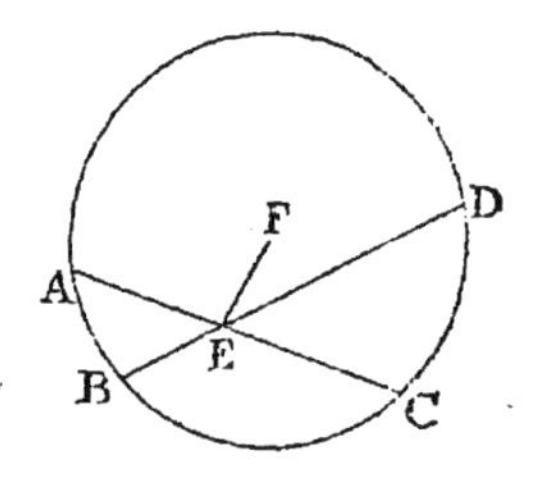

If one of the straight lines pass through the centre it is plain that it cannot be bisected by the other which does not pass through the centre.

But if neither of them pass through the centre, if possible, let AE be equal to EC, and BE equal to ED.

Take F the centre of the circle [III. 1.

and join EF.

Then, because FE, a straight line drawn through the centre, bisects another straight line AC which does not pass through the centre; [*Hypothesis.*

FE cuts AC at right angles; [III. 3.

therefore the angle FEA is a right angle.

Again, because the straight line FE bisects the straight line BD, which does not pass through the centre, [*Hyp.*

FE cuts BD at right angles; [III. 3.

therefore the angle FEB is a right angle.

But the angle FEA was shewn to be a right angle;

therefore the angle FEA is equal to the angle FEB, [*Ax.* 11.

the less to the greater; which is impossible.

Therefore AC, BD do not bisect each other.

Wherefore, *if in a circle* &c. Q.E.D.

PROPOSITION 5. *THEOREM.*

If two circles cut one another, they shall not have the same centre.

Let the two circles ABC, CDG cut one another at the

points B, C: they shall not have the same centre.

For, if it be possible, let E be their centre; join EC, and draw any straight line EFG meeting the circumferences at F and G.

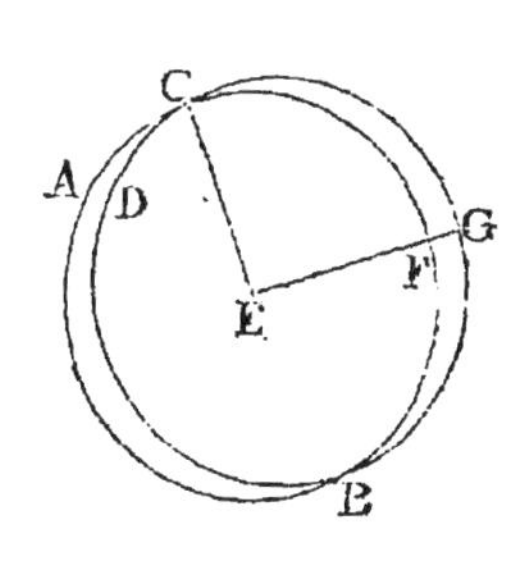

Then, because E is the centre of the circle ABC, EC is equal to EF. [I. *Definition* 15.
Again, because E is the centre of the circle CDG, EC is equal to EG. [I. *Definition* 15.
But EC was shewn to be equal to EF;
therefore EF is equal to EG, [*Axiom* 1.
the less to the greater; which is impossible.
Therefore E is not the centre of the circles ABC, CDG.

Wherefore, *if two circles* &c. Q.E.D.

PROPOSITION 6. *THEOREM.*

If two circles touch one another internally, they shall not have the same centre.

Let the two circles ABC, CDE touch one another internally at the point C: they shall not have the same centre.

For, if it be possible, let F be their centre; join FC, and draw any straight line FEB, meeting the circumferences at E and B.

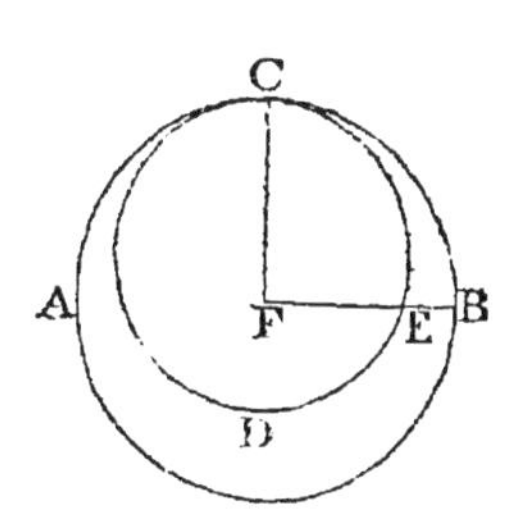

Then, because F is the centre of the circle ABC, FC is equal to FB. [I. *Def.* 15.
Again, because F is the centre of the circle CDE, FC is equal to FE. [I. *Definition* 15.
But FC was shewn to be equal to FB;
therefore FE is equal to FB, [*Axiom* 1.
the less to the greater; which is impossible.
Therefore F is not the centre of the circles ABC, CDE.

Wherefore, *if two circles* &c. Q.E.D.

PROPOSITION 7. *THEOREM.*

If any point be taken in the diameter of a circle which is not the centre, of all the straight lines which can be drawn from this point to the circumference, the greatest is that in which the centre is, and the other part of the diameter is the least; and, of any others, that which is nearer to the straight line which passes through the centre, is always greater than one more remote; and from the same point there can be drawn to the circumference two straight lines, and only two, which are equal to one another, one on each side of the shortest line.

Let *ABCD* be a circle and *AD* its diameter, in which let any point *F* be taken which is not the centre; let *E* be the centre: of all the straight lines *FB*, *FC*, *FG*, &c. that can be drawn from *F* to the circumference, *FA*, which passes through *E*, shall be the greatest, and *FD*, the other part of the diameter *AD*, shall be the least; and of the others *FB* shall be greater than *FC*, and *FC* than *FG*.

Join *BE*, *CE*, *GE*.

Then, because any two sides of a triangle are greater than the third side, [I. 20.

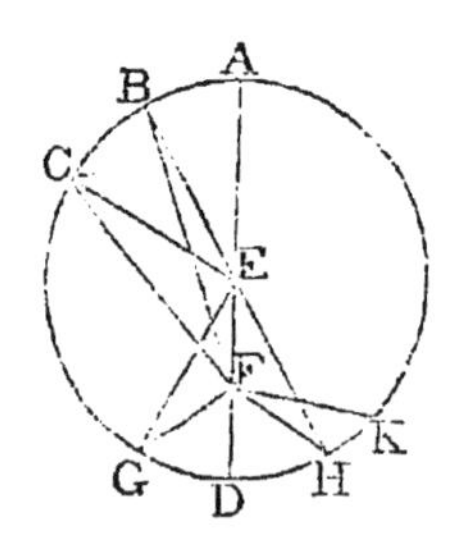

therefore *BE*, *EF* are greater than *BF*.

But *BE* is equal to *AE*; [I. *Def.* 15.

therefore *AE*, *EF* are greater than *BF*,

that is, *AF* is greater than *BF*.

Again, because *BE* is equal to *CE*, [I. *Definition* 15.

and *EF* is common to the two triangles *BEF*, *CEF*;

the two sides *BE*, *EF* are equal to the two sides *CE*, *EF*, each to each;

but the angle *BEF* is greater than the angle *CEF*;

therefore the base *FB* is greater than the base *FC*. [I. 24.

In the same manner it may be shewn that *FC* is greater than *FG*.

Again, because *GF*, *FE* are greater than *EG*, [I. 20.

and that EG is equal to ED; [I. *Definition* 15.
therefore GF, FE are greater than ED.
Take away the common part FE, and the remainder GF is greater than the remainder FD.

Therefore FA is the greatest, and FD the least of all the straight lines from F to the circumference; and FB is greater than FC, and FC than FG.

Also, there can be drawn two equal straight lines from the point F to the circumference, one on each side of the shortest line FD.

For, at the point E, in the straight line EF, make the angle FEH equal to the angle FEG, [I. 23.
and join FH.

Then, because EG is equal to EH, [I. *Definition* 15.
and EF is common to the two triangles GEF, HEF;
the two sides EG, EF are equal to the two sides EH, EF, each to each;
and the angle GEF is equal to the angle HEF; [*Constr.*
therefore the base FG is equal to the base FH. [I. 4.

But, besides FH, no other straight line can be drawn from F to the circumference, equal to FG.

For, if it be possible, let FK be equal to FG.
Then, because FK is equal to FG, [*Hypothesis.*
and FH is also equal to FG,
therefore FH is equal to FK; [*Axiom* 1.
that is, a line nearer to that which passes through the centre is equal to a line which is more remote;
which is impossible by what has been already shewn.

Wherefore, *if any point be taken* &c. Q.E.D.

PROPOSITION 8. *THEOREM.*

If any point be taken without a circle, and straight lines be drawn from it to the circumference, one of which passes through the centre; of those which fall on the concave circumference, the greatest is that which passes through the centre, and of the rest, that which is nearer to the one passing through the centre is always greater than one more remote; but of those which fall on the convex circumference, the least is that between the point without the circle and the diameter; and of the rest, that which is nearer to the least is always less than one more remote; and from the same point there can be drawn to the circumference two straight lines, and only two, which are equal to one another, one on each side of the shortest line.

Let ABC be a circle, and D any point without it, and from D let the straight lines DA, DE, DF, DC be drawn to the circumference, of which DA passes through the centre: of those which fall on the concave circumference $AEFC$, the greatest shall be DA which passes through the centre, and the nearer to it shall be greater than the more remote, namely, DE greater than DF, and DF greater than DC; but of those which fall on the convex circumference $GKLH$, the least shall be DG between the point D and the diameter AG, and the nearer to it shall be less than the more remote, namely, DK less than DL, and DL less than DH.

Take M, the centre of the circle ABC, [III. 1.

and join ME, MF, MC, MH, ML, MK.

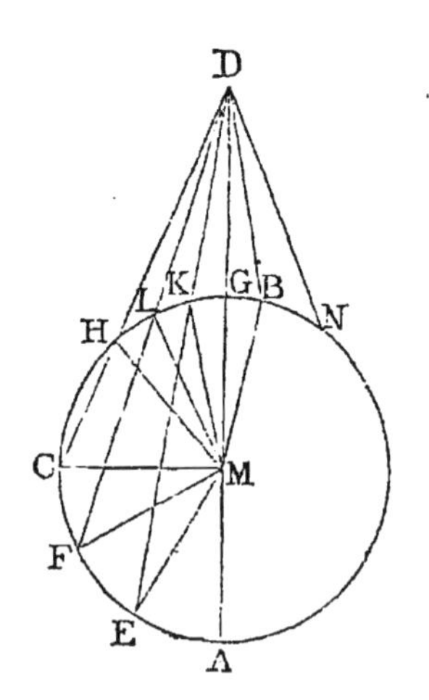

Then, because any two sides of a triangle are greater than the third side, [I. 20.

therefore EM, MD are greater than ED.

But EM is equal to AM; [I. *Def.* 15.

therefore AM, MD are greater than ED,

that is, AD is greater than ED.

Again, because EM is equal to FM,

and MD is common to the two triangles EMD, FMD;

the two sides EM, MD are equal to the two sides FM, MD, each to each;

but the angle EMD is greater than the angle FMD;

therefore the base ED is greater than the base FD. [I. 24.

In the same manner it may be shewn that FD is greater than CD.

Therefore DA is the greatest, and DE greater than DF, and DF greater than DC.

Again, because MK, KD are greater than MD, [I. 20.
and MK is equal to MG, [I. *Definition* 15.
the remainder KD is greater than the remainder GD,
that is, GD is less than KD.

And because MLD is a triangle, and from the points M, D, the extremities of its side MD, the straight lines MK, DK are drawn to the point K within the triangle, therefore MK, KD are less than ML, LD; [I. 21.
and MK is equal to ML; [I. *Definition* 15.
therefore the remainder KD is less than the remainder LD.

In the same manner it may be shewn that LD is less than HD.

Therefore DG is the least, and DK less than DL, and DL less than DH.

Also, there can be drawn two equal straight lines from the point D to the circumference, one on each side of the least line.

For, at the point M, in the straight line MD, make the angle DMB equal to the angle DMK, [I. 23.
and join DB.

Then, because MK is equal to MB,
and MD is common to the two triangles KMD, BMD;
the two sides KM, MD are equal to the two sides BM, MD, each to each;
and the angle DMK is equal to the angle DMB; [*Constr.*
therefore the base DK is equal to the base DB. [I. 4.

But, besides DB, no other straight line can be drawn from D to the circumference, equal to DK.

For, if it be possible, let DN be equal to DK.
Then, because DN is equal to DK,
and DB is also equal to DK,
therefore DB is equal to DN; [*Axiom* 1.
that is, a line nearer to the least is equal to one which is more remote;
which is impossible by what has been already shewn.

Wherefore, *if any point be taken* &c. Q.E.D.

PROPOSITION 9. *THEOREM.*

If a point be taken within a circle, from which there fall more than two equal straight lines to the circumference, that point is the centre of the circle.

Let the point D be taken within the circle ABC, from which to the circumference there fall more than two equal straight lines, namely DA, DB, DC: the point D shall be the centre of the circle.

For, if not, let E be the centre; join DE and produce it both ways to meet the circumference at F and G; then FG is a diameter of the circle.

Then, because in FG, a diameter of the circle ABC, the point D is taken, which is not the centre, DG is the greatest straight line from D to the circumference, and DC is greater than DB, and DB greater than DA; [III. 7.

but they are likewise equal, by hypothesis;

which is impossible.

Therefore E is not the centre of the circle ABC.

In the same manner it may be shewn that any other point than D is not the centre;

therefore D is the centre of the circle ABC.

Wherefore, *if a point be taken* &c. Q.E.D.

PROPOSITION 10. *THEOREM.*

One circumference of a circle cannot cut another at more than two points.

If it be possible, let the circumference ABC cut the circumference DEF at more than two points, namely, at the points B, G, F.

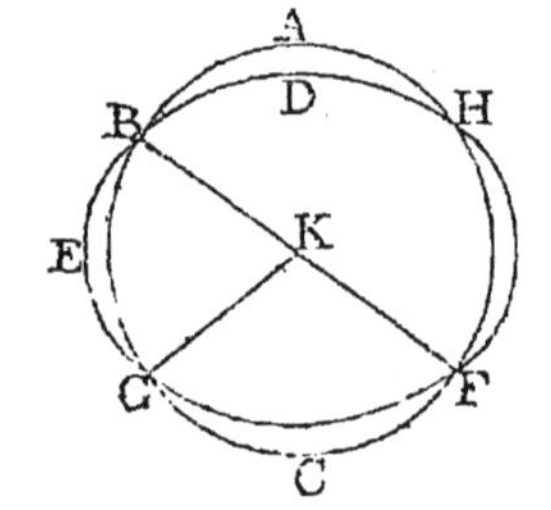

Take K, the centre of the circle ABC, [III. 1.

and join KB, KG, KF.

Then, because K is the centre of the circle ABC,

therefore KB, KG, KF are all equal to each other. [I. *Def.* 15.

And because within the circle DEF, the point K is taken, from which to the circumference DEF fall more than two equal straight lines KB, KG, KF, therefore K is the centre of the circle DEF. [III. 9.

But K is also the centre of the circle ABC. [*Construction.*

Therefore the same point is the centre of two circles which cut one another;

which is impossible. [III. 5.

Wherefore, *one circumference* &c. Q.E.D.

PROPOSITION 11. *THEOREM.*

If two circles touch one another internally, the straight line which joins their centres, being produced, shall pass through the point of contact.

Let the two circles ABC, ADE touch one another internally at the point A; and let F be the centre of the circle ABC, and G the centre of the circle ADE: the straight line which joins the centres F, G, being produced, shall pass through the point A.

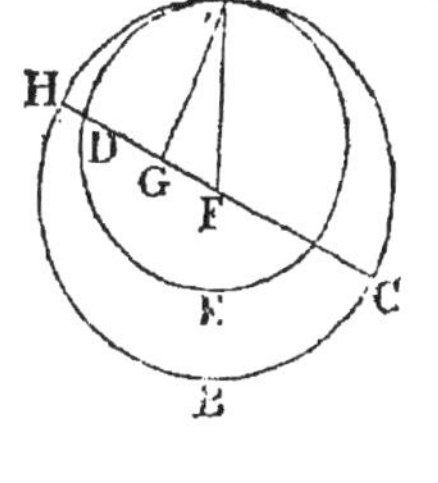

For, if not, let it pass otherwise, if possible, as $FGDH$, and join AF, AG.

Then, because AG, GF are greater than AF, [I. 20.

and AF is equal to HF, [I. *Def.* 15.

therefore AG, GF, are greater than HF.

Take away the common part GF;

therefore the remainder AG is greater than the remainder HG.

But AG is equal to DG. [I. *Definition* 15.

Therefore DG is greater than HG, the less than the greater; which is impossible.

Therefore the straight line which joins the points F, G, being produced, cannot pass otherwise than through the point A,

that is, it must pass through A.

Wherefore, *if two circles* &c. Q.E.D.

PROPOSITION 12. *THEOREM.*

If two circles touch one another externally, the straight line which joins their centres shall pass through the point of contact.

Let the two circles *ABC*, *ADE* touch one another externally at the point *A*; and let *F* be the centre of the circle *ABC*, and *G* the centre of the circle *ADE*: the straight line which joins the points *F*, *G*, shall pass through the point *A*.

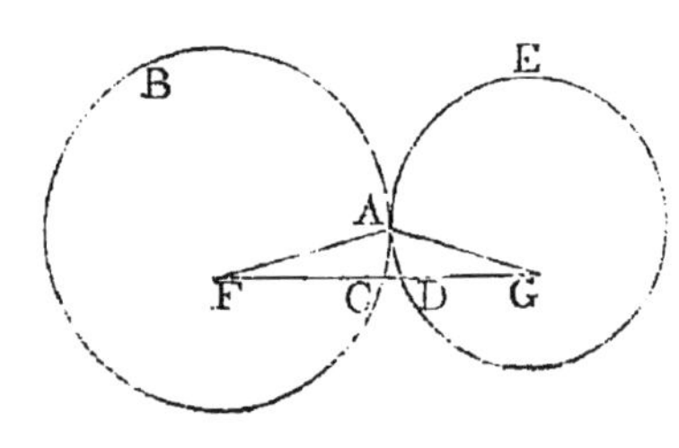

For, if not, let it pass otherwise, if possible, as *FCDG*, and join *FA*, *AG*.

Then, because *F* is the centre of the circle *ABC*, *FA* is equal to *FC*; [I. *Def.* 15.
and because *G* is the centre of the circle *ADE*, *GA* is equal to *GD*;
therefore *FA*, *AG* are equal to *FC*, *DG*. [*Axiom* 2.
Therefore the whole *FG* is greater than *FA*, *AG*.
But *FG* is also less than *FA*, *AG*; [I. 20.
which is impossible.
Therefore the straight line which joins the points *F*, *G*, cannot pass otherwise than through the point *A*,
that is, it must pass through *A*.

Wherefore, *if two circles* &c. Q.E.D.

PROPOSITION 13. *THEOREM.*

One circle cannot touch another at more points than one, whether it touches it on the inside or outside.

For, if it be possible, let the circle *EBF* touch the circle *ABC* at more points than one; and first on the inside, at the points *B*, *D*. Join *BD*, and draw *GH* bisecting *BD* at right angles. [I. 10, 11.

Then, because the two points *B*, *D* are in the circumference of each of the circles, the straight line *BD* falls within each of them; [III. 2.

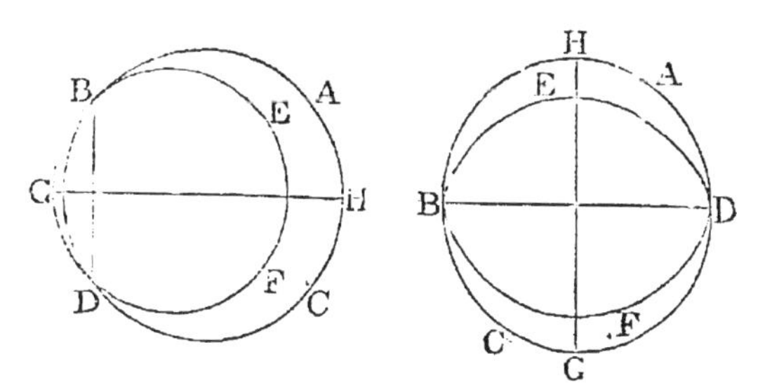

and therefore the centre of each circle is in the straight line GH which bisects BD at right angles; [III. 1, *Corol.*
therefore GH passes through the point of contact. [III. 11.
But GH does not pass through the point of contact, because the points B, D are out of the line GH;
which is absurd.

Therefore one circle cannot touch another on the inside at more points than one.

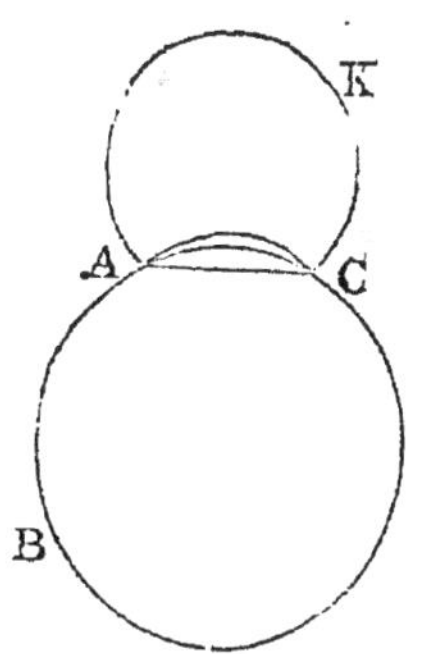

Nor can one circle touch another on the outside at more points than one.

For, if it be possible, let the circle ACK touch the circle ABC at the points A, C. Join AC.

Then, because the two points A, C are in the circumference of the circle ACK, the straight line AC which joins them, falls within the circle ACK; [III. 2.
but the circle ACK is without the circle ABC; [*Hypothesis.*
therefore the straight line AC is without the circle ABC.
But because the two points A, C are in the circumference of the circle ABC, the straight line AC falls within the circle ABC; [III. 2.
which is absurd.

Therefore one circle cannot touch another on the outside at more points than one.

And it has been shewn that one circle cannot touch another on the inside at more points than one.

Wherefore, *one circle* &c. Q.E.D.

PROPOSITION 14. *THEOREM.*

Equal straight lines in a circle are equally distant from the centre: and those which are equally distant from the centre are equal to one another.

Let the straight lines *AB*, *CD* in the circle *ABDC*, be equal to one another: they shall be equally distant from the centre.

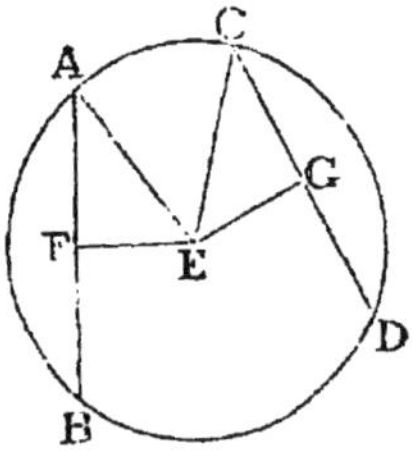

Take *E*, the centre of the circle *ABDC*; [III. 1.

and from *E* draw *EF*, *EG* perpendiculars to *AB*, *CD*; [I. 12.

and join *EA*, *EC*.

Then, because the straight line *EF*, passing through the centre, cuts the straight line *AB*, which does not pass through the centre, at right angles, it also bisects it; [III. 3.

therefore *AF* is equal to *FB*, and *AB* is double of *AF*.

For the like reason *CD* is double of *CG*.

But *AB* is equal to *CD*; [*Hypothesis*.

therefore *AF* is equal to *CG*. [*Axiom* 7.

And because *AE* is equal to *CE*, [I. *Definition* 15.

the square on *AE* is equal to the square on *CE*.

But the squares on *AF*, *FE* are equal to the square on *AE*, because the angle *AFE* is a right angle; [I. 47.

and for the like reason the squares on *CG*, *GE* are equal to the square on *CE*;

therefore the squares on *AF*, *FE* are equal to the squares on *CG*, *GE*. [*Axiom* 1.

But the square on *AF* is equal to the square on *CG*, because *AF* is equal to *CG*;

therefore the remaining square on *FE* is equal to the remaining square on *GE*; [*Axiom* 3.

and therefore the straight line *EF* is equal to the straight line *EG*.

But straight lines in a circle are said to be equally distant

from the centre, when the perpendiculars drawn to them from the centre are equal; [III. *Definition* 4.
therefore AB, CD are equally distant from the centre.

Next, let the straight lines AB, CD be equally distant from the centre, that is, let EF be equal to EG: AB shall be equal to CD.

For, the same construction being made, it may be shewn, as before, that AB is double of AF, and CD double of CG, and that the squares on EF, FA are equal to the squares on EG, GC;
but the square on EF is equal to the square on EG, because EF is equal to EG; [*Hypothesis*.
therefore the remaining square on FA is equal to the remaining square on GC, [*Axiom* 3.
and therefore the straight line AF is equal to the straight line CG.
But AB was shewn to be double of AF, and CD double of CG.
Therefore AB is equal to CD. [*Axiom* 6.

Wherefore, *equal straight lines* &c. Q.E.D.

PROPOSITION 15. *THEOREM.*

The diameter is the greatest straight line in a circle; and, of all others, that which is nearer to the centre is always greater than one more remote; and the greater is nearer to the centre than the less.

Let $ABCD$ be a circle, of which AD is a diameter, and E the centre; and let BC be nearer to the centre than FG: AD shall be greater than any straight line BC which is not a diameter, and BC shall be greater than FG.

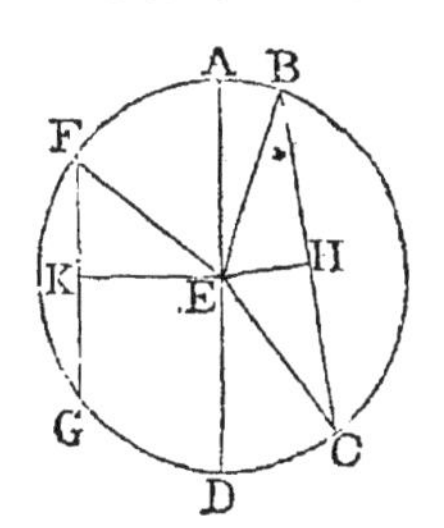

From the centre E draw EH, EK perpendiculars to BC, FG, [I. 12.
and join EB, EC, EF.

Then, because AE is equal to BE, and ED to EC, [I. *Def.* 15.
therefore AD is equal to BE, EC; [*Axiom* 2.

but *BE*, *EC* are greater than *BC*; [I. 20.
therefore also *AD* is greater than *BC*.

And, because *BC* is nearer to the centre than *FG*, [*Hypothesis.*
EH is less than *EK*. [III. *Def.* 5.
Now it may be shewn, as in the preceding proposition, that *BC* is double of *BH*, and *FG* double of *FK*, and that the squares on *EH*, *HB* are equal to the squares on *EK*, *KF*.

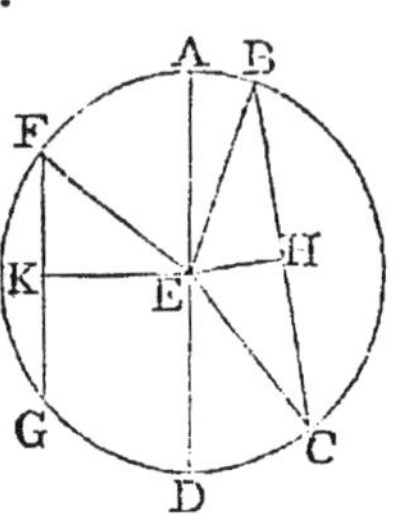

But the square on *EH* is less than the square on *EK*, because *EH* is less than *EK*;
therefore the square on *HB* is greater than the square on *KF*;
and therefore the straight line *BH* is greater than the straight line *FK*;
and therefore *BC* is greater than *FG*.

Next, let *BC* be greater than *FG*: *BC* shall be nearer to the centre than *FG*, that is, the same construction being made, *EH* shall be less than *EK*.

For, because *BC* is greater than *FG*, *BH* is greater than *FK*.
But the squares on *BH*, *HE* are equal to the squares on *FK*, *KE*;
and the square on *BH* is greater than the square on *FK*, because *BH* is greater than *FK*;
therefore the square on *HE* is less than the square on *KE*;
and therefore the straight line *EH* is less than the straight line *EK*.

Wherefore, *the diameter* &c. Q.E.D.

PROPOSITION 16. *THEOREM.*

The straight line drawn at right angles to the diameter of a circle from the extremity of it, falls without the circle; and no straight line can be drawn from the extremity, between that straight line and the circumference, so as not to cut the circle.

Let ABC be a circle, of which D is the centre and AB a diameter: the straight line drawn at right angles to AB, from its extremity A, shall fall without the circle.

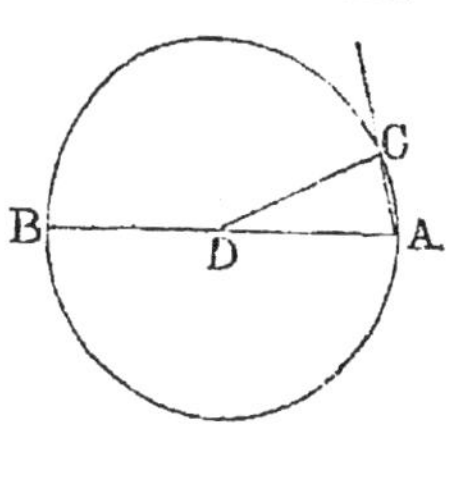

For, if not, let it fall, if possible, within the circle, as AC, and draw DC to the point C, where it meets the circumference.

Then, because DA is equal to DC, [I. *Definition* 15.
the angle DAC is equal to the angle DCA. [I. 5.
But the anglè DAC is a right angle; [*Hypothesis.*
therefore the angle DCA is a right angle;
and therefore the angles DAC, DCA are equal to two right angles; which is impossible. [I. 17.
Therefore the straight line drawn from A at right angles to AB does not fall within the circle.

And in the same manner it may be shewn that it does not fall on the circumference.

Therefore it must fall without the circle, as AE.

Also between the straight line AE and the circumference, no straight line can be drawn from the point A, which does not cut the circle.

For, if possible, let AF be between them; and from the centre D draw DG perpendicular to AF; [I. 12.
let DG meet the circumference at H.

Then, because the angle DGA is a right angle, [*Construction.*
the angle DAG is less than a right angle; [I. 17.
therefore DA is greater than DG. [I. 19.
But DA is equal to DH; [I. *Definition* 15.
therefore DH is greater than DG, the less than the greater; which is impossible.

Therefore no straight line can be drawn from the point A between AE and the circumference, so as not to cut the circle.

Wherefore, *the straight line* &c. Q.E.D.

COROLLARY. From this it is manifest, that the straight line which is drawn at right angles to the diameter of a circle from the extremity of it, touches the circle; [III. *Def.* 2.
and that it touches the circle at one point only,
because if it did meet the circle at two points it would fall within it. [III. 2.
Also it is evident, that there can be but one straight line which touches the circle at the same point.

PROPOSITION 17. *PROBLEM.*

To draw a straight line from a given point, either without or in the circumference, which shall touch a given circle.

First, let the given point A be without the given circle BCD: it is required to draw from A a straight line, which shall touch the given circle.

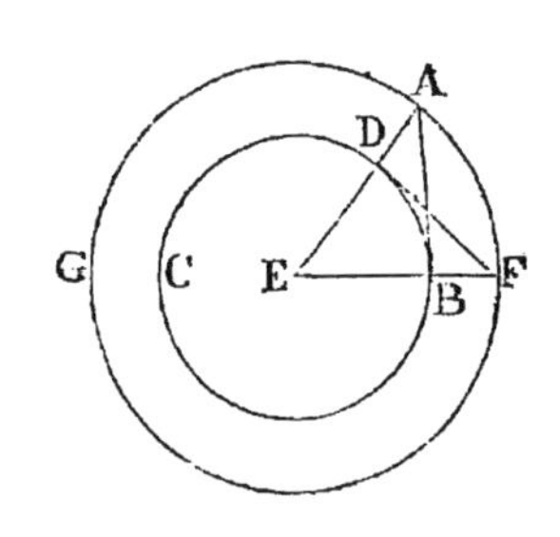

Take E, the centre of the circle, [III. 1.
and join AE cutting the circumference of the given circle at D;
and from the centre E, at the distance EA, describe the circle AFG; from the point D draw DF at right angles to EA, [I. 11.
and join EF cutting the circumference of the given circle at B;
join AB. AB shall touch the circle BCD.

For, because E is the centre of the circle AFG, EA is equal to EF. [I. *Definition* 15.
And because E is the centre of the circle BCD, EB is equal to ED. [I. *Definition* 15.
Therefore the two sides AE, EB are equal to the two sides FE, ED, each to each;
and the angle at E is common to the two triangles AEB, FED;
therefore the triangle AEB is equal to the triangle FED, and the other angles to the other angles, each to each, to which the equal sides are opposite; [I. 4.

therefore the angle ABE is equal to the angle FDE.
But the angle FDE is a right angle; [*Construction.*
therefore the angle ABE is a right angle. [*Axiom* 1.
And EB is drawn from the centre; but the straight line drawn at right angles to a diameter of a circle, from the extremity of it, touches the circle; [III. 16, *Corollary.*
therefore AB touches the circle.
And AB is drawn from the given point A. Q.E.F.

But if the given point be in the circumference of the circle, as the point D, draw DE to the centre E, and DF at right angles to DE; then DF touches the circle. [III. 16, *Cor.*

PROPOSITION 18. *THEOREM.*

If a straight line touch a circle the straight line drawn from the centre to the point of contact shall be perpendicular to the line touching the circle.

Let the straight line DE touch the circle ABC at the point C; take F, the centre of the circle ABC, and draw the straight line FC: FC shall be perpendicular to DE.

For if not, let FG be drawn from the point F perpendicular to DE, meeting the circumference at B.

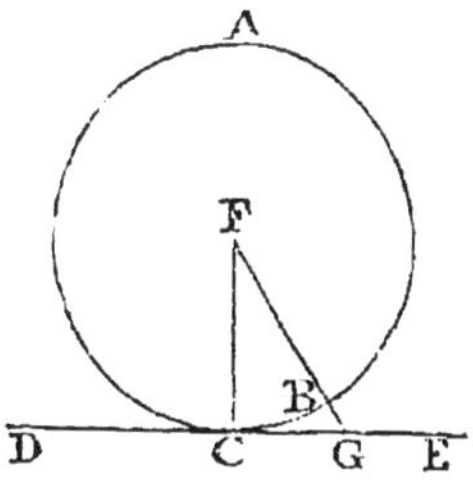

Then, because FGC is a right angle, [*Hypothesis.*
FCG is an acute angle; [I. 17.
and the greater angle of every triangle is subtended by the greater side; [I. 19.
therefore FC is greater than FG.
But FC is equal to FB; [I. *Definition* 15.
therefore FB is greater than FG, the less than the greater; which is impossible.
Therefore FG is not perpendicular to DE.

In the same manner it may be shewn that no other straight line from F is perpendicular to DE, but FC; therefore FC is perpendicular to DE.

Wherefore, *if a straight line* &c. Q.E.D.

PROPOSITION 19. *THEOREM.*

If a straight line touch a circle, and from the point of contact a straight line be drawn at right angles to the touching line, the centre of the circle shall be in that line.

Let the straight line DE touch the circle ABC at C, and from C let CA be drawn at right angles to DE: the centre of the circle shall be in CA.

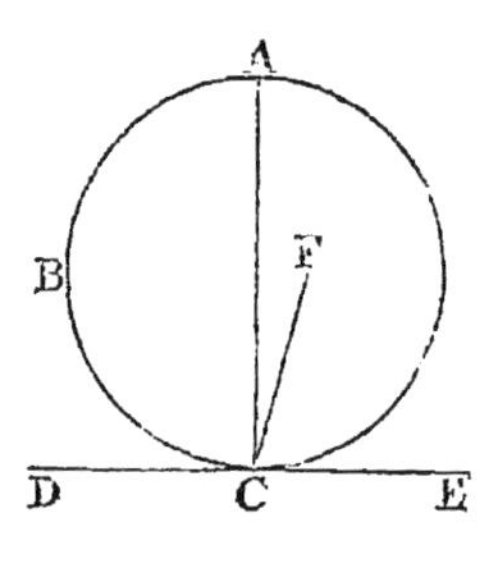

For, if not, if possible, let F be the centre, and join CF.

Then, because DE touches the circle ABC, and FC is drawn from the centre to the point of contact, FC is perpendicular to DE; [III. 18.

therefore the angle FCE is a right angle.

But the angle ACE is also a right angle; [*Construction.*

therefore the angle FCE is equal to the angle ACE, [*Ax.* 11.

the less to the greater; which is impossible.

Therefore F is not the centre of the circle ABC.

In the same manner it may be shewn that no other point out of CA is the centre; therefore the centre is in CA.

Wherefore, *if a straight line* &c. Q.E.D.

PROPOSITION 20. *THEOREM.*

The angle at the centre of a circle is double of the angle at the circumference on the same base, that is, on the same arc.

Let ABC be a circle, and BEC an angle at the centre, and BAC an angle at the circumference, which have the same arc, BC, for their base: the angle BEC shall be double of the angle BAC.

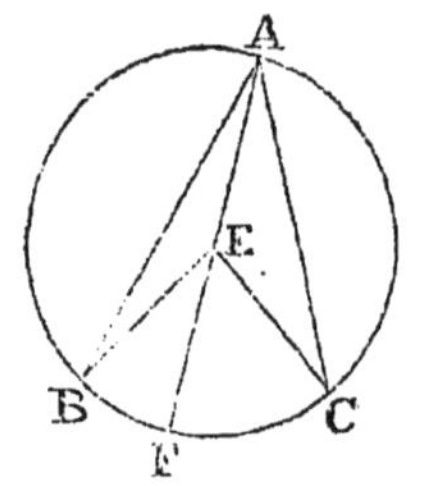

Join AE, and produce it to F.

First let the centre of the circle be within the angle BAC.

Then, because EA is equal to EB, the angle EAB is equal to the angle EBA; [I. 5.

therefore the angles EAB, EBA are double of the angle EAB.

But the angle BEF is equal to the angles EAB, EBA; [I. 32.
therefore the angle BEF is double of the angle EAB.

For the same reason the angle FEC is double of the angle EAC.

Therefore the whole angle BEC is double of the whole angle BAC.

Next, let the centre of the circle be without the angle BAC.

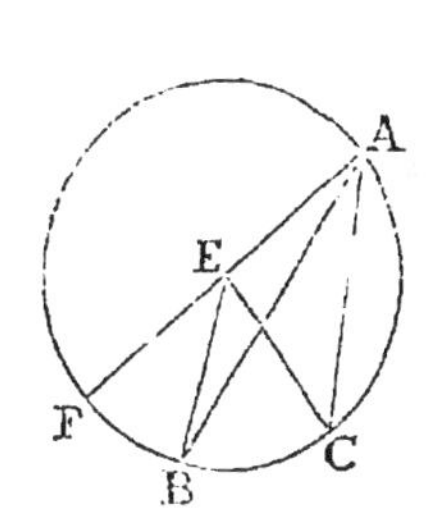

Then it may be shewn, as in the first case, that the angle FEC is double of the angle FAC, and that the angle FEB, a part of the first, is double of the angle FAB, a part of the other; therefore the remaining angle BEC is double of the remaining angle BAC.

Wherefore, *the angle at the centre* &c. Q.E.D.

PROPOSITION 21. *THEOREM.*

The angles in the same segment of a circle are equal to one another.

Let $ABCD$ be a circle, and BAD, BED angles in the same segment $BAED$: the angles BAD, BED shall be equal to one another.

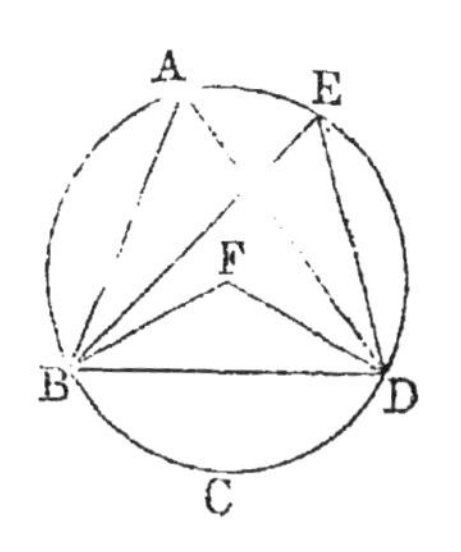

Take F the centre of the circle $ABCD$. [III. 1.

First let the segment $BAED$ be greater than a semicircle.

Join BF, DF.

Then, because the angle BFD is at the centre, and the angle BAD is at the circumference, and that they have the same arc for their base, namely, BCD;

therefore the angle BFD is double of the angle BAD. [III. 20.

For the same reason the angle BFD is double of the angle BED.

Therefore the angle BAD is equal to the angle BED. [*Ax.* 7.

Next, let the segment $BAED$ be not greater than a semicircle.

Draw AF to the centre, and produce it to meet the circumference at C, and join CE.

Then the segment $BAEC$ is greater than a semicircle, and therefore the angles BAC, BEC in it, are equal, by the first case.

For the same reason, because the segment $CAED$ is greater than a semicircle, the angles CAD, CED are equal.

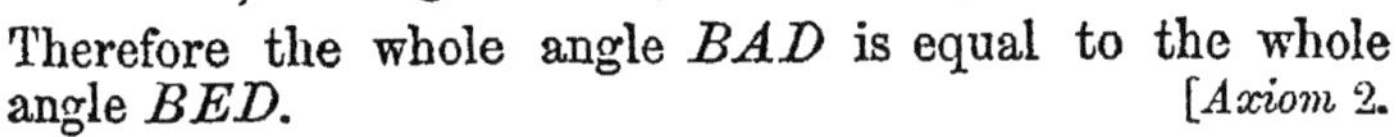

Therefore the whole angle BAD is equal to the whole angle BED. [*Axiom* 2.

Wherefore, *the angles in the same segment* &c. Q.E.D.

PROPOSITION 22. *THEOREM.*

The opposite angles of any quadrilateral figure inscribed in a circle are together equal to two right angles.

Let $ABCD$ be a quadrilateral figure inscribed in the circle $ABCD$: any two of its opposite angles shall be together equal to two right angles.

Join AC, BD.

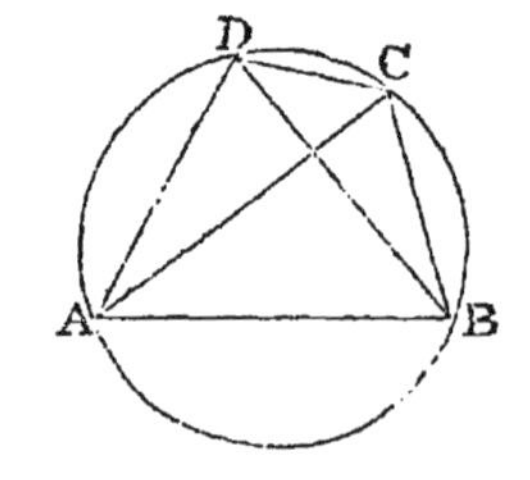

Then, because the three angles of every triangle are together equal to two right angles, [I. 32.

the three angles of the triangle CAB, namely, CAB, ACB, ABC are together equal to two right angles.

But the angle CAB is equal to the angle CDB, because they are in the same segment $CDAB$; [III. 21.

and the angle ACB is equal to the angle ADB, because they are in the same segment $ADCB$;

therefore the two angles CAB, ACB are together equal to the whole angle ADC. [*Axiom* 2.

To each of these equals add the angle ABC;

therefore the three angles CAB, ACB, ABC, are equal to the two angles ABC, ADC.

But the angles CAB, ACB, ABC are together equal to two right angles; [I. 32.

therefore also the angles ABC, ADC are together equal to two right angles.

In the same manner it may be shewn that the angles BAD, BCD are together equal to two right angles.

Wherefore, *the opposite angles* &c. Q.E.D.

PROPOSITION 23. *THEOREM.*

On the same straight line, and on the same side of it, there cannot be two similar segments of circles, not coinciding with one another.

If it be possible, on the same straight line AB, and on the same side of it, let there be two similar segments of circles ACB, ADB, not coinciding with one another.

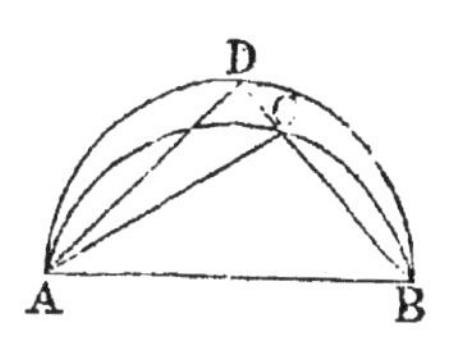

Then, because the circle ACB cuts the circle ADB at the two points A, B, they cannot cut one another at any other point; [III. 10.

therefore one of the segments must fall within the other; let ACB fall within ADB; draw the straight line BCD, and join AC, AD.

Then, because ACB, ADB are, by hypothesis, similar segments of circles, and that similar segments of circles contain equal angles, [III. *Definition* 11.

therefore the angle ACB is equal to the angle ADB;

that is, the exterior angle of the triangle ACD is equal to the interior and opposite angle;

which is impossible. [I. 16.

Wherefore, *on the same straight line* &c. Q.E.D.

PROPOSITION 24. *THEOREM.*

Similar segments of circles on equal straight lines are equal to one another.

Let AEB, CFD be similar segments of circles on the equal straight lines AB, CD: the segment AEB shall be equal to the segment CFD.

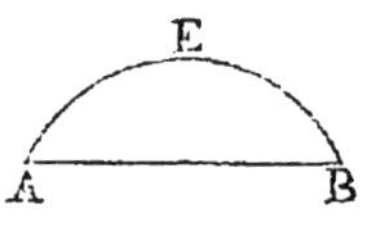

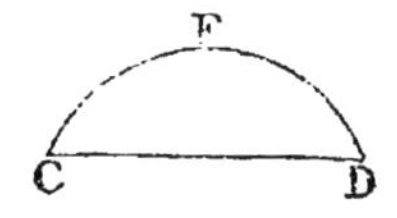

For if the segment AEB be applied to the segment CFD, so that the point A may be on the point C, and the straight line AB on the straight line CD, the point B will coincide with the point D, because AB is equal to CD.

Therefore, the straight line AB coinciding with the straight line CD, the segment AEB must coincide with the segment CFD; [III. 23.

and is therefore equal to it.

Wherefore, *similar segments* &c. Q.E.D.

PROPOSITION 25. *PROBLEM.*

A segment of a circle being given, to describe the circle of which it is a segment.

Let ABC be the given segment of a circle: it is required to describe the circle of which it is a segment.

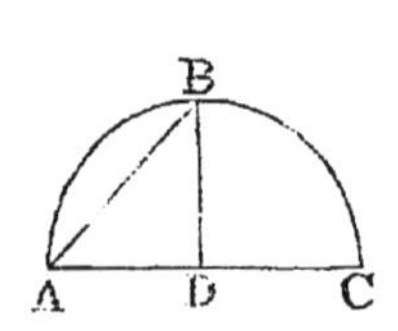

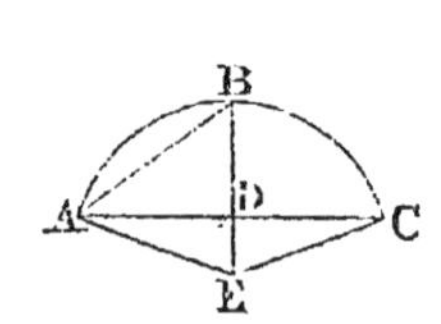

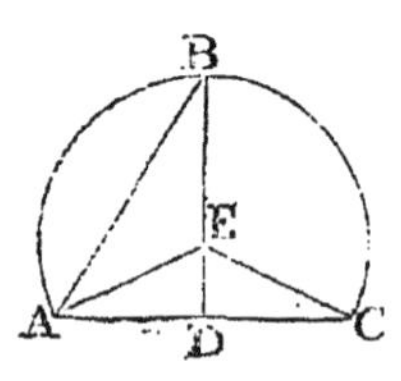

Bisect AC at D; [I. 10.

from the point D draw DB at right angles to AC; [I. 11.

and join AB.

First, let the angles ABD, BAD, be equal to one another.

Then DB is equal to DA; [I. 6.

but DA is equal to DC; [*Construction.*

therefore DB is equal to DC. [*Axiom* 1.

Therefore the three straight lines DA, DB, DC are all equal; and therefore D is the centre of the circle. [III. 9.

From the centre D, at the distance of any of the three DA, DB, DC, describe a circle; this will pass through the other points, and the circle of which ABC is a segment is described.

And because the centre D is in AC, the segment ABC is a semicircle.

Next, let the angles ABD, BAD be not equal to one another.

At the point A, in the straight line AB, make the angle BAE equal to the angle ABD; [I. 23.

produce BD, if necessary, to E, and join EC.

Then, because the angle BAE is equal to the angle ABE, [*Construction.*

EA is equal to EB. [I. 6.

And because AD is equal to CD, [*Construction.*

and DE is common to the two triangles ADE, CDE;

the two sides AD, DE are equal to the two sides CD, DE, each to each;

and the angle ADE is equal to the angle CDE, for each of them is a right angle; [*Construction.*

therefore the base EA is equal to the base EC. [I. 4.

But EA was shewn to be equal to EB;

therefore EB is equal to EC. [*Axiom* 1.

Therefore the three straight lines EA, EB, EC are all equal; and therefore E is the centre of the circle. [III. 9.

From the centre E, at the distance of any of the three EA, EB, EC, describe a circle; this will pass through the other points, and the circle of which ABC is a segment is described.

And it is evident, that if the angle ABD be greater than the angle BAD, the centre E falls without the segment ABC, which is therefore less than a semicircle; but if the angle ABD be less than the angle BAD, the centre E falls within the segment ABC, which is therefore greater than a semicircle.

Wherefore, *a segment of a circle being given, the circle has been described of which it is a segment.* Q.E.F.

PROPOSITION 26. *THEOREM.*

In equal circles, equal angles stand on equal arcs, whether they be at the centres or circumferences.

Let *ABC*, *DEF* be equal circles; and let *BGC*, *EHF* be equal angles in them at their centres, and *BAC*, *EDF* equal angles at their circumferences: the arc *BKC* shall be equal to the arc *ELF*.

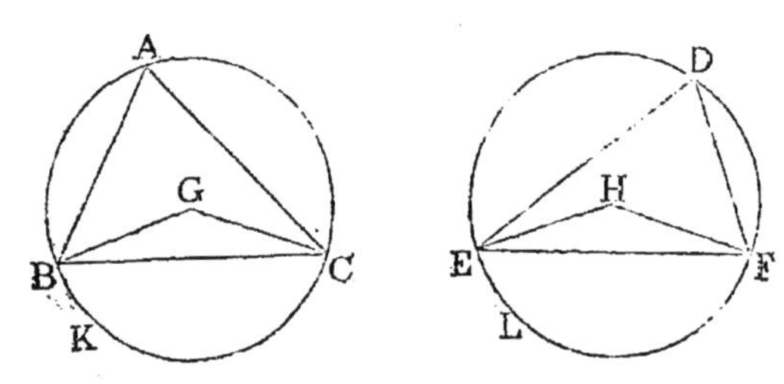

Join *BC*, *EF*.

Then, because the circles *ABC*, *DEF* are equal, [*Hyp.*
the straight lines from their centres are equal; [III. *Def.* 1.
therefore the two sides *BG*, *GC* are equal to the two sides *EH*, *HF*, each to each;
and the angle at *G* is equal to the angle at *H*; [*Hypothesis.*
therefore the base *BC* is equal to the base *EF*. [I. 4.

And because the angle at *A* is equal to the angle at *D*, [*Hyp.*
the segment *BAC* is similar to the segment *EDF*; [III. *Def.* 11.
and they are on equal straight lines *BC*, *EF*.
But similar segments of circles on equal straight lines are equal to one another; [III. 24.
therefore the segment *BAC* is equal to the segment *EDF*.

But the whole circle *ABC* is equal to the whole circle *DEF*; [*Hypothesis.*
therefore the remaining segment *BKC* is equal to the remaining segment *ELF*; [*Axiom* 3.
therefore the arc *BKC* is equal to the arc *ELF*.

Wherefore, *in equal circles* &c. Q.E.D.

PROPOSITION 27. *THEOREM.*

In equal circles, the angles which stand on equal arcs are equal to one another, whether they be at the centres or circumferences.

Let ABC, DEF be equal circles, and let the angles BGC, EHF at their centres, and the angles BAC, EDF at their circumferences, stand on equal arcs BC, EF: the angle BGC shall be equal to the angle EHF, and the angle BAC equal to the angle EDF.

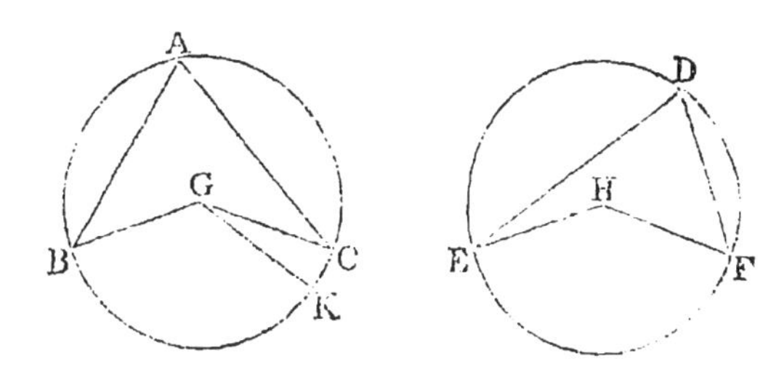

If the angle BGC be equal to the angle EHF, it is manifest that the angle BAC is also equal to the angle EDF. [III. 20, *Axiom* 7.

But, if not, one of them must be the greater. Let BGC be the greater, and at the point G, in the straight line BG, make the angle BGK equal to the angle EHF. [I. 23.

Then, because the angle BGK is equal to the angle EHF, and that in equal circles equal angles stand on equal arcs, when they are at the centres, [III. 26.

therefore the arc BK is equal to the arc EF.

But the arc EF is equal to the arc BC; [*Hypothesis.*

therefore the arc BK is equal to the arc BC, [*Axiom* 1.

the less to the greater; which is impossible.

Therefore the angle BGC is not unequal to the angle EHF, that is, it is equal to it.

And the angle at A is half of the angle BGC, and the angle at D is half of the angle EHF; [III. 20.

therefore the angle at A is equal to the angle at D. [*Ax.* 7.

Wherefore, *in equal circles* &c. Q.E.D.

PROPOSITION 28. *THEOREM.*

In equal circles, equal straight lines cut off equal arcs, the greater equal to the greater, and the less equal to the less.

Let *ABC, DEF* be equal circles, and *BC, EF* equal straight lines in them, which cut off the two greater arcs *BAC, EDF,* and the two less arcs *BGC, EHF*: the greater arc *BAC* shall be equal to the greater arc *EDF,* and the less arc *BGC* equal to the less arc *EHF.*

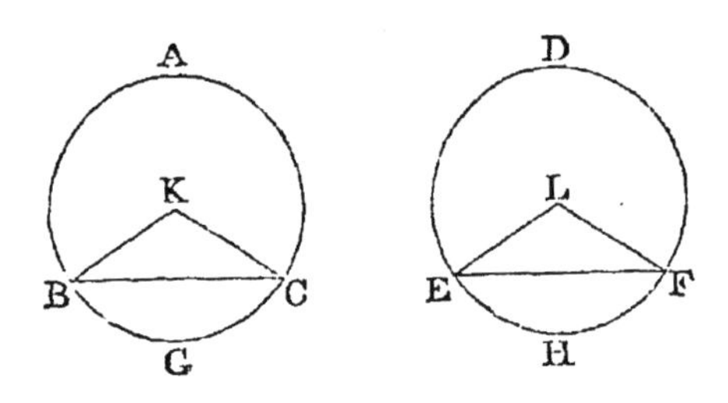

Take *K, L,* the centres of the circles, [III. 1.

and join *BK, KC, EL, LF.*

Then, because the circles are equal, [*Hypothesis.*

the straight lines from their centres are equal; [III. *Def.* 1.

therefore the two sides *BK, KC* are equal to the two sides *EL, LF,* each to each;

and the base *BC* is equal to the base *EF*; [*Hypothesis.*

therefore the angle *BKC* is equal to the angle *ELF.* [I. 8.

But in equal circles equal angles stand on equal arcs, when they are at the centres, [III. 26.

therefore the arc *BGC* is equal to the arc *EHF.*

But the circumference *ABGC* is equal to the circumference *DEHF*; [*Hypothesis.*

therefore the remaining arc *BAC* is equal to the remaining arc *EDF.* [*Axiom* 3.

Wherefore, *in equal circles* &c. Q.E.D.

PROPOSITION 29. *THEOREM.*

In equal circles, equal arcs are subtended by equal straight lines.

Let *ABC, DEF* be equal circles, and let *BGC, EHF* be equal arcs in them, and join *BC, EF*: the straight line *BC* shall be equal to the straight line *EF.*

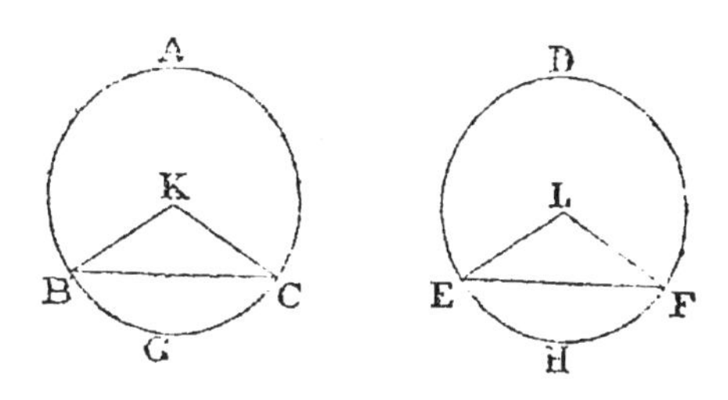

Take *K, L,* the centres of the circles, [III. 1.
and join *BK, KC, EL, LF.*

Then, because the arc *BGC* is equal to the arc *EHF,* [*Hypothesis.*
the angle *BKC* is equal to the angle *ELF.* [III. 27.
And because the circles *ABC, DEF* are equal, [*Hypothesis.*
the straight lines from their centres are equal; [III. *Def.* 1.
therefore the two sides *BK, KC* are equal to the two sides *EL, LF,* each to each;
and they contain equal angles;
therefore the base *BC* is equal to the base *EF.* [I. 4.

Wherefore, *in equal circles* &c. Q.E.D.

PROPOSITION 30. *PROBLEM.*

To bisect a given arc, that is, to divide it into two equal parts.

Let ADB be the given arc: it is required to bisect it.

Join AB;
bisect it at C; [I. 10.
from the point C draw CD at right angles to AB meeting the arc at D. [I. 11.
The arc ADB shall be bisected at the point D.

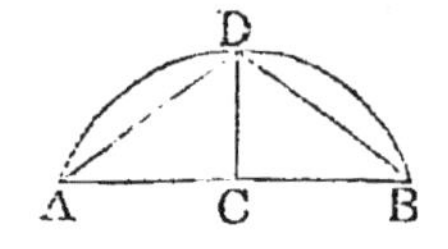

Join AD, DB.

Then, because AC is equal to CB, [*Construction.*

and CD is common to the two triangles ACD, BCD;

the two sides AC, CD are equal to the two sides BC, CD, each to each;

and the angle ACD is equal to the angle BCD, because each of them is a right angle; [*Construction.*

therefore the base AD is equal to the base BD. [I. 4.

But equal straight lines cut off equal arcs, the greater equal to the greater, and the less equal to the less; [III. 28.

and each of the arcs AD, DB is less than a semi-circumference, because DC, if produced, is a diameter; [III. 1. *Cor.*

therefore the arc AD is equal to the arc DB.

Wherefore *the given arc is bisected at* D. Q.E.F.

PROPOSITION 31. *THEOREM.*

In a circle the angle in a semicircle is a right angle; but the angle in a segment greater than a semicircle is less than a right angle; and the angle in a segment less than a semicircle is greater than a right angle.

Let $ABCD$ be a circle, of which BC is a diameter and E the centre; and draw CA, dividing the circle into the segments ABC, ADC, and join BA, AD, DC: the angle in the semicircle BAC shall be a right angle; but the angle in the segment ABC, which is greater than a semicircle, shall be less than a right angle; and the angle in the segment ADC, which is less than a semicircle, shall be greater than a right angle.

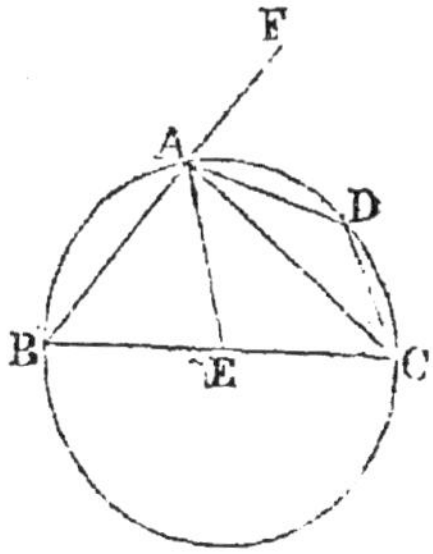

Join AE, and produce BA to F.

Then, because EA is equal to EB, [I. *Definition* 15.

the angle EAB is equal to the angle EBA; [I. 5.

and, because EA is equal to EC,
the angle EAC is equal to the angle ECA;

therefore the whole angle BAC is equal to the two angles, ABC, ACB. [*Axiom* 2.

But FAC, the exterior angle of the triangle ABC, is equal to the two angles ABC, ACB; [I. 32.

therefore the angle BAC is equal to the angle FAC, [*Ax.* 1.

and therefore each of them is a right angle. [I. *Def.* 10.

Therefore the angle in a semicircle BAC is a right angle.

And because the two angles ABC, BAC, of the triangle ABC, are together less than two right angles, [I. 17.

and that BAC has been shewn to be a right angle,

therefore the angle ABC is less than a right angle.

Therefore the angle in a segment ABC, greater than a semicircle, is less than a right angle.

And because $ABCD$ is a quadrilateral figure in a circle, any two of its opposite angles are together equal to two right angles; [III. 22.

therefore the angles ABC, ADC are together equal to two right angles.

But the angle ABC has been shewn to be less than a right angle;

therefore the angle ADC is greater than a right angle.

Therefore the angle in a segment ADC, less than a semicircle, is greater than a right angle.

Wherefore, *the angle* &c. Q.E.D.

COROLLARY. From the demonstration it is manifest that if one angle of a triangle be equal to the other two, it is a right angle.

For the angle adjacent to it is equal to the same two angles; [I. 32.

and when the adjacent angles are equal, they are right angles. [I. *Definition* 10.

PROPOSITION 32. *THEOREM.*

If a straight line touch a circle, and from the point of contact a straight line be drawn cutting the circle, the angles which this line makes with the line touching the circle shall be equal to the angles which are in the alternate segments of the circle.

Let the straight line *EF* touch the circle *ABCD* at the point *B*, and from the point *B* let the straight line *BD* be drawn, cutting the circle: the angles which *BD* makes with the touching line *EF*, shall be equal to the angles in the alternate segments of the circle; that is, the angle *DBF* shall be equal to the angle in the segment *BAD*, and the angle *DBE* shall be equal to the angle in the segment *BCD*.

From the point *B* draw *BA* at right angles to *EF*, [I. 11.
and take any point *C* in the arc *BD*, and join *AD*, *DC*, *CB*.

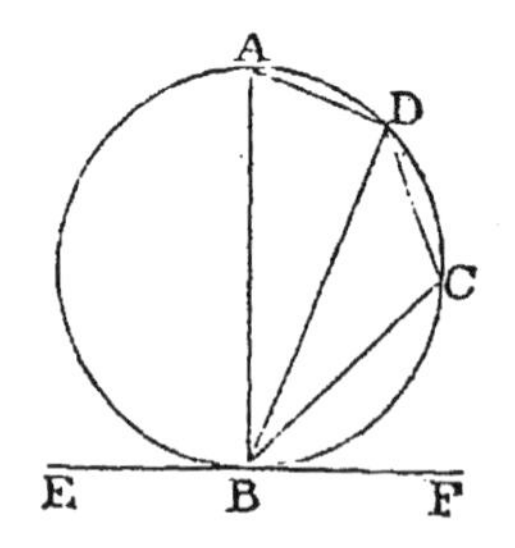

Then, because the straight line *EF* touches the circle *ABCD* at the point *B*, [*Hyp.*
and *BA* is drawn at right angles to the touching line from the point of contact *B*, [*Construction.*
therefore the centre of the circle is in *BA*. [III. 19.
Therefore the angle *ADB*, being in a semicircle, is a right angle. [III. 31.
Therefore the other two angles *BAD*, *ABD* are equal to a right angle. [I. 32.
But *ABF* is also a right angle. [*Construction.*

Therefore the angle ABF is equal to the angles BAD, ABD.

From each of these equals take away the common angle ABD;

therefore the remaining angle DBF is equal to the remaining angle BAD, [*Axiom* 3.

which is in the alternate segment of the circle.

And because $ABCD$ is a quadrilateral figure in a circle, the opposite angles BAD, BCD are together equal to two right angles. [III. 22.

But the angles DBF, DBE are together equal to two right angles. [I. 13.

Therefore the angles DBF, DBE are together equal to the angles BAD, BCD.

And the angle DBF has been shewn equal to the angle BAD;

therefore the remaining angle DBE is equal to the remaining angle BCD, [*Axiom* 3.

which is in the alternate segment of the circle.

Wherefore, *if a straight line* &c. Q.E.D.

PROPOSITION 33. *PROBLEM.*

On a given straight line to describe a segment of a circle, containing an angle equal to a given rectilineal angle.

Let AB be the given straight line, and C the given rectilineal angle: it is required to describe, on the given straight line AB, a segment of a circle containing an angle equal to the angle C.

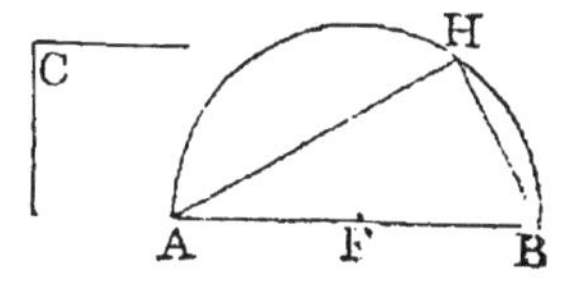

First, let the angle C be a right angle.

Bisect AB at F, [I. 10.

and from the centre F, at the distance FB, describe the semicircle AHB.

Then the angle AHB in a semicircle is equal to the right angle C. [III. 31.

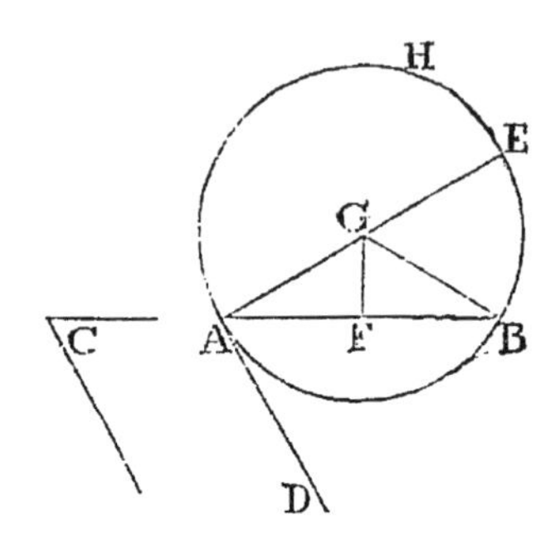

But if the angle C be not a right angle, at the point A, in the straight line AB, make the angle BAD equal to the angle C; [I. 23.
from the point A, draw AE at right angles to AD; [I. 11.
bisect AB at F; [I. 10.
from the point F, draw FG at right angles to AB; [I. 11.
and join GB.

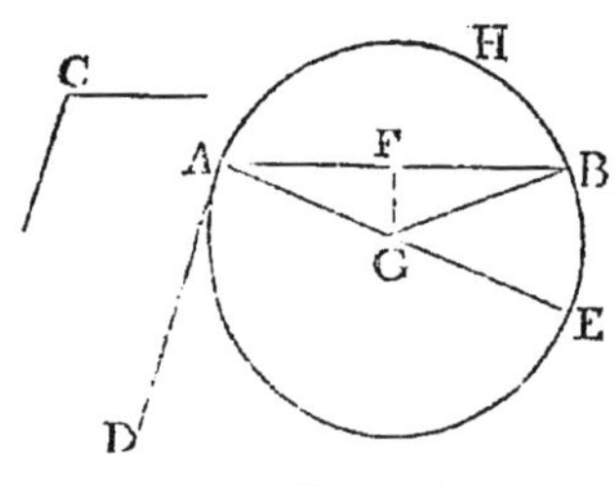

Then, because AF is equal to BF, [*Const.*
and FG is common to the two triangles AFG, BFG;
the two sides AF, FG are equal to the two sides BF, FG, each to each;
and the angle AFG is equal to the angle BFG; [I. *Definition* 10.
therefore the base AG is equal to the base BG; [I. 4.
and therefore the circle described from the centre G, at the distance GA, will pass through the point B.
Let this circle be described; and let it be AHB.
The segment AHB shall contain an angle equal to the given rectilineal angle C.

Because from the point A, the extremity of the diameter AE, AD is drawn at right angles to AE, [*Construction.*
therefore AD touches the circle. [III. 16. *Corollary.*
And because AB is drawn from the point of contact A, the angle DAB is equal to the angle in the alternate segment AHB. [III. 32.
But the angle DAB is equal to the angle C. [*Constr.*
Therefore the angle in the segment AHB is equal to the angle C. [*Axiom* 1.

Wherefore, *on the given straight line AB, the segment AHB of a circle has been described, containing an angle equal to the given angle C.* Q.E.F.

PROPOSITION 34. *PROBLEM.*

From a given circle to cut off a segment containing an angle equal to a given rectilineal angle.

Let ABC be the given circle, and D the given rectilineal angle: it is required to cut off from the circle ABC a segment containing an angle equal to the angle D.

Draw the straight line EF touching the circle ABC at the point B; [III. 17.
and at the point B, in the straight line BF, make the angle FBC equal to the angle D. [I. 23.
The segment BAC shall contain an angle equal to the angle D.

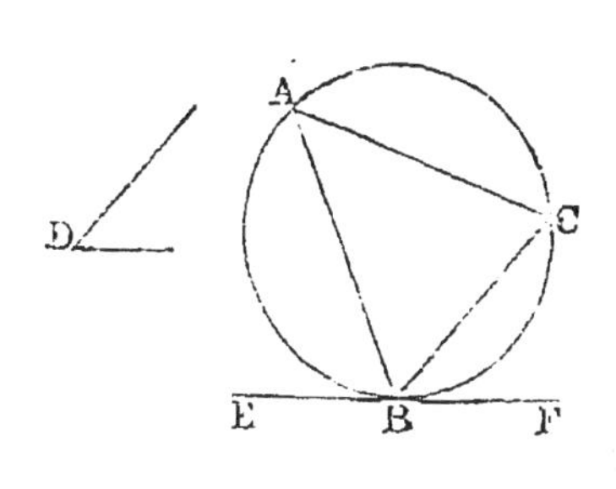

Because the straight line EF touches the circle ABC, and BC is drawn from the point of contact B, [*Constr.*
therefore the angle FBC is equal to the angle in the alternate segment BAC of the circle. [III. 32.

But the angle FBC is equal to the angle D. [*Construction.*

Therefore the angle in the segment BAC is equal to the angle D. [*Axiom* 1.

Wherefore, *from the given circle ABC, the segment BAC has been cut off, containing an angle equal to the given angle D.* Q.E.F.

PROPOSITION 35. *THEOREM.*

If two straight lines cut one another within a circle, the rectangle contained by the segments of one of them shall be equal to the rectangle contained by the segments of the other.

Let the two straight lines AC, BD cut one another at the point E, within the circle $ABCD$: the rectangle contained by AE, EC shall be equal to the rectangle contained by BE, ED.

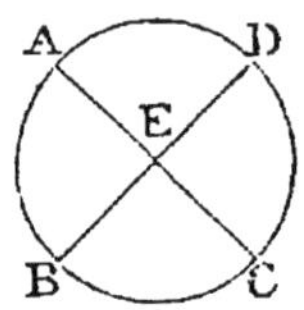

If AC and BD both pass through the centre, so that E is the centre, it is evident, since EA, EB, EC, ED are all equal, that the rectangle AE, EC is equal to the rectangle BE, ED.

But let one of them, BD, pass through the centre, and cut the other AC, which does not pass through the centre, at right angles, at the point E. Then, if BD be bisected at F, F is the centre of the circle $ABCD$; join AF.

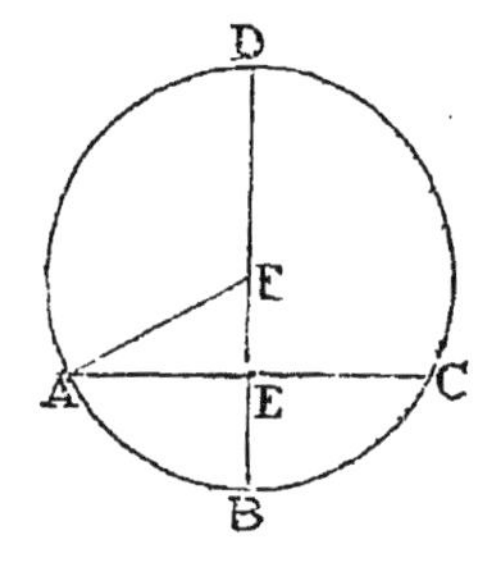

Then, because the straight line BD which passes through the centre, cuts the straight line AC, which does not pass through the centre, at right angles at the point E, [*Hypothesis.*
AE is equal to EC. [III. 3.

And because the straight line BD is divided into two equal parts at the point F, and into two unequal parts at the point E, the rectangle BE, ED, together with the square on EF, is equal to the square on FB, [II. 5.
that is, to the square on AF.

But the square on AF is equal to the squares on AE, EF. [I. 47.

Therefore the rectangle BE, ED, together with the square on EF, is equal to the squares on AE, EF. [*Axiom* 1.

Take away the common square on EF;
then the remaining rectangle BE, ED, is equal to the remaining square on AE,
that is, to the rectangle AE, EC.

Next, let BD, which passes through the centre, cut the other AC, which does not pass through the centre, at the point E, but not at right angles. Then, if BD be bisected at F, F is the centre of the circle $ABCD$; join AF, and from F draw FG perpendicular to AC. [I. 12.

Then AG is equal to GC; [III. 3.
therefore the rectangle AE, EC, together with the square on EG, is equal to the square on AG. [II. 5.
To each of these equals add the square on GF;
then the rectangle AE, EC, together with the squares on EG, GF, is equal to the squares on AG, GF. [*Axiom* 2.

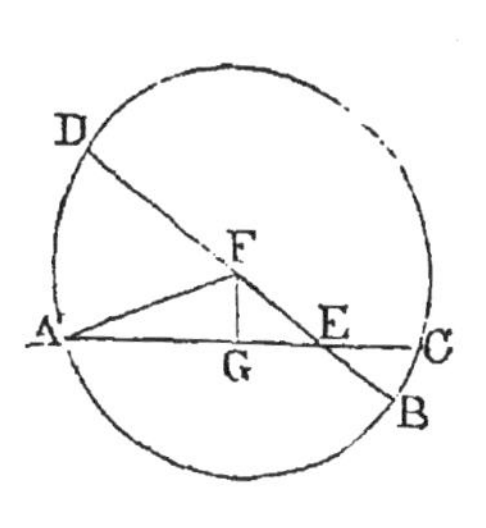

But the squares on EG, GF are equal to the square on EF;

and the squares on AG, GF are equal to the square on AF. [I. 47.

Therefore the rectangle AE, EC, together with the square on EF, is equal to the square on AF,

that is, to the square on FB.

But the square on FB is equal to the rectangle BE, ED, together with the square on EF. [II. 5.
Therefore the rectangle AE, EC, together with the square on EF, is equal to the rectangle BE, ED, together with the square on EF.

Take away the common square on EF;

then the remaining rectangle AE, EC is equal to the remaining rectangle BE, ED. [*Axiom* 3.

Lastly, let neither of the straight lines AC, BD pass through the centre.
Take the centre F, [III. 1.
and through E, the intersection of the straight lines AC, BD, draw the diameter $GEFH$.

Then, as has been shewn, the rectangle GE, EH is equal to the rectangle AE, EC, and also to the rectangle BE, ED;
therefore the rectangle AE, EC is equal to the rectangle BE, ED. [*Axiom* 1.

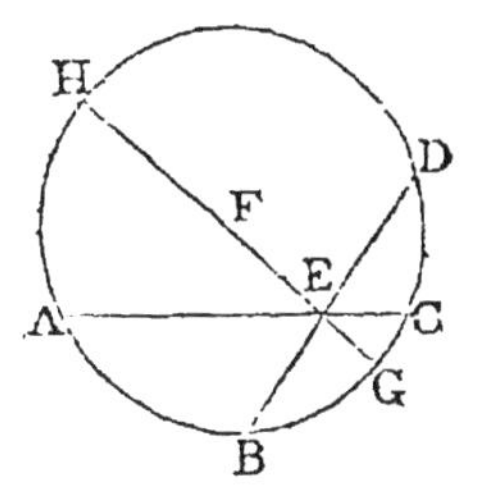

Wherefore, *if two straight lines* &c. Q.E.D.

PROPOSITION 36. *THEOREM.*

If from any point without a circle two straight lines be drawn, one of which cuts the circle, and the other touches it; the rectangle contained by the whole line which cuts the circle, and the part of it without the circle, shall be equal to the square on the line which touches it.

Let D be any point without the circle ABC, and let DCA, DB be two straight lines drawn from it, of which DCA cuts the circle and DB touches it: the rectangle AD, DC shall be equal to the square on DB.

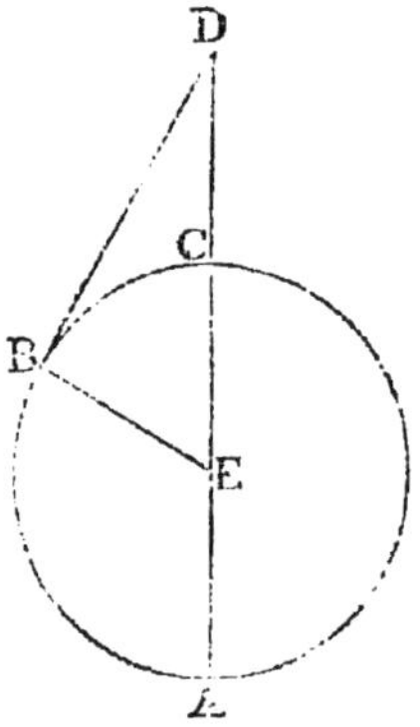

First, let DCA pass through the centre E, and join EB.

Then EBD is a right angle. [III. 18.

And because the straight line AC is bisected at E, and produced to D, the rectangle AD, DC together with the square on EC is equal to the square on ED. [II. 6.

But EC is equal to EB;

therefore the rectangle AD, DC together with the square on EB is equal to the square on ED.

But the square on ED is equal to the squares on EB, BD, because EBD is a right angle. [I. 47.

Therefore the rectangle AD, DC, together with the square on EB is equal to the squares on EB, BD.

Take away the common square on EB;

then the remaining rectangle AD, DC is equal to the square on DB. [*Axiom* 3.

Next let DCA not pass through the centre of the circle ABC; take the centre E; [III. 1.

from E draw EF perpendicular to AC; [I. 12.

and join EB, EC, ED.

Then, because the straight line EF which passes through the centre, cuts the straight line AC, which does not pass through the centre, at right angles, it also bisects it; [III. 3.

therefore AF is equal to FC.

And because the straight line AC is bisected at F, and produced to D, the rectangle AD, DC, together with the square on FC, is equal to the square on FD. [II. 6.

To each of these equals add the square on FE.

Therefore the rectangle AD, DC together with the squares on CF, FE, is equal to the squares on DF, FE. [*Axiom* 2.

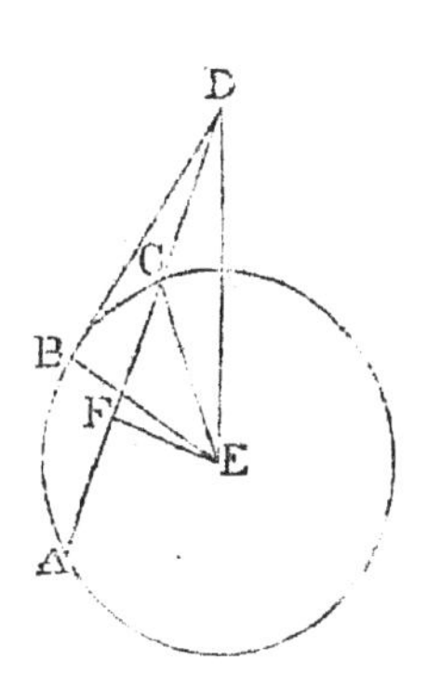

But the squares on CF, FE are equal to the square on CE, because CFE is a right angle; [I. 47.

and the squares on DF, FE are equal to the square on DE.

Therefore the rectangle AD, DC, together with the square on CE, is equal to the square on DE.

But CE is equal to BE;

therefore the rectangle AD, DC, together with the square on BE, is equal to the square on DE.

But the square on DE is equal to the squares on DB, BE, because EBD is a right angle. [I. 47.

Therefore the rectangle AD, DC, together with the square on BE, is equal to the squares on DB, BE.

Take away the common square on BE;

then the remaining rectangle AD, DC is equal to the square on DB. [*Axiom* 3.

Wherefore, *if from any point* &c. Q.E.D.

Corollary. If from any point without a circle, there be drawn two straight lines cutting it, as AB, AC, the rectangles contained by the whole lines and the parts of them without the circles are equal to one another; namely, the rectangle BA, AE is equal to the rectangle CA, AF; for each of them is equal to the square on the straight line AD, which touches the circle.

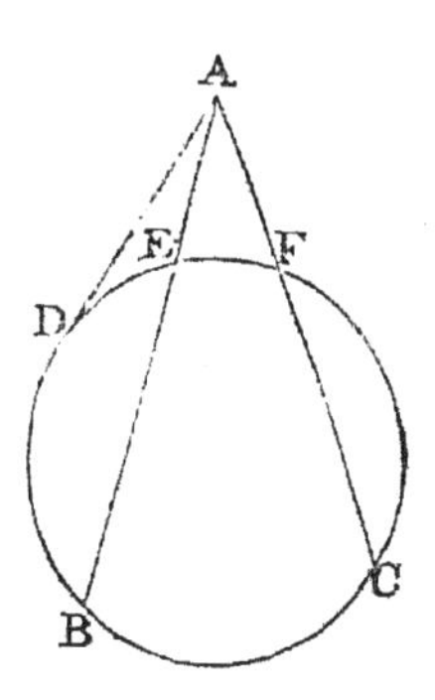

PROPOSITION 37. *THEOREM.*

If from any point without a circle there be drawn two straight lines, one of which cuts the circle, and the other meets it, and if the rectangle contained by the whole line which cuts the circle, and the part of it without the circle, be equal to the square on the line which meets the circle, the line which meets the circle shall touch it.

Let any point D be taken without the circle ABC, and from it let two straight lines DCA, DB be drawn, of which DCA cuts the circle, and DB meets it; and let the rectangle AD, DC be equal to the square on DB: DB shall touch the circle.

Draw the straight line DE, touching the circle ABC; [III. 17.
find F the centre, [III. 1.
and join FB, FD, FE.

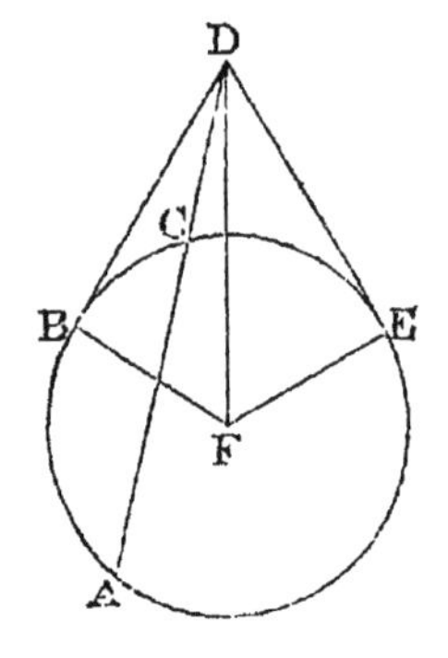

Then the angle FED is a right angle. [III. 18.
And because DE touches the circle ABC, and DCA cuts it, the rectangle AD, DC is equal to the square on DE. [III. 36.
But the rectangle AD, DC is equal to the square on DB. [*Hyp.*
Therefore the square on DE is equal to the square on DB; [*Ax.* 1.
therefore the straight line DE is equal to the straight line DB.

And EF is equal to BF; [I. *Definition* 15.
therefore the two sides DE, EF are equal to the two sides DB, BF each to each;
and the base DF is common to the two triangles DEF, DBF;
therefore the angle DEF is equal to the angle DBF. [I. 8.
But DEF is a right angle; [*Construction.*
therefore also DBF is a right angle.
And BF, if produced, is a diameter; and the straight line which is drawn at right angles to a diameter from the extremity of it touches the circle; [III. 16. *Corollary.*
therefore DB touches the circle ABC.

Wherefore, *if from a point* &c. Q.E.D.

BOOK IV.

DEFINITIONS.

1. A RECTILINEAL figure is said to be inscribed in another rectilineal figure, when all the angles of the inscribed figure are on the sides of the figure in which it is inscribed, each on each.

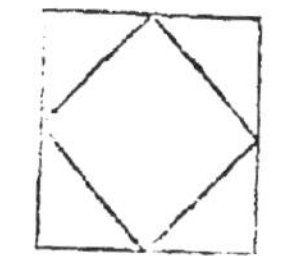

2. In like manner, a figure is said to be described about another figure, when all the sides of the circumscribed figure pass through the angular points of the figure about which it is described, each through each.

3. A rectilineal figure is said to be inscribed in a circle, when all the angles of the inscribed figure are on the circumference of the circle.

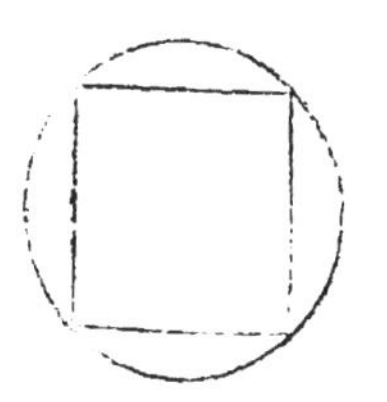

4. A rectilineal figure is said to be described about a circle, when each side of the circumscribed figure touches the circumference of the circle.

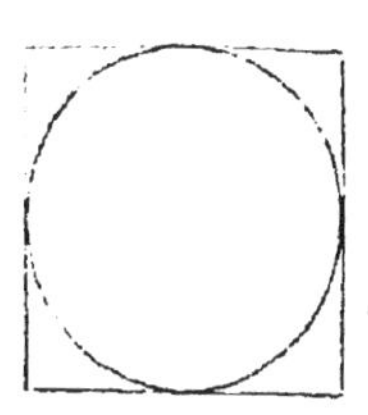

5. In like manner, a circle is said to be inscribed in a rectilineal figure, when the circumference of the circle touches each side of the figure.

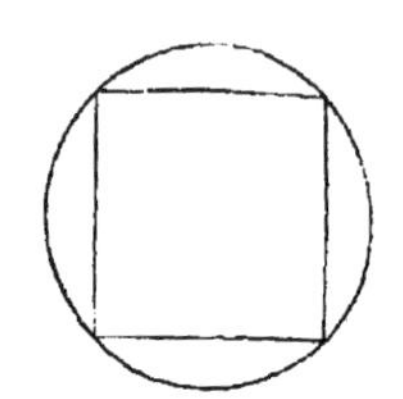

6. A circle is said to be described about a rectilineal figure, when the circumference of the circle passes through all the angular points of the figure about which it is described.

7. A straight line is said to be placed in a circle, when the extremities of it are in the circumference of the circle.

PROPOSITION 1. *PROBLEM.*

In a given circle, to place a straight line, equal to a given straight line, which is not greater than the diameter of the circle.

Let ABC be the given circle, and D the given straight line, not greater than the diameter of the circle: it is required to place in the circle ABC, a straight line equal to D.

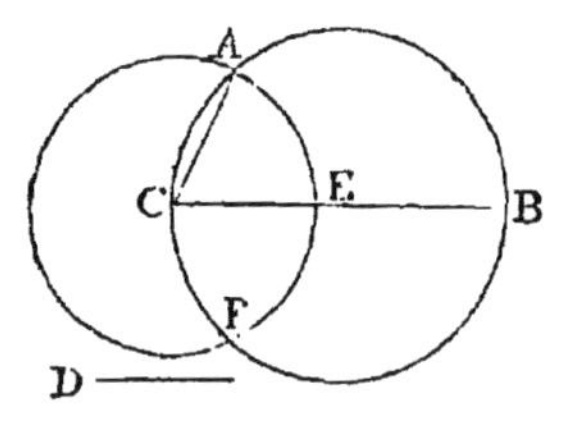

Draw BC, a diameter of the circle ABC.

Then, if BC is equal to D, the thing required is done; for in the circle ABC, a straight line is placed equal to D.

But, if it is not, BC is greater than D. [*Hypothesis.*

Make CE equal to D, [I. 3.

and from the centre C, at the distance CE, describe the circle AEF, and join CA.

Then, because C is the centre of the circle AEF,

CA is equal to CE; [I. *Definition* 15.

but CE is equal to D; [*Construction.*

therefore CA is equal to D. [*Axiom* 1.

Wherefore, *in the circle ABC, a straight line CA is placed equal to the given straight line D, which is not greater than the diameter of the circle.* Q.E.F.

PROPOSITION 2. *PROBLEM.*

In a given circle, to inscribe a triangle equiangular to a given triangle.

Let ABC be the given circle, and DEF the given triangle: it is required to inscribe in the circle ABC a triangle equiangular to the triangle DEF.

Draw the straight line GAH touching the circle at the point A; [III. 17.

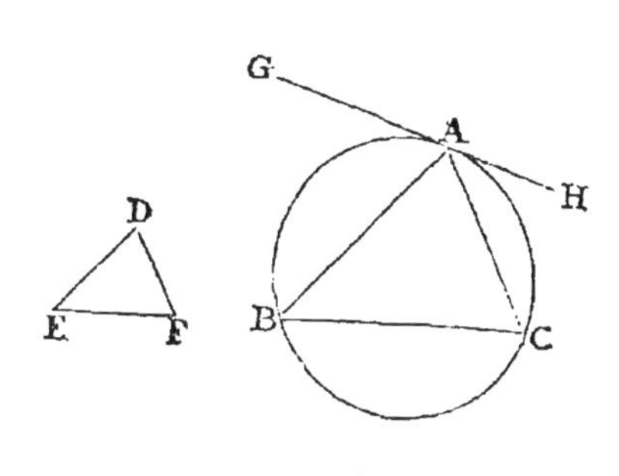

at the point A, in the straight line AH, make the angle HAC equal to the angle DEF; [I. 23.

and, at the point A, in the straight line AG, make the angle GAB equal to the angle DFE;

and join BC. ABC shall be the triangle required.

Because GAH touches the circle ABC, and AC is drawn from the point of contact A, [*Construction.*

therefore the angle HAC is equal to the angle ABC in the alternate segment of the circle. [III. 32.

But the angle HAC is equal to the angle DEF. [*Constr.*

Therefore the angle ABC is equal to the angle DEF. [*Ax.* 1.

For the same reason the angle ACB is equal to the angle DFE.

Therefore the remaining angle BAC is equal to the remaining angle EDF. [I. 32, *Axioms* 11 and 3.

Wherefore *the triangle ABC is equiangular to the triangle DEF, and it is inscribed in the circle ABC.* Q.E.F.

PROPOSITION 3. *PROBLEM.*

About a given circle, to describe a triangle equiangular to a given triangle.

Let ABC be the given circle, and DEF the given triangle: it is required to describe a triangle about the circle ABC, equiangular to the triangle DEF.

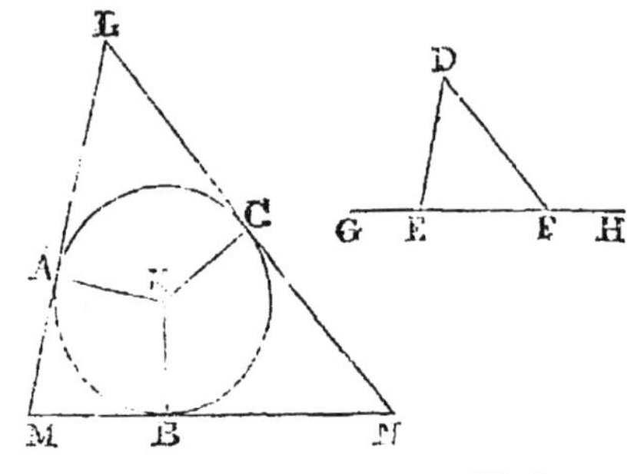

Produce EF both ways to the points G, H; take K the centre of the circle ABC; [III. 1.
from K draw any radius KB;
at the point K, in the straight line KB, make the angle BKA equal to the angle DEG, and the angle BKC equal to the angle DFH; [I. 23.
and through the points A, B, C, draw the straight lines LAM, MBN, NCL, touching the circle ABC. [III. 17.
LMN shall be the triangle required.

Because LM, MN, NL touch the circle ABC at the points A, B, C, [*Construction.*
to which from the centre are drawn KA, KB, KC,
therefore the angles at the points A, B, C are right angles. [III. 18.
And because the four angles of the quadrilateral figure $AMBK$ are together equal to four right angles,
for it can be divided into two triangles,
and that two of them KAM, KBM are right angles,
therefore the other two AKB, AMB are together equal to two right angles. [*Axiom* 3.
But the angles DEG, DEF are together equal to two right angles. [I. 13.
Therefore the angles AKB, AMB are equal to the angles DEG, DEF;
of which the angle AKB is equal to the angle DEG; [*Constr.*
therefore the remaining angle AMB is equal to the remaining angle DEF. [*Axiom* 3.

In the same manner the angle LNM may be shewn to be equal to the angle DFE.

Therefore the remaining angle MLN is equal to the remaining angle EDF. [I. 32, *Axioms* 11 and 3.

Wherefore *the triangle LMN is equiangular to the triangle DEF, and it is described about the circle ABC.* Q.E.F.

PROPOSITION 4. *PROBLEM.*

To inscribe a circle in a given triangle.

Let *ABC* be the given triangle: it is required to inscribe a circle in the triangle *ABC*.

Bisect the angles *ABC*, *ACB*, by the straight lines *BD*, *CD*, meeting one another at the point *D*; [I. 9.
and from *D* draw *DE*, *DF*, *DG* perpendiculars to *AB*, *BC*, *CA*. [I. 12.

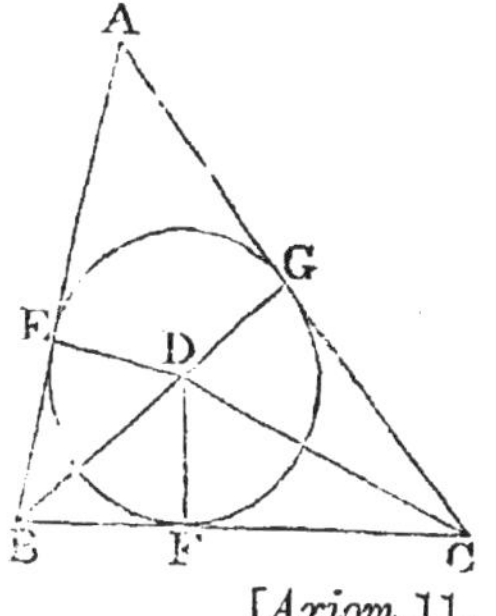

Then, because the angle *EBD* is equal to the angle *FBD*, for the angle *ABC* is bisected by *BD*, [Construction.
and that the right angle *BED* is equal to the right angle *BFD*; [Axiom 11.
therefore the two triangles *EBD*, *FBD* have two angles of the one equal to two angles of the other, each to each; and the side *BD*, which is opposite to one of the equal angles in each, is common to both;
therefore their other sides are equal; [I. 26.
therefore *DE* is equal to *DF*.

For the same reason *DG* is equal to *DF*.
Therefore *DE* is equal to *DG*. [Axiom 1.
Therefore the three straight lines *DE*, *DF*, *DG* are equal to one another, and the circle described from the centre *D*, at the distance of any one of them, will pass through the extremities of the other two;
and it will touch the straight lines *AB*, *BC*, *CA*, because the angles at the points *E*, *F*, *G* are right angles, and the straight line which is drawn from the extremity of a diameter, at right angles to it, touches the circle. [III. 16. *Cor.*
Therefore the straight lines *AB*, *BC*, *CA* do each of them touch the circle, and therefore the circle is inscribed in the triangle *ABC*.

Wherefore *a circle has been inscribed in the given triangle.* Q.E.F.

PROPOSITION 5. *PROBLEM.*

To describe a circle about a given triangle.

Let ABC be the given triangle: it is required to describe a circle about ABC.

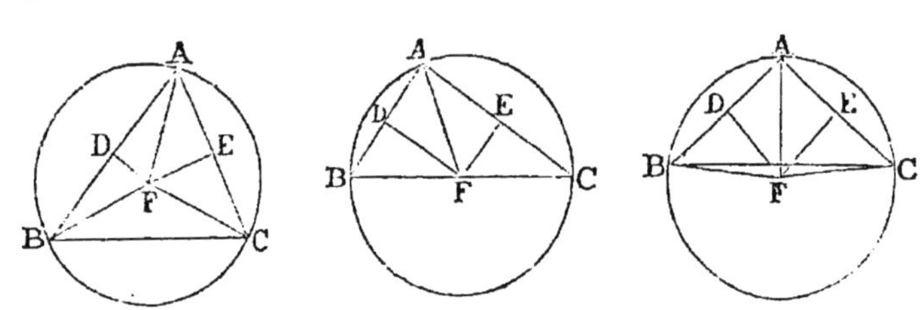

Bisect AB, AC at the points D, E; [I. 10.
from these points draw DF, EF, at right angles to AB, AC; [I. 11.
DF, EF, produced, will meet one another;
for if they do not meet they are parallel,
therefore AB, AC, which are at right angles to them are parallel; which is absurd:
let them meet at F, and join FA; also if the point F be not in BC, join BF, CF.

Then, because AD is equal to BD, [*Construction.*
and DF is common, and at right angles to AB,
therefore the base FA is equal to the base FB. [I. 4.
In the same manner it may be shewn that FC is equal to FA.
Therefore FB is equal to FC; [*Axiom* 1.
and FA, FB, FC are equal to one another.
Therefore the circle described from the centre F, at the distance of any one of them, will pass through the extremities of the other two, and will be described about the triangle ABC.

Wherefore *a triangle has been described about the given circle.* Q.E.F.

COROLLARY. And it is manifest, that when the centre of the circle falls within the triangle, each of its angles is less than a right angle, each of them being in a segment greater than a semicircle; and when the centre is in one of the sides of the triangle, the angle opposite to this side, being in a semicircle, is a right angle; and when the centre falls without the triangle, the angle opposite to the side beyond which it is, being in a segment less than a semicircle, is greater than a right angle. [III. 31.

Therefore, conversely, if the given triangle be acute-angled, the centre of the circle falls within it; if it be a right-angled triangle, the centre is in the side opposite to the right angle; and if it be an obtuse-angled triangle, the centre falls without the triangle, beyond the side opposite to the obtuse angle.

PROPOSITION 6. *PROBLEM.*

To inscribe a square in a given circle.

Let $ABCD$ be the given circle: it is required to inscribe a square in $ABCD$.

Draw two diameters AC, BD of the circle $ABCD$, at right angles to one another; [III. 1, I. 11.

and join AB, BC, CD, DA. The figure $ABCD$ shall be the square required.

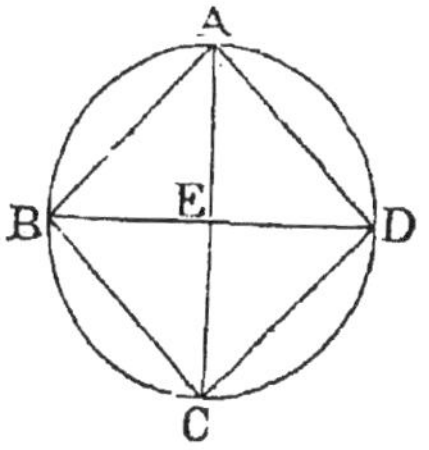

Because BE is equal to DE, for E is the centre;

and that EA is common, and at right angles to BD;

therefore the base BA is equal to the base DA. [I. 4.

And for the same reason BC, DC are each of them equal to BA, or DA.

Therefore the quadrilateral figure $ABCD$ is equilateral.

It is also rectangular.

For the straight line BD being a diameter of the circle $ABCD$, BAD is a semicircle; [*Construction.*

therefore the angle BAD is a right angle. [III. 31.

For the same reason each of the angles ABC, BCD, CDA is a right angle;

therefore the quadrilateral figure $ABCD$ is rectangular.

And it has been shewn to be equilateral; therefore it is a square.

Wherefore *a square has been inscribed in the given circle.* Q.E.F.

PROPOSITION 7. *PROBLEM.*

To describe a square about a given circle.

Let *ABCD* be the given circle: it is required to describe a square about it.

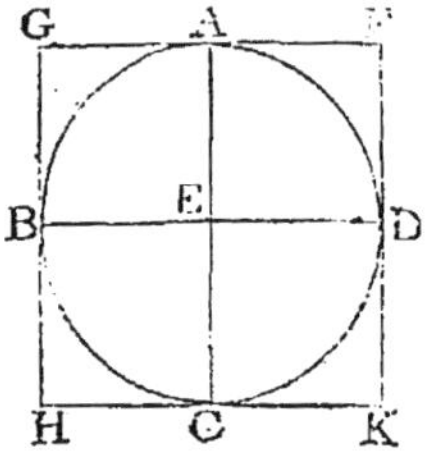

Draw two diameters *AC*, *BD* of the circle *ABCD*, at right angles to one another; [III. 1, I. 11.
and through the points *A*, *B*, *C*, *D*, draw *FG*, *GH*, *HK*, *KF* touching the circle. [III. 17.
The figure *GHKF* shall be the square required.

Because *FG* touches the circle *ABCD*, and *EA* is drawn from the centre *E* to the point of contact *A*, [*Construction.*
therefore the angles at *A* are right angles. [III. 18.
For the same reason the angles at the points *B*, *C*, *D* are right angles.
And because the angle *AEB* is a right angle, [*Construction.*
and also the angle *EBG* is a right angle,
therefore *GH* is parallel to *AC*. [I. 28.
For the same reason *AC* is parallel to *FK*.

In the same manner it may be shewn that each of the lines *GF*, *HK* is parallel to *BD*.
Therefore the figures *GK*, *GC*, *CF*, *FB*, *BK* are parallelograms;
and therefore *GF* is equal to *HK*, and *GH* to *FK*. [I. 34.

And because *AC* is equal to *BD*,
and that *AC* is equal to each of the two *GH*, *FK*,
and that *BD* is equal to each of the two *GF*, *HK*,
therefore *GH*, *FK* are each of them equal to *GF*, or *HK*;
therefore the quadrilateral figure *FGHK* is equilateral.

It is also rectangular.
For since *AEBG* is a parallelogram, and *AEB* a right angle,
therefore *AGB* is also a right angle. [I. 34.
In the same manner it may be shewn that the angles at *H*, *K*, *F* are right angles;

therefore the quadrilateral figure $FGHK$ is rectangular.

And it has been shewn to be equilateral; therefore it is a square.

Wherefore *a square has been described about the given circle.* Q.E.F.

PROPOSITION 8. *PROBLEM.*

To inscribe a circle in a given square.

Let $ABCD$ be the given square: it is required to inscribe a circle in $ABCD$.

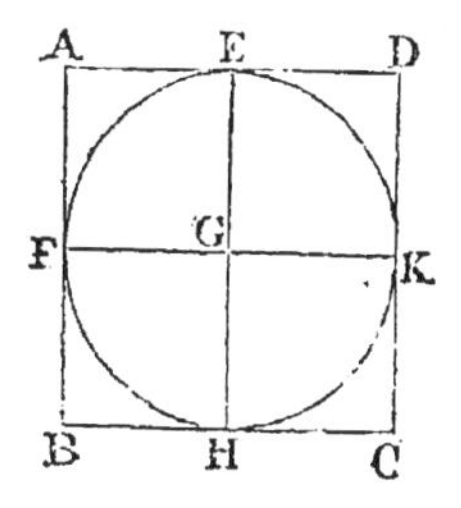

Bisect each of the sides AB, AD at the points F, E; [I. 10.
through E draw EH parallel to AB or DC, and through F draw FK parallel to AD or BC. [I. 31.

Then each of the figures AK, KB, AH, HD, AG, GC, BG, GD is a right-angled parallelogram;
and their opposite sides are equal. [I. 34.
And because AD is equal to AB, [I. *Definition* 30.
and that AE is half of AD, and AF half of AB, [*Constr.*
therefore AE is equal to AF. [*Axiom* 7.
Therefore the sides opposite to these are equal, namely, FG equal to GE. [I. 34.

In the same manner it may be shewn that the straight lines GH, GK are each of them equal to FG or GE.
Therefore the four straight lines GE, GF, GH, GK are equal to one another, and the circle described from the centre G, at the distance of any one of them, will pass through the extremities of the other three;
and it will touch the straight lines AB, BC, CD, DA, because the angles at the points E, F, H, K are right angles, and the straight line which is drawn from the extremity of a diameter, at right angles to it, touches the circle. [III. 16. *Corollary.*

Therefore the straight lines AB, BC, CD, DA do each of them touch the circle.

Wherefore *a circle has been inscribed in the given square.* Q.E.F.

PROPOSITION 9. *PROBLEM.*

To describe a circle about a given square.

Let $ABCD$ be the given square: it is required to describe a circle about $ABCD$.

Join AC, BD, cutting one another at E.

Then, because AB is equal to AD,
and AC is common to the two triangles BAC, DAC;
the two sides BA, AC are equal to the two sides DA, AC each to each;

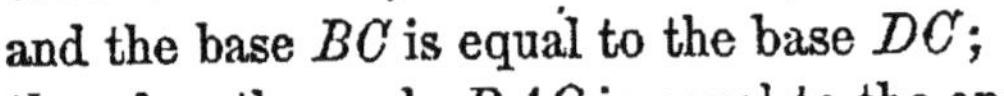

and the base BC is equal to the base DC;
therefore the angle BAC is equal to the angle DAC, [I. 8.
and the angle BAD is bisected by the straight line AC.

In the same manner it may be shewn that the angles ABC, BCD, CDA are severally bisected by the straight lines BD, AC.

Then, because the angle DAB is equal to the angle ABC,
and that the angle EAB is half the angle DAB,
and the angle EBA is half the angle ABC,
therefore the angle EAB is equal to the angle EBA; [*Ax.* 7.
and therefore the side EA is equal to the side EB. [I. 6.

In the same manner it may be shewn that the straight lines EC, ED are each of them equal to EA or EB.
Wherefore the four straight lines EA, EB, EC, ED are equal to one another, and the circle described from the centre E, at the distance of any one of them, will pass through the extremities of the other three, and will be described about the square $ABCD$.

Wherefore *a circle has been described about the given square.* Q.E.F.

PROPOSITION 10. *PROBLEM.*

To describe an isosceles triangle, having each of the angles at the base double of the third angle.

Take any straight line AB, and divide it at the point C, so that the rectangle AB, BC may be equal to the square on AC; [II. 11.

from the centre A, at the distance AB, describe the circle BDE, in which place the straight line BD equal to AC, which is not greater than the diameter of the circle BDE; [IV. 1.

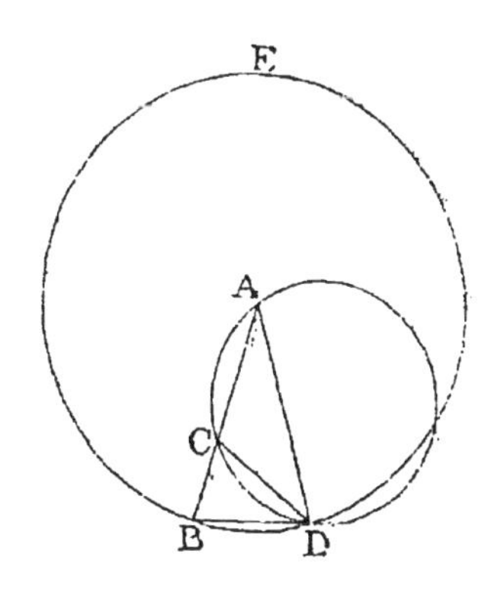

and join DA. The triangle ABD shall be such as is required; that is, each of the angles ABD, ADB shall be double of the third angle BAD.

Join DC; and about the triangle ACD describe the circle ACD. [IV. 5.

Then, because the rectangle AB, BC is equal to the square on AC, [*Construction.*

and that AC is equal to BD, [*Construction.*

therefore the rectangle AB, BC is equal to the square on BD.

And, because from the point B, without the circle ACD, two straight lines BCA, BD are drawn to the circumference, one of which cuts the circle, and the other meets it,

and that the rectangle AB, BC, contained by the whole of the cutting line, and the part of it without the circle, is equal to the square on BD which meets it;

therefore the straight line BD touches the circle ACD. [III. 37.

And, because BD touches the circle ACD, and DC is drawn from the point of contact D,

therefore the angle BDC is equal to the angle DAC in the alternate segment of the circle. [III. 32.

To each of these add the angle CDA;

therefore the whole angle BDA is equal to the two angles CDA, DAC. [*Axiom* 2.

But the exterior angle BCD is equal to the angles CDA, DAC. [I. 32.

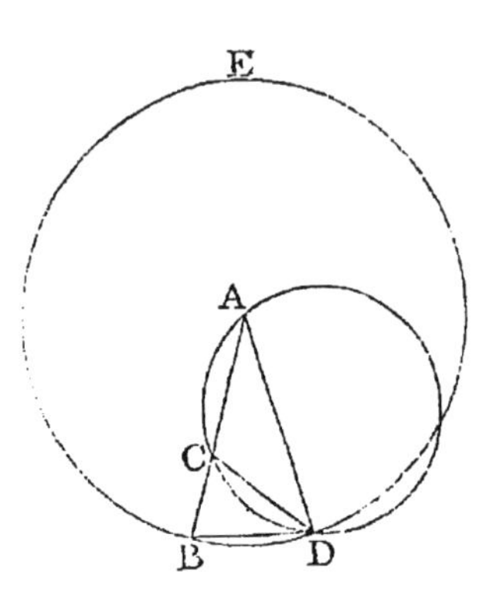

Therefore the angle BDA is equal to the angle BCD. [*Ax.* 1.
But the angle BDA is equal to the angle DBA, [I. 5.
because AD is equal to AB.
Therefore each of the angles BDA, DBA, is equal to the angle BCD. [*Axiom* 6.

And, because the angle DBC is equal to the angle BCD, the side DB is equal to the side DC; [I. 6.
but DB was made equal to CA;
therefore CA is equal to CD, [*Axiom* 6.
and therefore the angle CAD is equal to the angle CDA. [I. 5.
Therefore the angles CAD, CDA are together double of the angle CAD.
But the angle BCD is equal to the angles CAD, CDA. [I. 32.
Therefore the angle BCD is double of the angle CAD.
And the angle BCD has been shewn to be equal to each of the angles BDA, DBA;
therefore each of the angles BDA, DBA is double of the angle BAD.

Wherefore *an isosceles triangle has been described, having each of the angles at the base double of the third angle.* Q.E.F.

PROPOSITION 11. *PROBLEM.*

To inscribe an equilateral and equiangular pentagon in a given circle.

Let $ABCDE$ be the given circle: it is required to inscribe an equilateral and equiangular pentagon in the circle $ABCDE$.

Describe an isosceles triangle, FGH, having each of the angles at G, H, double of the angle at F; [IV. 10.
in the circle $ABCDE$, inscribe the triangle ACD, equiangular to the triangle FGH, so that the angle CAD may

be equal to the angle at F, and each of the angles ACD, ADC equal to the angle at G or H; [IV. 2.

and therefore each of the angles ACD, ADC is double of the angle CAD; bisect the angles ACD, ADC by the straight lines CE, DB; [I. 9. and join AB, BC, AE, ED.

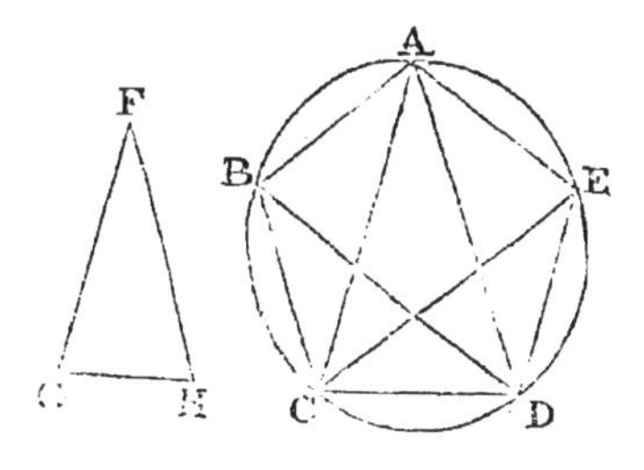

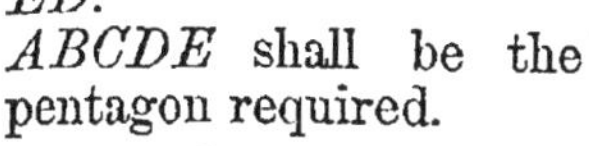

$ABCDE$ shall be the pentagon required.

For because each of the angles ACD, ADC is double of the angle CAD, and that they are bisected by the straight lines CE, DB, therefore the five angles ADB, BDC, CAD, DCE, ECA are equal to one another.

But equal angles stand on equal arcs; [III. 26.

therefore the five arcs AB, BC, CD, DE, EA are equal to one another.

And equal arcs are subtended by equal straight lines; [III. 29.

therefore the five straight lines AB, BC, CD, DE, EA are equal to one another;

and therefore the pentagon $ABCDE$ is equilateral.

It is also equiangular.

For, the arc AB is equal to the arc DE; to each of these add the arc BCD;

therefore the whole arc $ABCD$ is equal to the whole arc $BCDE$. [*Axiom* 2.

And the angle AED stands on the arc $ABCD$, and the angle BAE on the arc $BCDE$.

Therefore the angle AED is equal to the angle BAE. [III. 27.

For the same reason each of the angles ABC, BCD, CDE is equal to the angle AED or BAE;

therefore the pentagon $ABCDE$ is equiangular.

And it has been shewn to be equilateral.

Wherefore *an equilateral and equiangular pentagon has been inscribed in the given circle.* Q.E.F.

PROPOSITION 12. *PROBLEM.*

To describe an equilateral and equiangular pentagon about a given circle.

Let $ABCDE$ be the given circle: it is required to describe an equilateral and equiangular pentagon about the circle $ABCDE$.

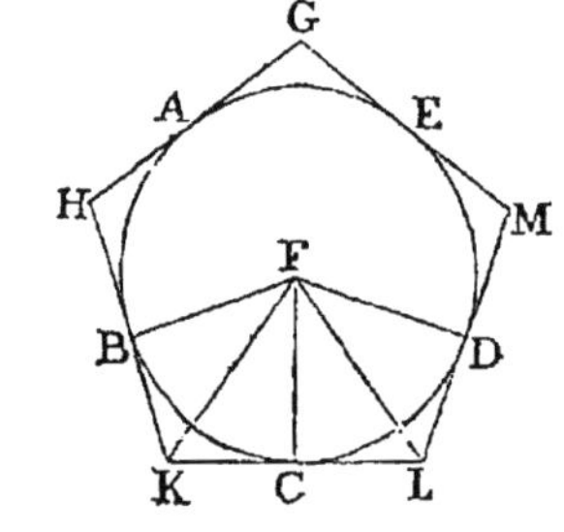

Let the angles of a pentagon, inscribed in the circle, by the last proposition, be at the points A, B, C, D, E, so that the arcs AB, BC, CD, DE, EA are equal; and through the points A, B, C, D, E, draw GH, HK, KL, LM, MG, touching the circle. [III. 17.

The figure $GHKLM$ shall be the pentagon required.

Take the centre F, and join FB, FK, FC, FL, FD.

Then, because the straight line KL touches the circle $ABCDE$ at the point C to which FC is drawn from the centre,

therefore FC is perpendicular to KL, [III. 18.

therefore each of the angles at C is a right angle.

For the same reason the angles at the points B, D are right angles.

And because the angle FCK is a right angle, the square on FK is equal to the squares on FC, CK. [I. 47.

For the same reason the square on FK is equal to the squares on FB, BK.

Therefore the squares on FC, CK are equal to the squares on FB, BK; [*Axiom* 1.

of which the square on FC is equal to the square on FB;

therefore the remaining square on CK is equal to the remaining square on BK, [*Axiom* 3.

and therefore the straight line CK is equal to the straight line BK.

And because FB is equal to FC,
and FK is common to the two triangles BFK, CFK;
the two sides BF, FK are equal to the two sides CF, FK, each to each;
and the base BK was shewn equal to the base CK;
therefore the angle BFK is equal to the angle CFK, [I. 8.
and the angle BKF to the angle CKF. [I. 4.
Therefore the angle BFC is double of the angle CFK, and the angle BKC is double of the angle CKF.

For the same reason the angle CFD is double of the angle CFL, and the angle CLD is double of the angle CLF.

And because the arc BC is equal to the arc CD,
the angle BFC is equal to the angle CFD; [III. 27.
and the angle BFC is double of the angle CFK, and the angle CFD is double of the angle CFL;
therefore the angle CFK is equal to the angle CFL. [*Ax.* 7.
And the right angle FCK is equal to the right angle FCL.
Therefore in the two triangles FCK, FCL, there are two angles of the one equal to two angles of the other, each to each;
and the side FC, which is adjacent to the equal angles in each, is common to both;
therefore their other sides are equal, each to each, and the third angle of the one equal to the third angle of the other;
therefore the straight line CK is equal to the straight line CL, and the angle FKC to the angle FLC. [I. 26.
And because CK is equal to CL, LK is double of CK.

In the same manner it may be shewn that HK is double of BK.

And because BK is equal to CK, as was shewn,
and that HK is double of BK, and LK double of CK,
therefore HK is equal to LK. [*Axiom* 6.

In the same manner it may be shewn that GH, GM, ML are each of them equal to HK or LK;
therefore the pentagon $GHKLM$ is equilateral.

It is also equiangular.

For, since the angle FKC is equal to the angle FLC, and that the angle HKL is double of the angle FKC, and the angle KLM double of the angle FLC, as was shewn, 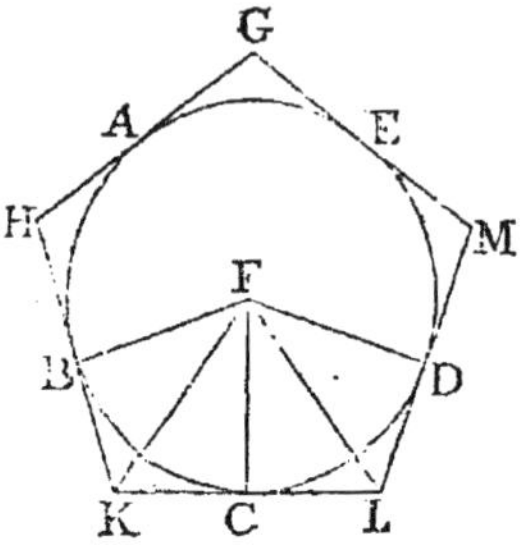
therefore the angle HKL is equal to the angle KLM. [*Axiom* 6.

In the same manner it may be shewn that each of the angles KHG, HGM, GML is equal to the angle HKL or KLM;

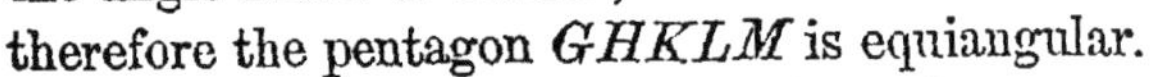

therefore the pentagon $GHKLM$ is equiangular.

And it has been shewn to be equilateral.

Wherefore, *an equilateral and equiangular pentagon has been described about the given circle.* Q.E.F.

PROPOSITION 13. *PROBLEM.*

To inscribe a circle in a given equilateral and equiangular pentagon.

Let $ABCDE$ be the given equilateral and equiangular pentagon: it is required to inscribe a circle in the pentagon $ABCDE$.

Bisect the angles BCD, CDE by the straight lines CF, DF; [I. 9.

and from the point F, at which they meet, draw the straight lines FB, FA, FE.

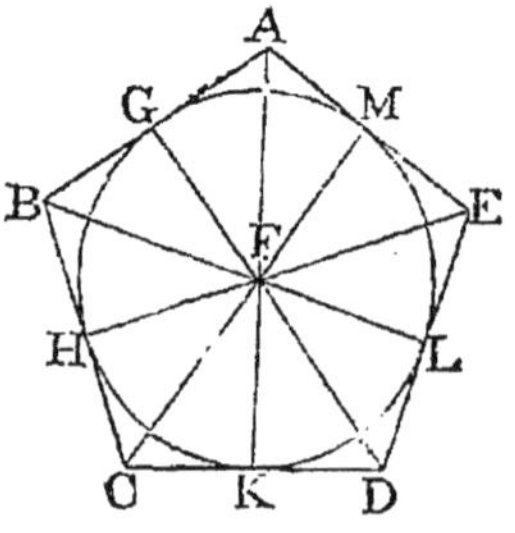

Then, because BC is equal to DC, [*Hypothesis.*

and CF is common to the two triangles BCF, DCF;

the two sides BC, CF are equal to the two sides DC, CF, each to each;

and the angle BCF is equal to the angle DCF; [*Constr.*

therefore the base BF is equal to the base DF, and the

other angles to the other angles to which the equal sides are opposite; [I. 4.
therefore the angle *CBF* is equal to the angle *CDF*.

And because the angle *CDE* is double of the angle *CDF*, and that the angle *CDE* is equal to the angle *CBA*, and the angle *CDF* is equal to the angle *CBF*,
therefore the angle *CBA* is double of the angle *CBF*;
therefore the angle *ABF* is equal to the angle *CBF*;
therefore the angle *ABC* is bisected by the straight line *BF*.

In the same manner it may be shewn that the angles *BAE*, *AED* are bisected by the straight lines *AF*, *EF*.

From the point *F* draw *FG*, *FH*, *FK*, *FL*, *FM* perpendiculars to the straight lines *AB*, *BC*, *CD*, *DE*, *EA*. [I. 12.

Then, because the angle *FCH* is equal to the angle *FCK*,
and the right angle *FHC* equal to the right angle *FKC*;
therefore in the two triangles *FHC*, *FKC*, there are two angles of the one equal to two angles of the other, each to each;
and the side *FC*, which is opposite to one of the equal angles in each, is common to both;
therefore their other sides are equal, each to each, and therefore the perpendicular *FH* is equal to the perpendicular *FK*. [I. 26.

In the same manner it may be shewn that *FL*, *FM*, *FG* are each of them equal to *FH* or *FK*.
Therefore the five straight lines *FG*, *FH*, *FK*, *FL*, *FM* are equal to one another, and the circle described from the centre *F*, at the distance of any one of them will pass through the extremities of the other four;
and it will touch the straight lines *AB*, *BC*, *CD*, *DE*, *EA*, because the angles at the points *G*, *H*, *K*, *L*, *M* are right angles, [*Construction.*
and the straight line drawn from the extremity of a diameter, at right angles to it, touches the circle; [III. 16.
Therefore each of the straight lines *AB*, *BC*, *CD*, *DE*, *EA* touches the circle.

Wherefore *a circle has been inscribed in the given equilateral and equiangular pentagon.* Q.E.F.

PROPOSITION 14. *PROBLEM.*

To describe a circle about a given equilateral and equiangular pentagon.

Let $ABCDE$ be the given equilateral and equiangular pentagon: it is required to describe a circle about it.

Bisect the angles BCD, CDE by the straight lines CF, DF; [I. 9. and from the point F, at which they meet, draw the straight lines FB, FA, FE.

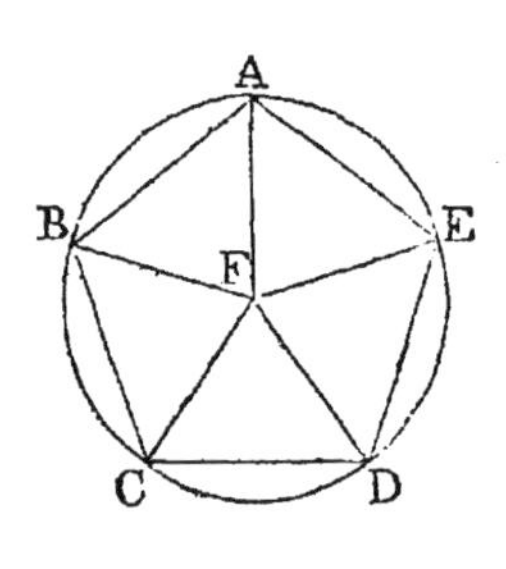

Then it may be shewn, as in the preceding proposition, that the angles CBA, BAE, AED are bisected by the straight lines BF, AF, EF.

And, because the angle BCD is equal to the angle CDE, and that the angle FCD is half of the angle BCD, and the angle FDC is half of the angle CDE, therefore the angle FCD is equal to the angle FDC; [*Ax.* 7. therefore the side FC is equal to the side FD. [I. 6.

In the same manner it may be shewn that FB, FA, FE are each of them equal to FC or FD; therefore the five straight lines FA, FB, FC, FD, FE are equal to one another, and the circle described from the centre F, at the distance of any one of them, will pass through the extremities of the other four, and will be described about the equilateral and equiangular pentagon $ABCDE$.

Wherefore *a circle has been described about the given equilateral and equiangular pentagon.* Q.E.F.

PROPOSITION 15. *PROBLEM.*

To inscribe an equilateral and equiangular hexagon in a given circle.

Let $ABCDEF$ be the given circle: it is required to inscribe an equilateral and equiangular hexagon in it.

Find the centre G of the circle $ABCDEF$, [III. 1.

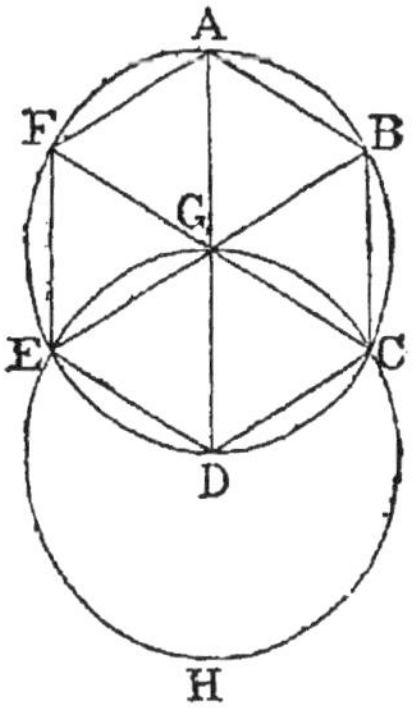

and draw the diameter AGD;

from the centre D, at the distance DG, describe the circle $EGCH$;

join EG, CG, and produce them to the points B, F; and join AB, BC, CD, DE, EF, FA.

The hexagon $ABCDEF$ shall be equilateral and equiangular.

For, because G is the centre of the circle $ABCDEF$, GE is equal to GD;

and because D is the centre of the circle $EGCH$, DE is equal to DG;

therefore GE is equal to DE, [*Axiom* 1.

and the triangle EGD is equilateral;

therefore the three angles EGD, GDE, DEG are equal to one another. [I. 5. *Corollary.*

But the three angles of a triangle are together equal to two right angles; [I. 32.

therefore the angle EGD is the third part of two right angles.

In the same manner it may be shewn, that the angle DGC is the third part of two right angles.

And because the straight line GC makes with the straight line EB the adjacent angles EGC, CGB together equal to two right angles, [I. 13.

therefore the remaining angle CGB is the third part of two right angles;

therefore the angles EGD, DGC, CGB are equal to one another.

And to these are equal the vertical opposite angles BGA, AGF, FGE. [I. 15.

Therefore the six angles EGD, DGC, CGB, BGA, AGF, FGE are equal to one another.

But equal angles stand on equal arcs; [III. 26.
therefore the six arcs AB, BC, CD, DE, EF, FA are equal to one another.

And equal arcs are subtended by equal straight lines; [III. 29.
therefore the six straight lines are equal to one another, and the hexagon is equilateral.

It is also equiangular.

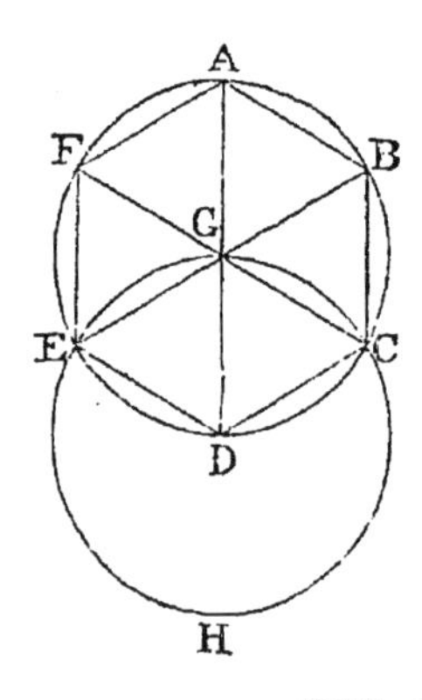

For, the arc AF is equal to the arc ED;
to each of these add the arc $ABCD$;
therefore the whole arc $FABCD$ is equal to the whole arc $ABCDE$;
and the angle FED stands on the arc $FABCD$,
and the angle AFE stands on the arc $ABCDE$;
therefore the angle FED is equal to the angle AFE. [III. 27.

In the same manner it may be shewn that the other angles of the hexagon $ABCDEF$ are each of them equal to the angle AFE or FED;
therefore the hexagon is equiangular.

And it has been shewn to be equilateral; and it is inscribed in the circle $ABCDEF$.

Wherefore *an equilateral and equiangular hexagon has been inscribed in the given circle.* Q.E.F.

COROLLARY. From this it is manifest that the side of the hexagon is equal to the straight line from the centre, that is, to the semidiameter of the circle.

Also, if through the points A, B, C, D, E, F, there be drawn straight lines touching the circle, an equilateral and equiangular hexagon will be described about the circle, as may be shewn from what was said of the pentagon; and a circle may be inscribed in a given equilateral and equiangular hexagon, and circumscribed about it, by a method like that used for the pentagon.

PROPOSITION 16. *PROBLEM.*

To inscribe an equilateral and equiangular quindecagon in a given circle.

Let $ABCD$ be the given circle: it is required to inscribe an equilateral and equiangular quindecagon in the circle $ABCD$.

Let AC be the side of an equilateral triangle inscribed in the circle; [IV. 2.

and let AB be the side of an equilateral and equiangular pentagon inscribed in the circle. [IV. 11.

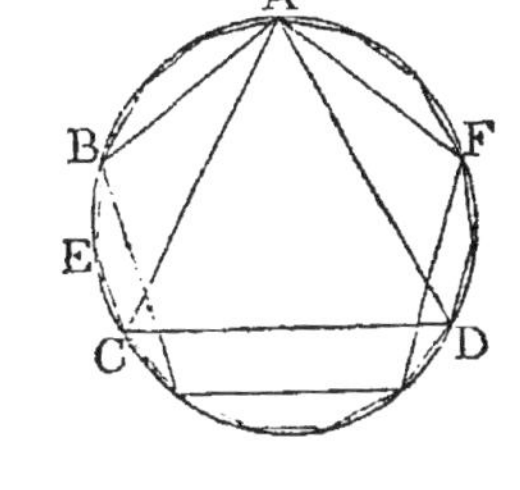

Then, of such equal parts as the whole circumference $ABCDF$ contains fifteen, the arc ABC, which is the third part of the whole, contains five, and the arc AB, which is the fifth part of the whole, contains three;

therefore their difference, the arc BC, contains two of the same parts.

Bisect the arc BC at E; [III. 30.

therefore each of the arcs BE, EC is the fifteenth part of the whole circumference $ABCDF$.

Therefore if the straight lines BE, EC be drawn, and straight lines equal to them be placed round in the whole circle, [IV. 1.

an equilateral and equiangular quindecagon will be inscribed in it. Q.E.F.

And, in the same manner as was done for the pentagon, if through the points of division made by inscribing the quindecagon, straight lines be drawn touching the circle, an equilateral and equiangular quindecagon will be described about it; and also, as for the pentagon, a circle may be inscribed in a given equilateral and equiangular quindecagon, and circumscribed about it.

BOOK V.

DEFINITIONS.

1. A LESS magnitude is said to be a part of a greater magnitude, when the less measures the greater; that is, when the less is contained a certain number of times exactly in the greater.

2. A greater magnitude is said to be a multiple of a less, when the greater is measured by the less; that is, when the greater contains the less a certain number of times exactly.

3. Ratio is a mutual relation of two magnitudes of the same kind to one another in respect of quantity.

4. Magnitudes are said to have a ratio to one another, when the less can be multiplied so as to exceed the other.

5. The first of four magnitudes is said to have the same ratio to the second, that the third has to the fourth, when any equimultiples whatever of the first and the third being taken, and any equimultiples whatever of the second and the fourth, if the multiple of the first be less than that of the second, the multiple of the third is also less than that of the fourth, and if the multiple of the first be equal to that of the second, the multiple of the third is also equal to that of the fourth, and if the multiple of the first be greater than that of the second, the multiple of the third is also greater than that of the fourth.

6. Magnitudes which have the same ratio are called proportionals.

When four magnitudes are proportionals it is usually expressed by saying, the first is to the second as the third is to the fourth.

7. When of the equimultiples of four magnitudes, taken as in the fifth definition, the multiple of the first is greater than the multiple of the second, but the multiple of the third is not greater than the multiple of the fourth, then the first is said to have to the second a greater ratio than the third has to the fourth; and the third is said to have to the fourth a less ratio than the first has to the second.

8. Analogy, or proportion, is the similitude of ratios.

9. Proportion consists in three terms at least.

10. When three magnitudes are proportionals, the first is said to have to the third the duplicate ratio of that which it has to the second.

[The second magnitude is said to be a *mean proportional* between the first and the third.]

11. When four magnitudes are continued proportionals, the first is said to have to the fourth, the triplicate ratio of that which it has to the second, and so on, quadruplicate, &c. increasing the denomination still by unity, in any number of proportionals.

Definition of compound ratio. When there are any number of magnitudes of the same kind, the first is said to have to the last of them, the ratio which is compounded of the ratio which the first has to the second, and of the ratio which the second has to the third, and of the ratio which the third has to the fourth, and so on unto the last magnitude.

For example, if A, B, C, D be four magnitudes of the same kind, the first A is said to have to the last D, the ratio compounded of the ratio of A to B, and of the ratio of B to C, and of the ratio of C to D; or, the ratio of A to D is said to be compounded of the ratios of A to B, B to C, and C to D.

And if A has to B the same ratio that E has to F; and B to C the same ratio that G has to H; and C to D the same ratio that K has to L; then, by this definition, A is said to have to D the ratio compounded of ratios which are the same with the ratios of E to F, G to H, and K to L.

And the same thing is to be understood when it is more briefly expressed by saying, *A* has to *D* the ratio compounded of the ratios of *E* to *F*, *G* to *H*, and *K* to *L*.

In like manner, the same things being supposed, if *M* has to *N* the same ratio that *A* has to *D*; then, for the sake of shortness, *M* is said to have to *N* the ratio compounded of the ratios of *E* to *F*, *G* to *H*, and *K* to *L*.

12. In proportionals, the antecedent terms are said to be homologous to one another; as also the consequents to one another.

Geometers make use of the following technical words, to signify certain ways of changing either the order or the magnitude of proportionals, so that they continue still to be proportionals.

13. *Permutando*, or *alternando*, by permutation or alternately; when there are four proportionals, and it is inferred that the first is to the third, as the second is to the fourth. V. 16.

14. *Invertendo*, by inversion; when there are four proportionals, and it is inferred, that the second is to the first as the fourth is to the third. V. *B*.

15. *Componendo*, by composition; when there are four proportionals, and it is inferred, that the first together with the second, is to the second, as the third together with the fourth, is to the fourth. V. 18.

16. *Dividendo*, by division; when there are four proportionals, and it is inferred, that the excess of the first above the second, is to the second, as the excess of the third above the fourth, is to the fourth. V. 17.

17. *Convertendo*, by conversion; when there are four proportionals, and it is inferred, that the first is to its excess above the second, as the third is to its excess above the fourth. V. *E*.

18. *Ex æquali distantia*, or *ex æquo*, from equality of distance; when there is any number of magnitudes more than two, and as many others, such that they are proportionals when taken two and two of each rank, and it is inferred, that the first is to the last of the first rank of magnitudes, as the first is to the last of the others.

Of this there are the two following kinds, which arise from the different order in which the magnitudes are taken, two and two.

19. *Ex æquali.* This term is used simply by itself, when the first magnitude is to the second of the first rank, as the first is to the second of the other rank; and the second is to the third of the first rank, as the second is to the third of the other; and so on in order; and the inference is that mentioned in the preceding definition. V. 22.

20. *Ex æquali in proportione perturbatâ seu inordinatâ,* from equality in perturbate or disorderly proportion. This term is used when the first magnitude is to the second of the first rank, as the last but one is to the last of the second rank; and the second is to the third of the first rank, as the last but two is to the last but one of the second rank; and the third is to the fourth of the first rank, as the last but three is to the last but two of the second rank; and so on in a cross order; and the inference is that mentioned in the eighteenth definition. V. 23.

AXIOMS.

1. Equimultiples of the same, or of equal magnitudes, are equal to one another.

2. Those magnitudes, of which the same or equal magnitudes are equimultiples, are equal to one another.

3. A multiple of a greater magnitude is greater than the same multiple of a less.

4. That magnitude, of which a multiple is greater than the same multiple of another, is greater than that other magnitude.

PROPOSITION 1. *THEOREM.*

If any number of magnitudes be equimultiples of as many, each of each; whatever multiple any one of them is of its part, the same multiple shall all the first magnitudes be of all the other.

Let any number of magnitudes *AB*, *CD* be equimultiples of as many others *E*, *F*, each of each: whatever multiple *AB* is of *E*, the same multiple shall *AB* and *CD* together, be of *E* and *F* together.

For, because *AB* is the same multiple of *E*, that *CD* is of *F*, as many magnitudes as there are in *AB* equal to *E*, so many are there in *CD* equal to *F*.

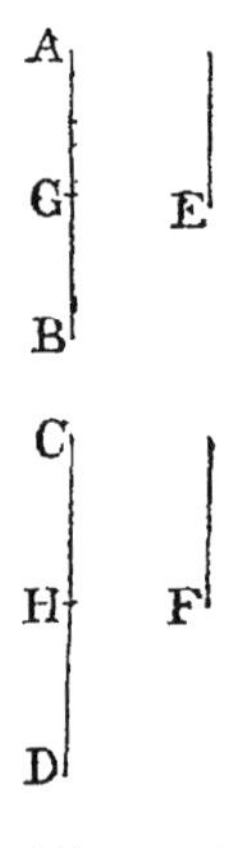

Divide *AB* into the magnitudes *AG*, *GB*, each equal to *E*; and *CD* into the magnitudes *CH*, *HD*, each equal to *F*.

Therefore the number of the magnitudes *CH*, *HD*, will be equal to the number of the magnitudes *AG*, *GB*.

And, because *AG* is equal to *E*, and *CH* equal to *F*, therefore *AG* and *CH* together are equal to *E* and *F* together; and because *GB* is equal to *E*, and *HD* equal to *F*, therefore *GB* and *HD* together are equal to *E* and *F* together. [*Axiom* 2.

Therefore as many magnitudes as there are in *AB* equal to *E*, so many are there in *AB* and *CD* together equal to *E* and *F* together.

Therefore whatever multiple *AB* is of *E*, the same multiple is *AB* and *CD* together, of *E* and *F* together.

Wherefore, *if any number of magnitudes* &c. Q.E.D.

PROPOSITION 2. *THEOREM.*

If the first be the same multiple of the second that the third is of the fourth, and the fifth the same multiple of the second that the sixth is of the fourth; the first together with the fifth shall be the same multiple of the second, that the third together with the sixth is of the fourth.

Let AB the first be the same multiple of C the second, that DE the third is of F the fourth, and let BG the fifth be the same multiple of C the second, that EH the sixth is of F the fourth: AG, the first together with the fifth, shall be the same multiple of C the second, that DH, the third together with the sixth, is of F the fourth.

For, because AB is the same multiple of C that DE is of F, as many magnitudes as there are in AB equal to C, so many are there in DE equal to F.

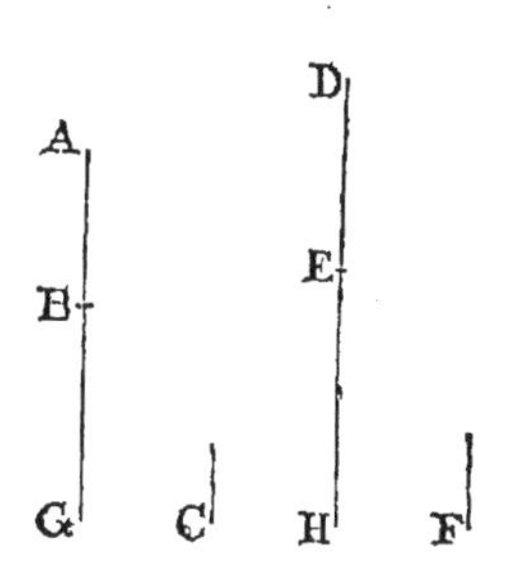

For the same reason, as many magnitudes as there are in BG equal to C, so many are there in EH equal to F.

Therefore as many magnitudes as there are in the whole AG equal to C, so many are there in the whole DH equal to F.

Therefore AG is the same multiple of C that DH is of F.

Wherefore, *if the first be the same multiple* &c. Q.E.D.

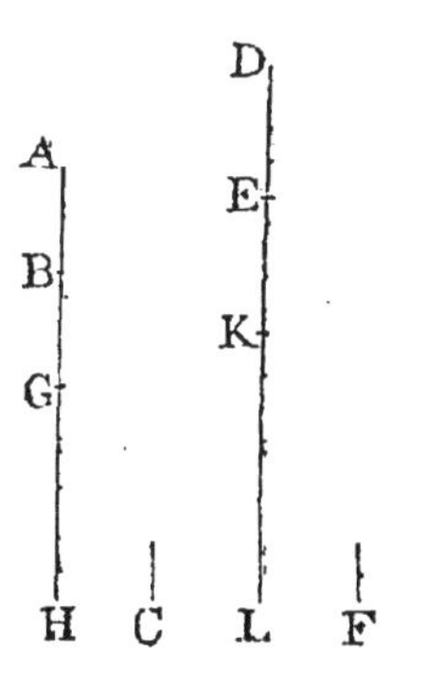

COROLLARY. From this it is plain, that if any number of magnitudes AB, BG, GH be multiples of another C; and as many DE, EK, KL be the same multiples of F, each of each; then the whole of the first, namely, AH, is the same multiple of C, that the whole of the last, namely, DL, is of F.

PROPOSITION 3. *THEOREM.*

If the first be the same multiple of the second that the third is of the fourth, and if of the first and the third there be taken equimultiples, these shall be equimultiples, the one of the second, and the other of the fourth.

Let A the first be the same multiple of B the second, that C the third is of D the fourth; and of A and C let the equimultiples EF and GH be taken: EF shall be the same multiple of B that GH is of D.

For, because EF is the same multiple of A that GH is of C, [*Hypothesis.*

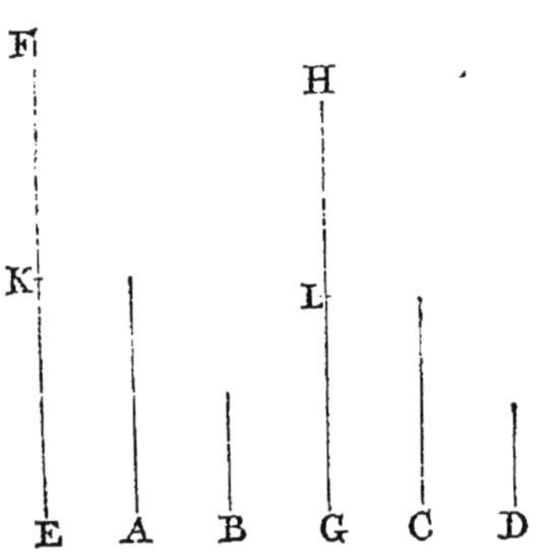

as many magnitudes as there are in EF equal to A, so many are there in GH equal to C.

Divide EF into the magnitudes EK, KF, each equal to A; and GH into the magnitudes GL, LH, each equal to C.

Therefore the number of the magnitudes EK, KF, will be equal to the number of the magnitudes GL, LH.

And because A is the same multiple of B that C is of D, [*Hypothesis.*

and that EK is equal to A, and GL is equal to C; [*Constr.*

therefore EK is the same multiple of B that GL is of D.

For the same reason KF is the same multiple of B that LH is of D.

Therefore because EK the first is the same multiple of B the second, that GL the third is of D the fourth,

and that KF the fifth is the same multiple of B the second, that LH the sixth is of D the fourth;

EF the first together with the fifth, is the same multiple of B the second, that GH the third together with the sixth, is of D the fourth. [V. 2.

In the same manner, if there be more parts in EF equal to A and in GH equal to C, it may be shewn that EF is the same multiple of B that GH is of D. [V. 2, *Cor.*

Wherefore, *if the first* &c. Q.E.D.

PROPOSITION 4. *THEOREM.*

If the first have the same ratio to the second that the third has to the fourth, and if there be taken any equi-

multiples whatever of the first and the third, and also any equimultiples whatever of the second and the fourth, then the multiple of the first shall have the same ratio to the multiple of the second, that the multiple of the third has to the multiple of the fourth.

Let A the first have to B the second, the same ratio that C the third has to D the fourth; and of A and C let there be taken any equimultiples whatever E and F, and of B and D any equimultiples whatever G and H: E shall have the same ratio to G that F has to H.

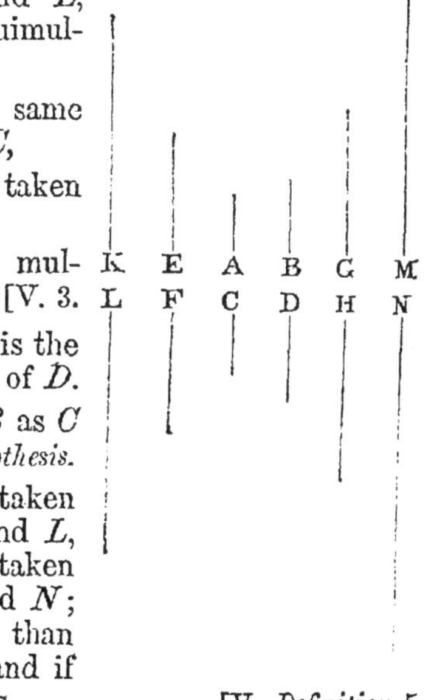

Take of E and F any equimultiples whatever K and L, and of G and H any equimultiples whatever M and N.

Then, because E is the same multiple of A that F is of C,

and of E and F have been taken equimultiples K and L;

therefore K is the same multiple of A that L is of C. [V. 3.

For the same reason, M is the same multiple of B that N is of D.

And because A is to B as C is to D, [*Hypothesis.*

and of A and C have been taken certain equimultiples K and L, and of B and D have been taken certain equimultiples M and N; therefore if K be greater than M, L is greater than N; and if equal, equal; and if less, less. [V. *Definition* 5.

But K and L are any equimultiples whatever of E and F, and M and N are any equimultiples whatever of G and H; therefore E is to G as F is to H. [V. *Definition* 5.

Wherefore, *if the first* &c. Q.E.D.

COROLLARY. Also if the first have the same ratio to the second that the third has to the fourth, then any equimultiples whatever of the first and third shall have the same ratio to the second and fourth; and the first and third shall have the same ratio to any equimultiples whatever of the second and fourth.

Let A the first have the same ratio to B the second, that C the third has to D the fourth; and of A and C let there be taken any equimultiples whatever E and F: E shall be to B as F is to D.

Take of E and F any equimultiples whatever K and L, and of B and D any equimultiples whatever G and H.

Then it may be shewn, as before, that K is the same multiple of A that L is of C.

And because A is to B as C is to D, [*Hypothesis.*

and of A and C have been taken certain equimultiples K and L, and of B and D have been taken certain equimultiples G and H;

therefore if K be greater than G, L is greater than H; and if equal, equal; and if less, less. [V. *Definition* 5.

But K and L are any equimultiples whatever of E and F, and G and H are any equimultiples whatever of B and D;

therefore E is to B as F is to D. [V. *Definition* 5.

In the same way the other case may be demonstrated.

PROPOSITION 5. *THEOREM.*

If one magnitude be the same multiple of another that a magnitude taken from the first is of a magnitude taken from the other, the remainder shall be the same multiple of the remainder that the whole is of the whole.

Let AB be the same multiple of CD, that AE taken from the first, is of CF taken from the other: the remainder EB shall be the same multiple of the remainder FD, that the whole AB is of the whole CD.

Take AG the same multiple of FD, that AE is of CF; therefore AE is the same multiple of CF that EG is of CD. [V. 1.

But AE is the same multiple of CF that AB is of CD;

therefore EG is the same multiple of CD that AB is of CD;

therefore EG is equal to AB. [V. *Axiom* 1.

From each of these take the common magnitude AE; then the remainder AG is equal to the remainder EB.

Then, because AE is the same multiple of CF that AG is of FD, [*Construction.*

and that AG is equal to EB;

therefore AE is the same multiple of CF that EB is of FD.

But AE is the same multiple of CF that AB is of CD; [*Hypothesis.*

therefore EB is the same multiple of FD that AB is of CD.

Wherefore, *if one magnitude* &c. Q.E.D.

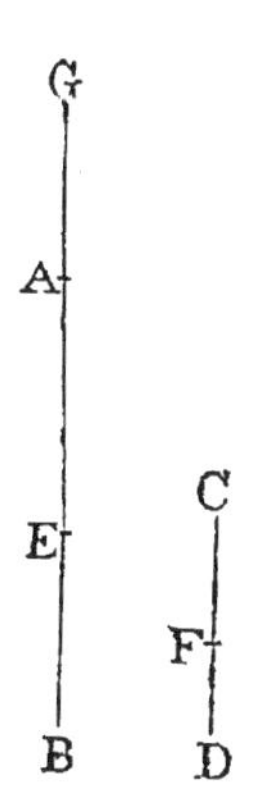

PROPOSITION 6. *THEOREM.*

If two magnitudes be equimultiples of two others, and if equimultiples of these be taken from the first two, the remainders shall be either equal to these others, or equimultiples of them.

Let the two magnitudes AB, CD be equimultiples of the two E, F; and let AG, CH, taken from the first two, be equimultiples of the same E, F: the remainders GB, HD shall be either equal to E, F, or equimultiples of them.

First, let GB be equal to E: HD shall be equal to F. Make CK equal to F.

Then, because AG is the same multiple of E that CH is of F, [*Hyp.*

and that GB is equal to E, and CK is equal to F;

therefore AB is the same multiple of E that KH is of F.

But AB is the same multiple of E that CD is of F; [*Hypothesis.*

therefore KH is the same multiple of F that CD is of F;

therefore KH is equal to CD. [V. *Axiom* 1.

From each of these take the common magnitude CH; then the remainder CK is equal to the remainder HD.

But CK is equal to F; [*Construction.*

therefore HD is equal to F.

Next let GB be a multiple of E: HD shall be the same multiple of F.

Make CK the same multiple of F that GB is of E.

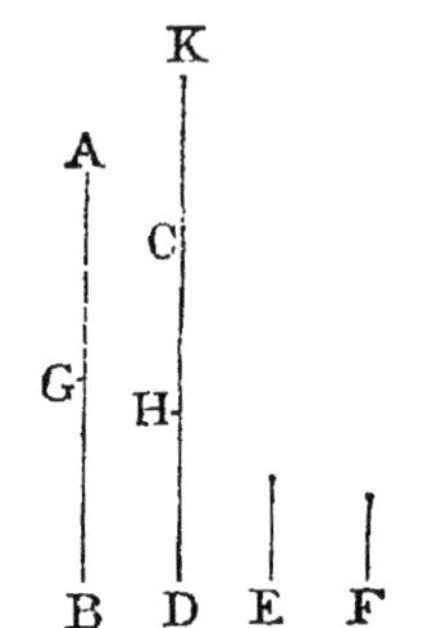

Then, because AG is the same multiple of E that CH is of F, [*Hypothesis.*

and GB is the same multiple of E that CK is of F; [*Constr.*

therefore AB is the same multiple of E that KH is of F. [V. 2.

But AB is the same multiple of E that CD is of F; [*Hyp.*

therefore KH is the same multiple of F that CD is of F; therefore KH is equal to CD. [V. *Axiom* 1.

From each of these take the common magnitude CH; then the remainder CK is equal to the remainder HD.

And because CK is the same multiple of F that GB is of E, [*Construction.*

and that CK is equal to HD;

therefore HD is the same multiple of F that GB is of E.

Wherefore, *if two magnitudes* &c. Q.E.D.

PROPOSITION *A*. *THEOREM.*

If the first of four magnitudes have the same ratio to the second that the third has to the fourth, then, if the first be greater than the second, the third shall also be greater than the fourth, and if equal equal, and if less less.

Take any equimultiples of each of them, as the doubles of each.

Then if the double of the first be greater than the double of the second, the double of the third is greater than the double of the fourth. [V. *Definition* 5.

But if the first be greater than the second, the double of the first is greater than the double of the second;

therefore the double of the third is greater than the double of the fourth,

and therefore the third is greater than the fourth.

In the same manner, if the first be equal to the second, or less than it, the third may be shewn to be equal to the fourth, or less than it.

Wherefore, *if the first* &c. Q.E.D.

PROPOSITION *B. THEOREM.*

If four magnitudes be proportionals, they shall also be proportionals when taken inversely.

Let A be to B as C is to D: then also, inversely, B shall be to A as D is to C.

Take of B and D any equimultiples whatever E and F;

and of A and C any equimultiples whatever G and H.

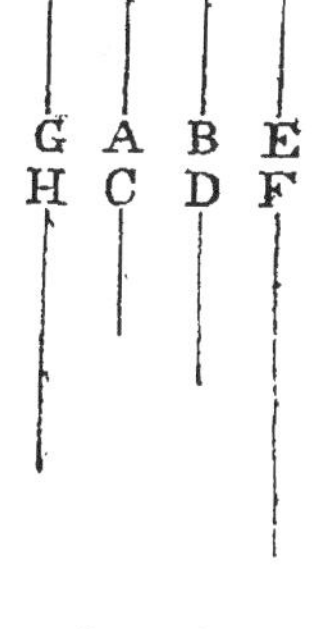

First, let E be greater than G, then G is less than E.

Then, because A is to B as C is to D; [*Hypothesis.*

and of A and C the first and third, G and H are equimultiples;

and of B and D the second and fourth, E and F are equimultiples;

and that G is less than E;

therefore H is less than F; [V. *Def.* 5.

that is, F is greater than H.

Therefore, if E be greater than G, F is greater than H.

In the same manner, if E be equal to G, F may be shewn to be equal to H; and if less, less.

But E and F are any equimultiples whatever of B and D, and G and H are any equimultiples whatever of A and C; [*Construction.*

therefore B is to A as D is to C. [V. *Definition* 5.

Wherefore, *if four magnitudes* &c. Q.E.D.

PROPOSITION C. *THEOREM.*

If the first be the same multiple of the second, or the same part of it, that the third is of the fourth, the first shall be to the second as the third is to the fourth.

First, let A be the same multiple of B that C is of D: A shall be to B as C is to D.

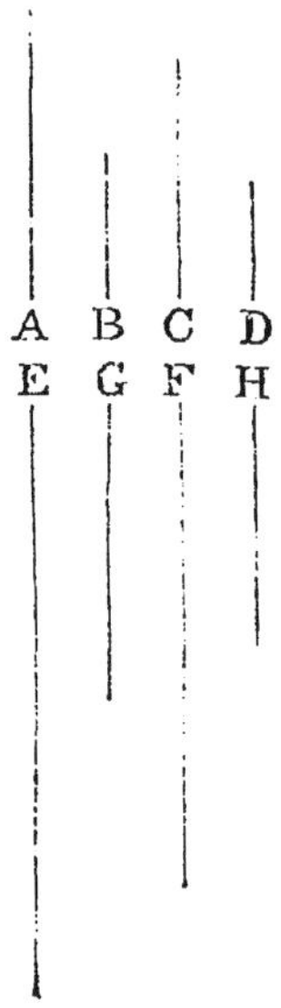

Take of A and C any equimultiples whatever E and F; and of B and D any equimultiples whatever G and H.

A B C D
E G F H

Then, because A is the same multiple of B that C is of D; [*Hypothesis.*
and that E is the same multiple of A that F is of C; [*Construction.*
therefore E is the same multiple of B that F is of D; [V. 3.
that is, E and F are equimultiples of B and D.

But G and H are equimultiples of B and D; [*Construction.*
therefore if E be a greater multiple of B than G is of B, F is a greater multiple of D than H is of D;
that is, if E be greater than G, F is greater than H.

In the same manner, if E be equal to G, F may be shewn to be equal to H; and if less, less.

But E and F are any equimultiples whatever of A and C, and G and H are any equimultiples whatever of B and D; [*Construction.*
therefore A is to B as C is to D. [V. *Definition* 5.

Next, let A be the same part of B that C is of D: A shall be to B as C is to D.

For, since A is the same part of B that C is of D,
therefore B is the same multiple of A that D is of C;
therefore, by the preceding case, B is to A as D is to C;

A B C D

therefore, inversely, A is to B as C is to D. [V. *B.*

Wherefore, *if the first* &c. Q.E.D.

PROPOSITION D. THEOREM.

If the first be to the second as the third is to the fourth, and if the first be a multiple, or a part, of the second, the third shall be the same multiple, or the same part, of the fourth.

Let A be to B as C is to D.

And first, let A be a multiple of B: C shall be the same multiple of D.

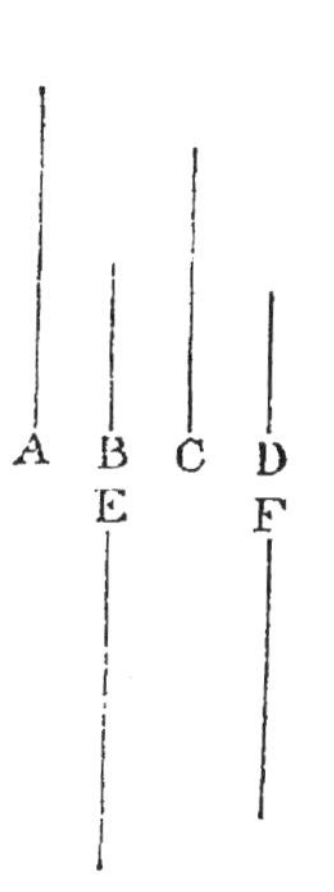

Take E equal to A; and whatever multiple A or E is of B, make F the same multiple of D.

Then, because A is to B as C is to D, [*Hypothesis.*

and of B the second and D the fourth have been taken equimultiples E and F; [*Construction.*

therefore A is to E as C is to F. [V. 4, *Corollary.*

But A is equal to E; [*Construction.*

therefore C is equal to F. [V. *A.*

And F is the same multiple of D that A is of B; [*Construction.*

therefore C is the same multiple of D that A is of B.

Next, let A be a part of B: C shall be the same part of D.

For, because A is to B as C is to D; [*Hypothesis.*

therefore, inversely, B is to A as D is to C. [V. *B.*

But A is a part of B; [*Hypothesis.*

that is, B is a multiple of A;

therefore, by the preceding case, D is the same multiple of C;

that is, C is the same part of D that A is of B.

Wherefore, *if the first* &c. Q.E.D.

PROPOSITION 7. THEOREM.

Equal magnitudes have the same ratio to the same magnitude; and the same has the same ratio to equal magnitudes.

Let A and B be equal magnitudes, and C any other magnitude: each of the magnitudes A and B shall have the same ratio to C; and C shall have the same ratio to each of the magnitudes A and B.

Take of A and B any equimultiples whatever D and E; and of C any multiple whatever F.

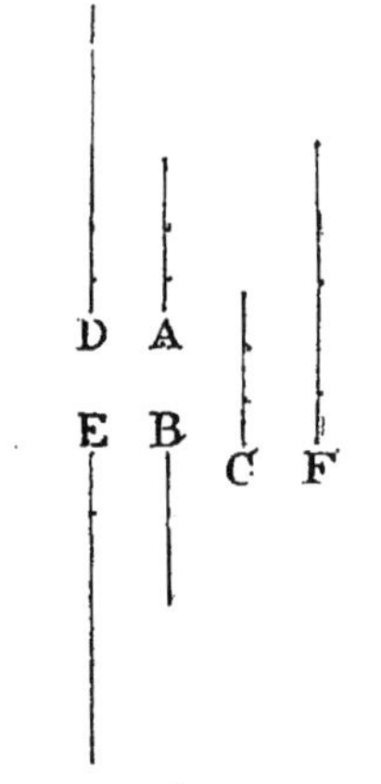

Then, because D is the same multiple of A that E is of B, [*Construction.*
and that A is equal to B; [*Hypothesis.*
therefore D is equal to E. [V. *Axiom* 1.
Therefore if D be greater than F, E is greater than F; and if equal, equal; and if less, less.

But D and E are any equimultiples whatever of A and B, and F is any multiple whatever of C; [*Construction.*
therefore A is to C as B is to C. [V. *Def.* 5.

Also C shall have the same ratio to A that it has to B.

For the same construction being made, it may be shewn, as before, that D is equal to E.
Therefore if F be greater than D, F is greater than E; and if equal, equal; and if less, less.

But F is any multiple whatever of C, and D and E are any equimultiples whatever of A and B; [*Construction.*
therefore C is to A as C is to B. [V. *Definition* 5.

Wherefore, *equal magnitudes* &c. Q.E.D.

PROPOSITION 8. *THEOREM.*

Of unequal magnitudes, the greater has a greater ratio to the same than the less has; and the same magnitude has a greater ratio to the less than it has to the greater.

Let AB and BC be unequal magnitudes, of which AB is the greater; and let D be any other magnitude whatever: AB shall have a greater ratio to D than BC has to D; and D shall have a greater ratio to BC than it has to AB.

If the magnitude which is not the greater of the two AC, CB, be not less than D, take EF, FG the doubles of AC, CB (Figure 1).

Fig. 1.

But if that which is not the greater of the two AC, CB, be less than D (Figures 2 and 3), this magnitude can be multiplied, so as to become greater than D, whether it be AC or CB.

Let it be multiplied until it becomes greater than D, and let the other be multiplied as often.

Let EF be the multiple thus taken of AC, and FG the same multiple of CB;

therefore EF and FG are each of them greater than D.

And in all the cases, take H the double of D, K its triple, and so on, until the multiple of D taken is the first which is greater than FG. Let L be that multiple of D, namely, the first which is greater than FG; and let K be the multiple of D which is next less than L.

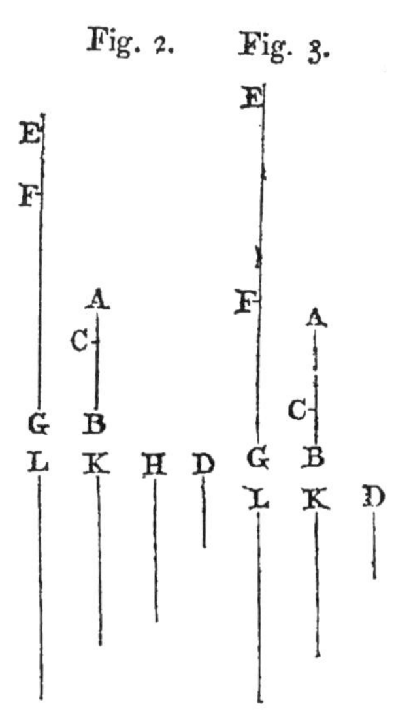

Then, because L is the first multiple of D which is greater than FG, [*Construction.*

the next preceding multiple K is not greater than FG;

that is, FG is not less than K.

And because EF is the same multiple of AC that FG is of CB, [*Construction.*

therefore EG is the same multiple of AB that FG is of CB; [V. 1.

that is, EG and FG are equimultiples of AB and CB.

And it was shewn that FG is not less than K,
and EF is greater than D; [*Construction.*
therefore the whole EG is greater than K and D together.
But K and D together are equal to L; [*Construction.*
therefore EG is greater than L.
But FG is not greater than L.
And EG and FG were shewn to be equimultiples of AB and BC;
and L is a multiple of D. [*Construction.*
Therefore AB has to D a greater ratio than BC has to D. [V. *Definition* 7.

Also, D shall have to BC a greater ratio than it has to AB.
For, the same construction being made, it may be shewn, that L is greater than FG but not greater than EG.
And L is a multiple of D, [*Construction.*
and EG and FG were shewn to be equimultiples of AB and CB.
Therefore D has to BC a greater ratio than it has to AB. [V. *Definition* 7.

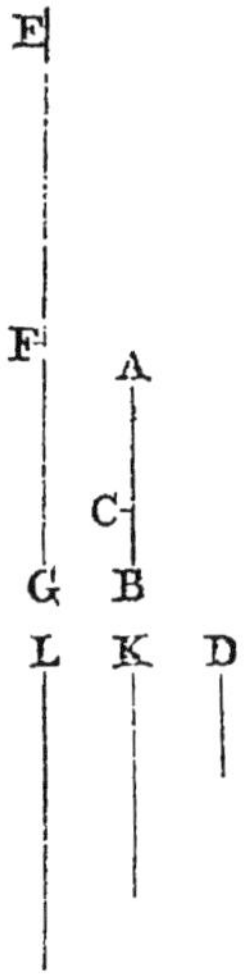

Wherefore, *of unequal magnitudes* &c. Q.E.D.

PROPOSITION 9. *THEOREM.*

Magnitudes which have the same ratio to the same magnitude, are equal to one another; and those to which the same magnitude has the same ratio, are equal to one another.

First, let A and B have the same ratio to C: A shall be equal to B.

For, if A is not equal to B, one of them must be greater than the other; let A be the greater.

Then, by what was shewn in Proposition 8, there are

some equimultiples of A and B, and some multiple of C, such that the multiple of A is greater than the multiple of C, but the multiple of B is not greater than the multiple of C.

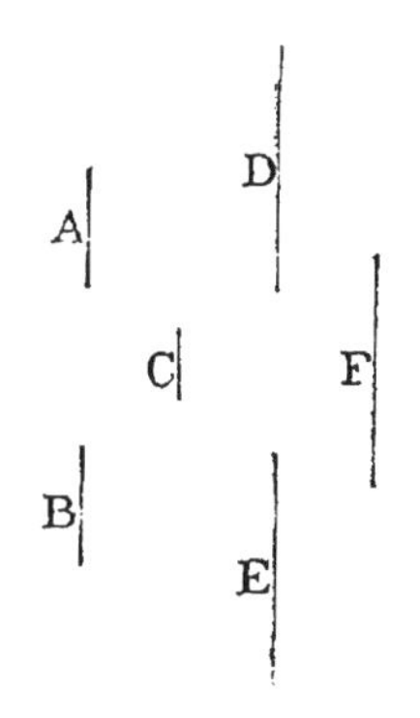

Let such multiples be taken; and let D and E be the equimultiples of A and B, and F the multiple of C; so that D is greater than F, but E is not greater than F.

Then, because A is to C as B is to C; and of A and B are taken equimultiples D and E, and of C is taken a multiple F;

and that D is greater than F; [*Construction.*

therefore E is also greater than F. [V. *Definition* 5.

But E is not greater than F; [*Construction.*

which is impossible.

Therefore A and B are not unequal; that is, they are equal.

Next, let C have the same ratio to A and B: A shall be equal to B.

For, if A is not equal to B, one of them must be greater than the other; let A be the greater.

Then, by what was shewn in Proposition 8, there is some multiple F of C, and some equimultiples E and D of B and A, such that F is greater than E, but not greater than D.

And, because C is to B as C is to A, [*Hypothesis.*

and that F the multiple of the first is greater than E the multiple of the second, [*Construction.*

therefore F the multiple of the third is greater than D the multiple of the fourth. [V. *Definition* 5.

But F is not greater than D; [*Construction.*

which is impossible.

Therefore A and B are not unequal; that is, they are equal.

Wherefore, *magnitudes which* &c. Q.E.D.

PROPOSITION 10. *THEOREM.*

That magnitude which has a greater ratio than another has to the same magnitude is the greater of the two; and that magnitude to which the same has a greater ratio than it has to another magnitude is the less of the two.

First, let A have to C a greater ratio than B has to C: A shall be greater than B.

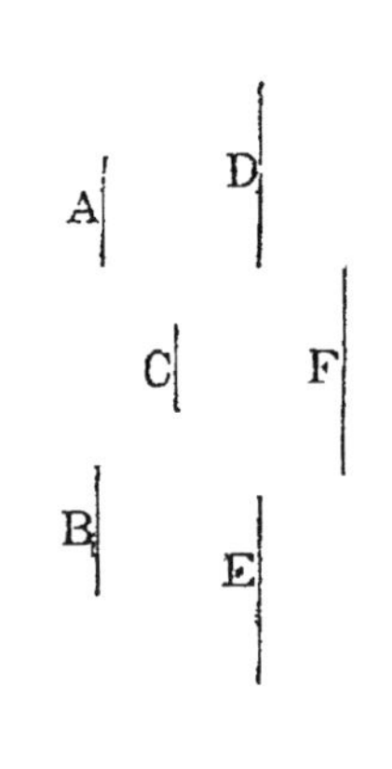

For, because A has a greater ratio to C than B has to C, there are some equimultiples of A and B, and some multiple of C, such that the multiple of A is greater than the multiple of C, but the multiple of B is not greater than the multiple of C. [V. *Def.* 7.
Let such multiples be taken; and let D and E be the equimultiples of A and B, and F the multiple of C; so that D is greater than F, but E is not greater than F;
therefore D is greater than E.
And because D and E are equimultiples of A and B, and that D is greater than E,
therefore A is greater than B. [V. *Axiom* 4.

Next, let C have to B a greater ratio than it has to A: B shall be less than A.

For there is some multiple F of C, and some equimultiples E and D of B and A, such that F is greater than E, but not greater than D; [V. *Definition* 7.
therefore E is less than D.
And because E and D are equimultiples of B and A, and that E is less than D,
therefore B is less than A. [V. *Axiom* 4.

Wherefore, *that magnitude* &c. Q.E.D.

PROPOSITION 11. *THEOREM.*

Ratios that are the same to the same ratio, are the same to one another.

Let A be to B as C is to D, and let C be to D as E is to F: A shall be to B as E is to F.

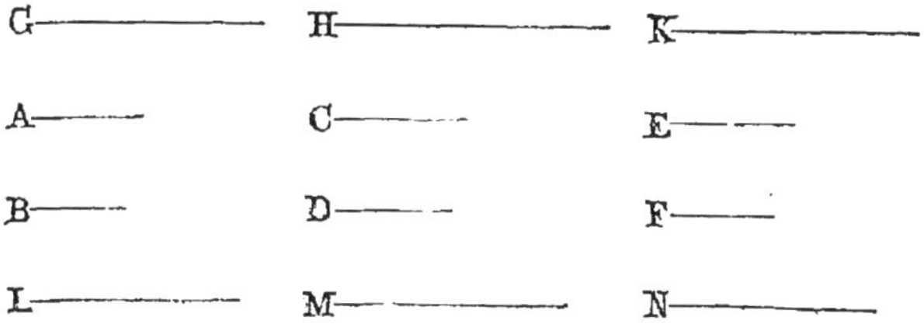

G H K

A C E

B D F

L M N

Take of A, C, E any equimultiples whatever G, H, K; and of B, D, F any equimultiples whatever L, M, N.

Then, because A is to B as C is to D, [*Hypothesis.*
and that G and H are equimultiples of A and C, and L and M are equimultiples of B and D; [*Construction.*
therefore if G be greater than L, H is greater than M; and if equal, equal; and if less, less. [V. *Definition* 5.

Again, because C is to D as E is to F, [*Hypothesis.*
and that H and K are equimultiples of C and E, and M and N are equimultiples of D and F; [*Construction.*
therefore if H be greater than M, K is greater than N; and if equal, equal; and if less, less. [V. *Definition* 5.

But it has been shewn that if G be greater than L, H is greater than M; and if equal, equal; and if less, less.
Therefore if G be greater than L, K is greater than N; and if equal, equal; and if less, less.
And G and K are any equimultiples whatever of A and E, and L and N are any equimultiples whatever of B and F.
Therefore A is to B as E is to F. [V. *Definition* 5.

Wherefore, *ratios that are the same* &c. Q.E.D.

PROPOSITION 12. *THEOREM.*

If any number of magnitudes be proportionals, as one of the antecedents is to its consequent, so shall all the antecedents be to all the consequents.

Let any number of magnitudes A, B, C, D, E, F be proportionals; namely, as A is to B, so let C be to D, and E to F: as A is to B, so shall A, C, E together be to B, D, F together.

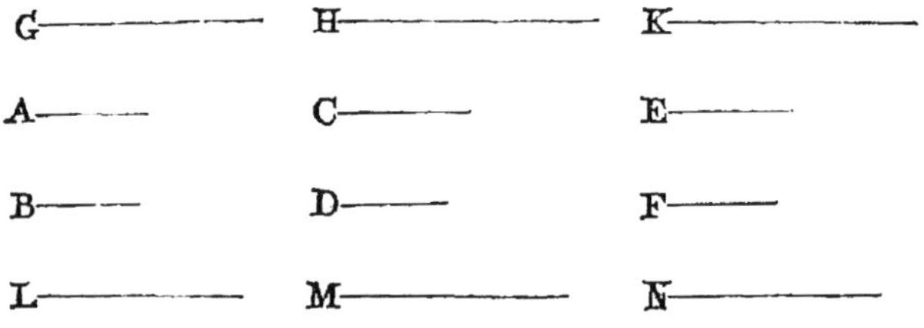

Take of A, C, E any equimultiples whatever G, H, K; and of B, D, F any equimultiples whatever L, M, N.

Then, because A is to B as C is to D and as E is to F, and that G, H, K are equimultiples of A, C, E, and L, M, N equimultiples of B, D, F; [*Construction.*

therefore if G be greater than L, H is greater than M, and K is greater than N; and if equal, equal; and if less, less. [V. *Definition* 5.

Therefore, if G be greater than L, then G, H, K together are greater than L, M, N together; and if equal, equal; and if less, less.

But G, and G, H, K together, are any equimultiples whatever of A, and A, C, E together; [V. 1.

and L, and L, M, N together are any equimultiples whatever of B, and B, D, F together. [V. 1.

Therefore as A is to B, so are A, C, E together to B, D, F together. [V. *Definition* 5.

Wherefore, *if any number* &c. Q.E.D.

PROPOSITION 13. *THEOREM.*

If the first have the same ratio to the second which the third has to the fourth, but the third to the fourth a greater ratio than the fifth to the sixth, the first shall have to the second a greater ratio than the fifth has to the sixth.

Let A the first have the same ratio to B the second that C the third has to D the fourth, but C the third a greater ratio to D the fourth than E the fifth to F the sixth: A the first shall have to B the second a greater ratio than E the fifth has to F the sixth.

M———— G———— H————

A—— C—— E——

B—— D—— F——

N———— K———— L————

For, because C has a greater ratio to D than E has to F, there are some equimultiples of C and E, and some equimultiples of D and F, such that the multiple of C is greater than the multiple of D, but the multiple of E is not greater than the multiple of F. [V. *Definition* 7.

Let such multiples be taken, and let G and H be the equimultiples of C and E, and K and L the equimultiples of D and F;

so that G is greater than K, but H is not greater than L.

And whatever multiple G is of C, take M the same multiple of A; and whatever multiple K is of D, take N the same multiple of B.

Then, because A is to B as C is to D, [*Hypothesis.*

and M and G are equimultiples of A and C, and N and K are equimultiples of B and D; [*Construction.*

therefore if M be greater than N, G is greater than K; and if equal, equal; and if less, less. [V. *Definition* 5.

But G is greater than K; [*Construction.*

therefore M is greater than N.

But H is not greater than L; [*Construction.*

and M and H are equimultiples of A and E, and N and L are equimultiples of B and F; [*Construction.*

therefore A has a greater ratio to B than E has to F.

Wherefore, *if the first* &c. Q.E.D.

COROLLARY. And if the first have a greater ratio to the second than the third has to the fourth, but the third the same ratio to the fourth that the fifth has to the sixth, it may be shewn, in the same manner, that the first has a greater ratio to the second than the fifth has to the sixth.

PROPOSITION 14. *THEOREM.*

If the first have the same ratio to the second that the third has to the fourth, then if the first be greater than the third the second shall be greater than the fourth; and if equal, equal; and if less, less.

Let A the first have the same ratio to B the second that C the third has to D the fourth: if A be greater than C, B shall be greater than D; if equal, equal; and if less, less.

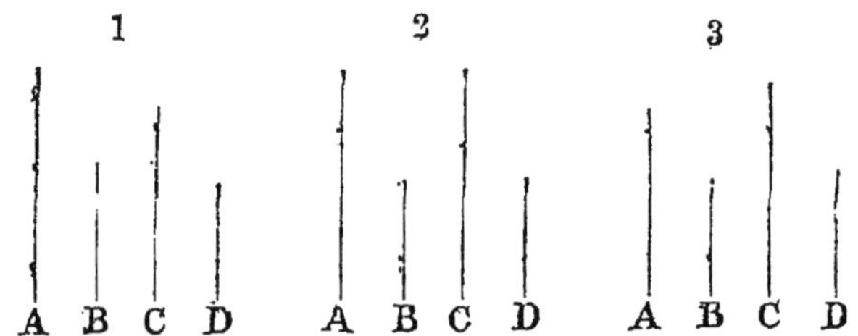

First, let A be greater than C: B shall be greater than D.

For, because A is greater than C, [*Hypothesis.*
and B is any other magnitude;
therefore A has to B a greater ratio than C has to B. [V. 8.
But A is to B as C is to D. [*Hypothesis.*
Therefore C has to D a greater ratio than C has to B. [V. 13.

But of two magnitudes, that to which the same has the greater ratio is the less. [V. 10.
Therefore D is less than B; that is, B is greater than D.

Secondly, let A be equal to C: B shall be equal to D.

For, A is to B as C, that is A, is to D. [*Hypothesis.*
Therefore B is equal to D. [V. 9.

Thirdly, let A be less than C: B shall be less than D.
For, C is greater than A.
And because C is to D as A is to B; [*Hypothesis.*
and C is greater than A;
therefore, by the first case, D is greater than B;
that is, B is less than D.

Wherefore, *if the first* &c. Q.E.D.

PROPOSITION 15. *THEOREM.*

Magnitudes have the same ratio to one another that their equimultiples have.

Let AB be the same multiple of C that DE is of F: C shall be to F as AB is to DE.

For, because AB is the same multiple of C that DE is of F, [*Hypothesis.*
therefore as many magnitudes as there are in AB equal to C, so many are there in DE equal to F.
Divide AB into the magnitudes AG, GH, HB, each equal to C; and DE into the magnitudes DK, KL, LE, each equal to F.
Therefore the number of the magnitudes AG, GH, HB will be equal to the number of the magnitudes DK, KL, LE.

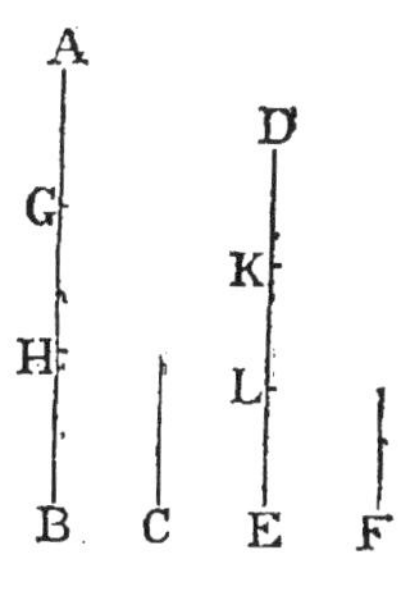

And because AG, GH, HB are all equal; [*Construction.*
and that DK, KL, LE are also all equal;
therefore AG is to DK as GH is to KL, and as HB is to LE. [V. 7.
But as one of the antecedents is to its consequent, so are all the antecedents to all the consequents. [V. 12.
Therefore as AG is to DK so is AB to DE.
But AG is equal to C, and DK is equal to F.
Therefore as C is to F so is AB to DE.

Wherefore, *magnitudes* &c. Q.E.D.

PROPOSITION 16. *THEOREM.*

If four magnitudes of the same kind be proportionals, they shall also be proportionals when taken alternately.

Let A, B, C, D be four magnitudes of the same kind which are proportionals; namely, as A is to B so let C be to D: they shall also be proportionals when taken alternately, that is, A shall be to C as B is to D.

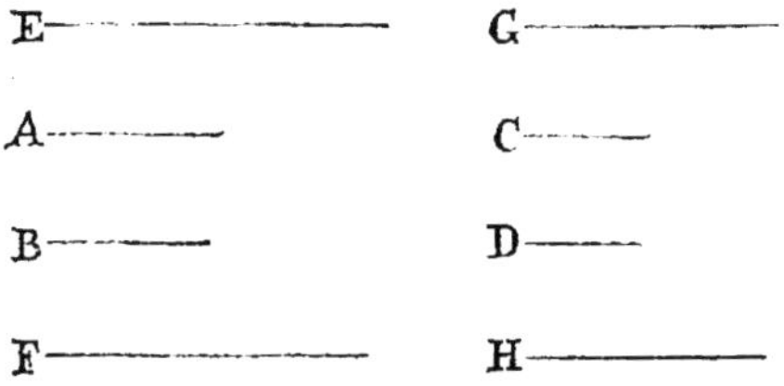

E G

A C

B D

F H

Take of A and B any equimultiples whatever E and F, and of C and D any equimultiples whatever G and H.

Then, because E is the same multiple of A that F is of B, and that magnitudes have the same ratio to one another that their equimultiples have; [V. 15.

therefore A is to B as E is to F.

But A is to B as C is to D. [*Hypothesis.*

Therefore C is to D as E is to F. [V. 11.

Again, because G and H are equimultiples of C and D, therefore C is to D as G is to H. [V. 15.

But it was shewn that C is to D as E is to F.

Therefore E is to F as G is to H. [V. 11.

But when four magnitudes are proportionals, if the first be greater than the third, the second is greater than the fourth; and if equal, equal; and if less, less. [V. 14.

Therefore if E be greater than G, F is greater than H; and if equal, equal; and if less, less.

But E and F are any equimultiples whatever of A and B, and G and H are any equimultiples whatever of C and D. [*Construction.*

Therefore A is to C as B is to D. [V. *Definition* 5.

Wherefore, *if four magnitudes* &c. Q.E.D.

PROPOSITION 17. *THEOREM.*

If magnitudes, taken jointly, be proportionals, they shall also be proportionals when taken separately; that is, if two magnitudes taken together have to one of them the same ratio which two others have to one of these, the remaining one of the first two shall have to the other the same ratio which the remaining one of the last two has to the other of these.

Let AB, BE, CD, DF be the magnitudes which, taken jointly, are proportionals; that is, let AB be to BE as CD is to DF: they shall also be proportionals when taken separately; that is, AE shall be to EB as CF is to FD.

Take of AE, EB, CF, FD any equimultiples whatever GH, HK, LM, MN;

and, again, of EB, FD take any equimultiples whatever KX, NP.

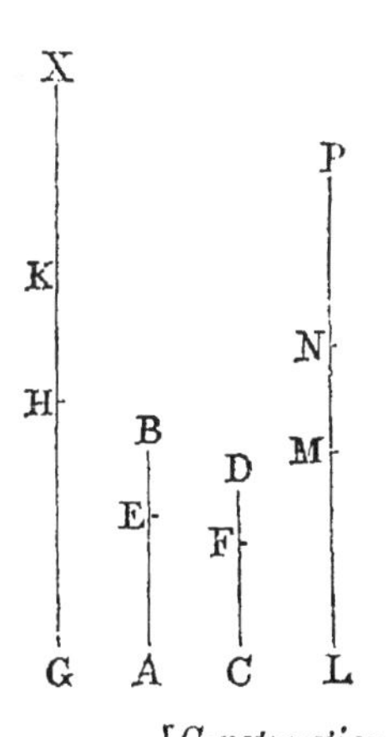

Then, because GH is the same multiple of AE that HK is of EB; therefore GH is the same multiple of AE that GK is of AB. [V. 1.

But GH is the same multiple of AE that LM is of CF, [*Constr.*

therefore GK is the same multiple of AB that LM is of CF.

Again, because LM is the same multiple of CF that MN is of FD, [*Construction.*

therefore LM is the same multiple of CF that LN is of CD. [V. 1.

But LM was shewn to be the same multiple of CF that GK is of AB.

Therefore GK is the same multiple of AB that LN is of CD;

that is, GK and LN are equimultiples of AB and CD.

Again, because HK is the same multiple of EB that MN is of FD, and that KX is the same multiple of EB that NP is of FD, [*Construction.*

therefore HX is the same multiple of EB that MP is of FD; [V. 2.

that is, HX and MP are equimultiples of EB and FD.

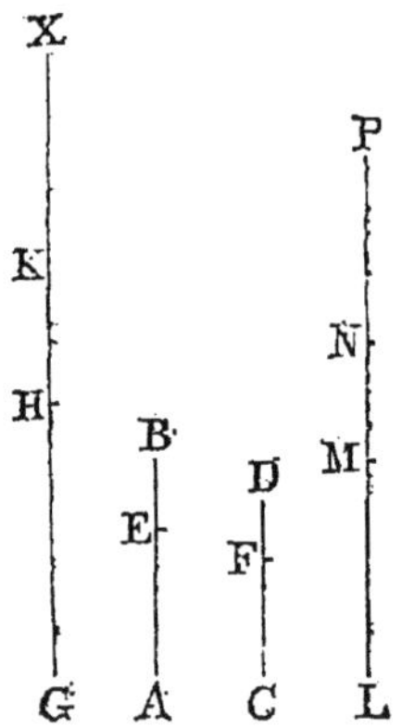

And because AB is to BE as CD is to DF, [*Hypothesis.*

and that GK and LN are equimultiples of AB and CD, and HX and MP are equimultiples of EB and FD,

therefore if GK be greater than HX, LN is greater than MP; and if equal, equal; and if less, less. [V. *Def.* 5.

But if GH be greater than KX, then, by adding the common magnitude HK to both, GK is greater than HX;

therefore also LN is greater than MP;

and, by taking away the common magnitude MN from both, LM is greater than NP.

Thus if GH be greater than KX, LM is greater than NP.

In like manner it may be shewn that, if GH be equal to KX, LM is equal to NP; and if less, less.

But GH and LM are any equimultiples whatever of AE and CF, and KX and NP are any equimultiples whatever of EB and FD; [*Construction.*

therefore AE is to EB as CF is to FD. [V. *Definition* 5.

Wherefore, *if four magnitudes* &c. Q.E.D.

PROPOSITION 18. *THEOREM.*

If magnitudes, taken separately, be proportionals, they shall also be proportionals when taken jointly; that is, if the first be to the second as the third to the fourth, the first and second together shall be to the second as the third and fourth together to the fourth.

Let *AE*, *EB*, *CF*, *FD* be proportionals; that is, let *AE* be to *EB* as *CF* is to *FD*: they shall also be proportionals when taken jointly; that is, *AB* shall be to *BE* as *CD* is to *DF*.

Take of *AB*, *BE*, *CD*, *DF* any equimultiples whatever *GH*, *HK*, *LM*, *MN*;

and, again, of *BE*, *DF* take any equimultiples whatever *KO*, *NP*.

Then, because *KO* and *NP* are equimultiples of *BE* and *DF*, and that *KH* and *NM* are also equimultiples of *BE* and *DF*; [*Construction.*

therefore if *KO*, the multiple of *BE*, be greater than *KH*, which is a multiple of the same *BE*, then *NP* the multiple of *DF* is also greater than *NM* the multiple of the same *DF*; and if *KO* be equal to *KH*, *NP* is equal to *NM*; and if less, less.

First, let *KO* be not greater than *KH*;

therefore *NP* is not greater than *NM*.

And because *GH* and *HK* are equimultiples of *AB* and *BE*, [*Construction.*

and that *AB* is greater than *BE*,

therefore *GH* is greater than *HK*; [V. *Axiom* 3.

but *KO* is not greater than *KH*; [*Hypothesis.*

therefore *GH* is greater than *KO*.

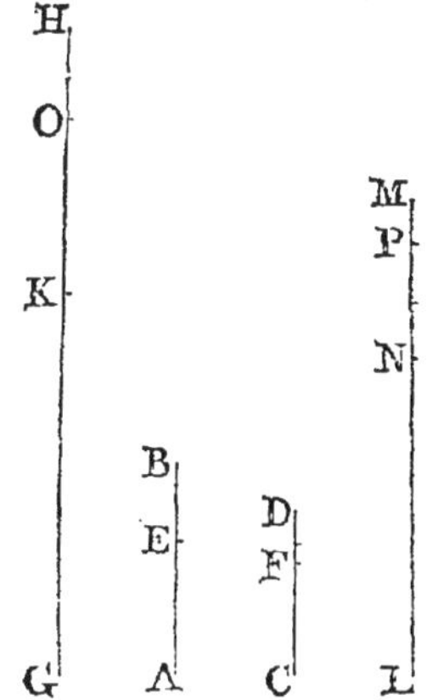

In like manner it may be shewn that *LM* is greater than *NP*.

Thus if *KO* be not greater than *KH*, then *GH*, the multiple of *AB*, is always greater than *KO*, the multiple of *BE*;

and likewise *LM*, the multiple of *CD*, is greater than *NP*, the multiple of *DF*.

Next, let KO be greater than KH;

therefore, as has been shewn, NP is greater than NM.

And because the whole GH is the same multiple of the whole AB that HK is of BE, [*Construction.*

therefore the remainder GK is the same multiple of the remainder AE that GH is of AB; [V. 5.

which is the same that LM is of CD. [*Construction.*

In like manner, because the whole LM is the same multiple of the whole CD that MN is of DF, [*Construction.*

therefore the remainder LN is the same multiple of the remainder CF that LM is of CD. [V. 5.

But it was shewn that LM is the same multiple of CD that GK is of AE.

Therefore GK is the same multiple of AE that LN is of CF;

that is, GK and LN are equimultiples of AE and CF.

And because KO and NP are equimultiples of BE and DF; [*Construction.*

therefore, if from KO and NP there be taken KH and NM, which are also equimultiples of BE and DF, [*Constr.*

the remainders HO and MP are either equal to BE and DF, or are equimultiples of them. [V. 6.

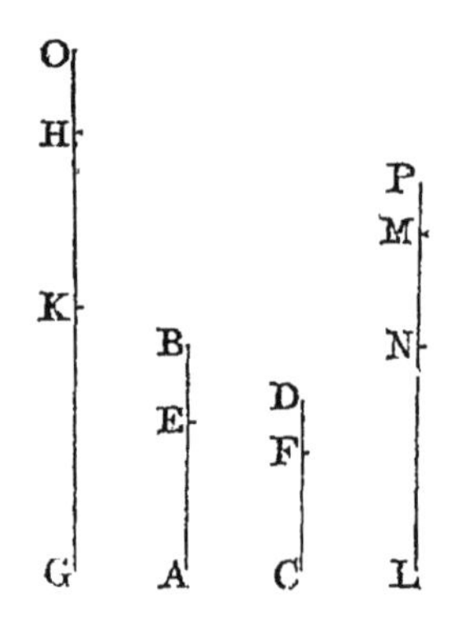

Suppose that HO and MP are equal to BE and DF. Then, because AE is to EB as CF is to FD, [*Hypothesis.* and that GK and LN are equimultiples of AE and CF; therefore GK is to EB as LN is to FD. [V. 4, *Cor.* But HO is equal to BE, and MP is equal to DF; [*Hyp.* therefore GK is to HO as LN is to MP.

Therefore if GK be greater than HO, LN is greater than MP; and if equal, equal; and if less, less. [V. *A*.

Again, suppose that HO and MP are equimultiples of EB and FD.

Then, because AE is to EB as CF is to FD; [*Hypothesis.*
and that GK and LN are equimultiples of AE and CF, and HO and MP are equimultiples of EB and FD;
therefore if GK be greater than HO, LN is greater than MP; and if equal, equal; and if less, less; [V. *Definition* 5.
which was likewise shewn on the preceding supposition.

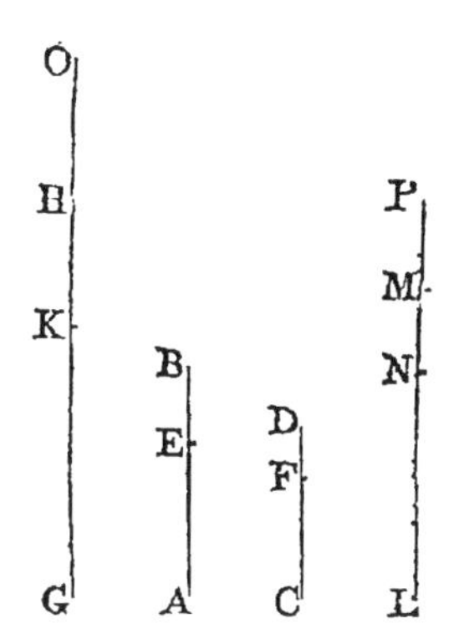

But if GH be greater than KO, then by taking the common magnitude KH from both, GK is greater than HO;

therefore also LN is greater than MP;

and, by adding the common magnitude NM to both, LM is greater than NP.

Thus if GH be greater than KO, LM is greater than NP.

In like manner it may be shewn, that if GH be equal to KO, LM is equal to NP; and if less, less.

And in the case in which KO is not greater than KH, it has been shewn that GH is always greater than KO, and also LM greater than NP.

But GH and LM are any equimultiples whatever of AB and CD, and KO and NP are any equimultiples whatever of BE and DF, [*Construction.*

therefore AB is to BE as CD is to DF. [V. *Definition* 5.

Wherefore, *if magnitudes* &c. Q.E.D.

PROPOSITION 19. *THEOREM.*

If a whole magnitude be to a whole as a magnitude taken from the first is to a magnitude taken from the other, the remainder shall be to the remainder as the whole is to the whole.

Let the whole AB be to the whole CD as AE, a magnitude taken from AB, is to CF, a magnitude taken from CD: the remainder EB shall be to the remainder FD as the whole AB is to the whole CD.

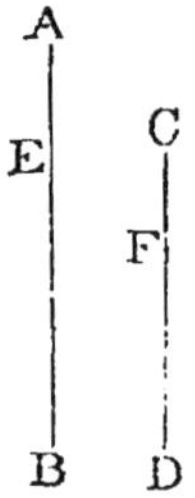

For, because AB is to CD as AE is to CF, [*Hypothesis.*
therefore, alternately, AB is to AE as CD is to CF. [V. 16.
And if magnitudes taken jointly be proportionals, they are also proportionals when taken separately; [V. 17.
therefore EB is to AE as FD is to CF;
therefore, alternately, EB is to FD as AE is to CF. [V. 16.
But AE is to CF as AB is to CD; [*Hyp.*
therefore EB is to FD as AB is to CD. [V. 11.

Wherefore, *if a whole* &c. Q.E.D.

COROLLARY. If the whole be to the whole as a magnitude taken from the first is to a magnitude taken from the other, the remainder shall be to the remainder as the magnitude taken from the first is to the magnitude taken from the other. The demonstration is contained in the preceding.

PROPOSITION *E.* *THEOREM.*

If four magnitudes be proportionals, they shall also be proportionals by conversion; that is, the first shall be to its excess above the second as the third is to its excess above the fourth.

Let AB be to BE as CD is to DF: AB shall be to AE as CD is to CF.

For, because AB is to BE as CD is to DF; [*Hypothesis.*

therefore, by division, AE is to EB as CF is to FD; [V. 17.

and, by inversion, EB is to AE as FD is to CF. [V. *B.*

Therefore, by composition, AB is to AE as CD is to CF. [V. 18.

Wherefore, *if four magnitudes* &c. Q.E.D.

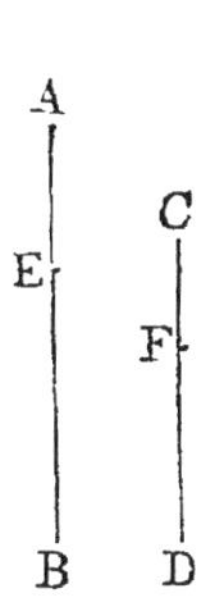

PROPOSITION 20. *THEOREM.*

If there be three magnitudes, and other three, which have the same ratio, taken two and two, then, if the first be greater than the third, the fourth shall be greater than the sixth; and if equal, equal; and if less, less.

Let A, B, C be three magnitudes, and D, E, F other three, which have the same ratio taken two and two; that is, let A be to B as D is to E, and let B be to C as E is to F: if A be greater than C, D shall be greater than F; and if equal, equal; and if less, less.

First, let A be greater than C: D shall be greater than F.

For, because A is greater than C, and B is any other magnitude,

therefore A has to B a greater ratio than C has to B. [V. 8.

But A is to B as D is to E; [*Hypothesis.*

therefore D has to E a greater ratio than C has to B. [V. 13.

And because B is to C as E is to F, [*Hyp.*

therefore, by inversion, C is to B as F is to E. [V. *B.*

And it was shewn that D has to E a greater ratio than C has to B;

therefore D has to E a greater ratio than F has to E; [V. 13, *Cor.*

therefore D is greater than F. [V. 10.

Secondly, let A be equal to C: D shall be equal to F.

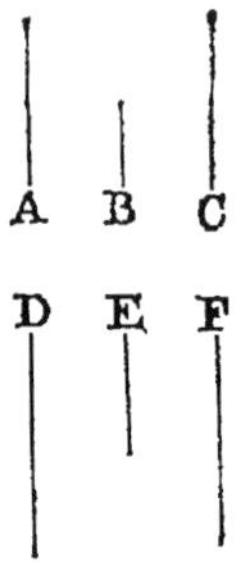

For, because A is equal to C, and B is any other magnitude,

therefore A is to B as C is to B. [V. 7.

But A is to B as D is to E, [*Hypothesis.*

and C is to B as F is to E, [*Hyp.* V. *B.*

therefore D is to E as F is to E; [V. 11.

and therefore D is equal to F. [V. 9.

Lastly, let A be less than C: D shall be less than F.

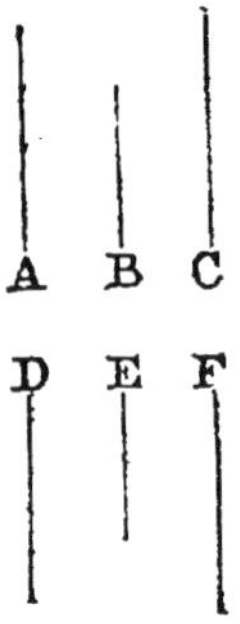

For C is greater than A;

and, as was shewn in the first case, C is to B as F is to E;

and, in the same manner, B is to A as E is to D;

therefore, by the first case, F is greater than D;

that is, D is less than F.

Wherefore, *if there be three* &c. Q.E.D.

PROPOSITION 21. *THEOREM.*

If there be three magnitudes, and other three, which have the same ratio, taken two and two, but in a cross order, then if the first be greater than the third, the fourth shall be greater than the sixth; and if equal, equal; and if less, less.

Let A, B, C be three magnitudes, and D, E, F other three, which have the same ratio, taken two and two, but in a cross order; that is, let A be to B as E is to F, and let B be to C as D is to E: if A be greater than C, D shall be greater than F; and if equal, equal; and if less, less.

First, let A be greater than C: D shall be greater than F.

For, because A is greater than C, and B is any other magnitude,

therefore A has to B a greater ratio than C has to B. [V. 8.

But A is to B as E is to F; [*Hypothesis.*

therefore E has to F a greater ratio than C has to B. [V. 13.

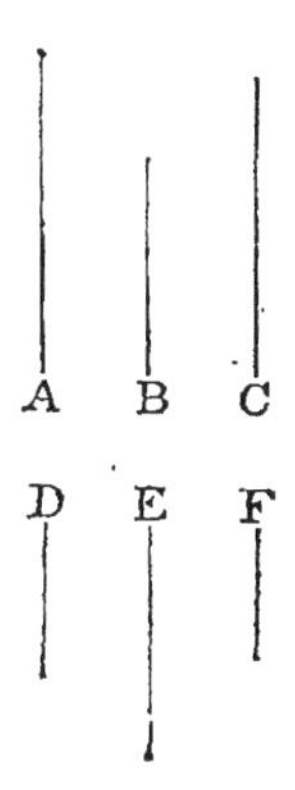

And because B is to C as D is to E, [*Hypothesis.*

therefore, by inversion, C is to B as E is to D. [V. *B.*

And it was shewn that E has to F a greater ratio than C has to B;

therefore E has to F a greater ratio than E has to D; [V. 13, *Cor.*

therefore F is less than D; [V. 10.

that is, D is greater than F.

Secondly, let A be equal to C: D shall be equal to F.

For, because A is equal to C, and B is any other magnitude,

therefore A is to B as C is to B. [V. 7.

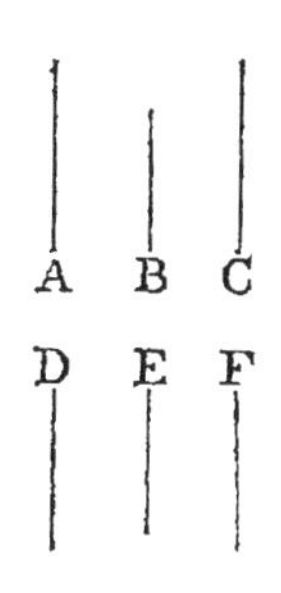

But A is to B as E is to F; [*Hyp.*

and C is to B as E is to D; [*Hyp.* V. *B.*

therefore E is to F as E is to D; [V. 11.

and therefore D is equal to F. [V. 9.

Lastly, let A be less than C: D shall be less than F.

For C is greater than A;

and, as was shewn in the first case, C is to B as E is to D;

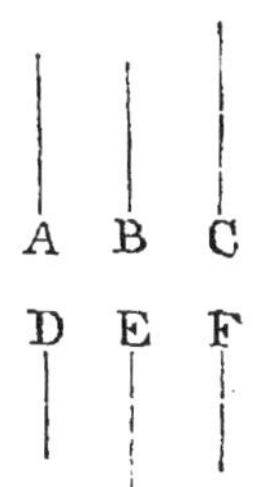

and, in the same manner, B is to A as F is to E;

therefore, by the first case, F is greater than D;

that is, D is less than F.

Wherefore, *if there be three* &c. Q.E.D.

PROPOSITION 22. *THEOREM.*

If there be any number of magnitudes, and as many others, which have the same ratio, taken two and two in order, the first shall have to the last of the first magnitudes the same ratio which the first of the others has to the last.

[This proposition is usually cited by the words *ex æquali.*]

First, let there be three magnitudes A, B, C, and other three D, E, F, which have the same ratio, taken two and two in order; that is, let A be to B as D is to E, and let B be to C as E is to F: A shall be to C as D is to F.

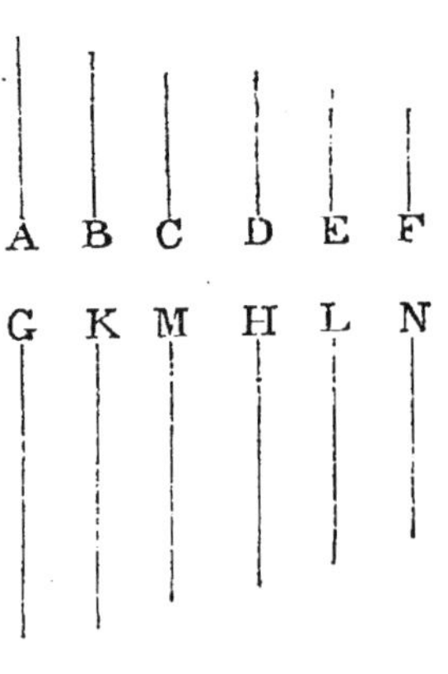

Take of A and D any equimultiples whatever G and H;

and of B and E any equimultiples whatever K and L;

and of C and F any equimultiples whatever M and N.

Then, because A is to B as D is to E; [*Hypothesis.*

and that G and H are equimultiples of A and D,

and K and L equimultiples of B and E; [*Construction.*

therefore G is to K as H is to L. [V. 4.

For the same reason, K is to M as L is to N.

And because there are three magnitudes G, K, M, and other three H, L, N, which have the same ratio taken two and two,

therefore if G be greater than M, H is greater than N; and if equal, equal; and if less, less. [V. 20.

But G and H are any equimultiples whatever of A and D, and M and N are any equimultiples whatever of C and F.

Therefore A is to C as D is to F. [V. *Definition* 5.

Next, let there be four magnitudes, A, B, C, D, and

A. B. C. D.
E. F. G. H.

other four E, F, G, H, which have the same ratio taken two and two in order; namely, let A be to B as E is to F, and B to C as F is to G, and C to D as G is to H: A shall be to D as E is to H.

For, because A, B, C are three magnitudes, and E, F, G other three, which have the same ratio, taken two and two in order, [*Hypothesis.*

therefore, by the first case, A is to C as E is to G.

But C is to D as G is to H; [*Hypothesis.*

therefore also, by the first case, A is to D as E is to H.

And so on, whatever be the number of magnitudes.

Wherefore, *if there be any number* &c. Q.E.D.

PROPOSITION 23. *THEOREM.*

If there be any number of magnitudes, and as many others, which have the same ratio, taken two and two in a cross order, the first shall have to the last of the first magnitudes the same ratio which the first of the others has to the last.

First, let there be three magnitudes, A, B, C, and other three D, E, F, which have the same ratio, taken two and two in a cross order; namely, let A be to B as E is to F, and B to C as D is to E: A shall be to C as D is to F.

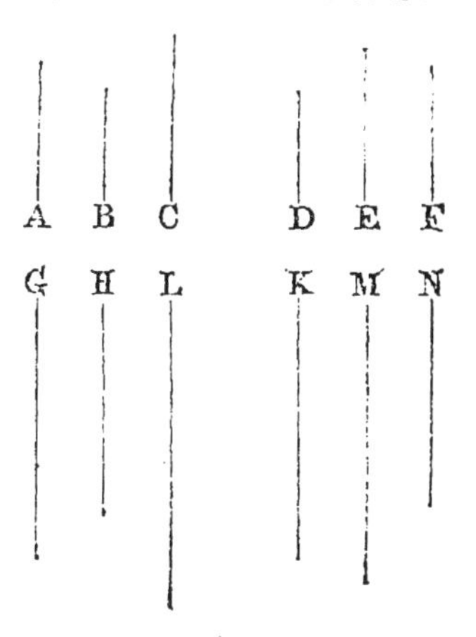

Take of A, B, D any equimultiples whatever G, H, K; and of C, E, F any equimultiples whatever L, M, N.

Then because G and H are equimultiples of A and B, and that magnitudes have the same ratio which their equimultiples have; [V. 15.

therefore A is to B as G is to H.

And, for the same reason, E is to F as M is to N.

But A is to B as E is to F. [*Hypothesis.*

Therefore G is to H as M is to N. [V. 11.

And because B is to C as D is to E, [*Hypothesis.*

and that H and K are equimultiples of B and D, and L and M are equimultiples of C and E; [*Constr.*

therefore H is to L as K is to M. [V. 4.

And it has been shewn that G is to H as M is to N.

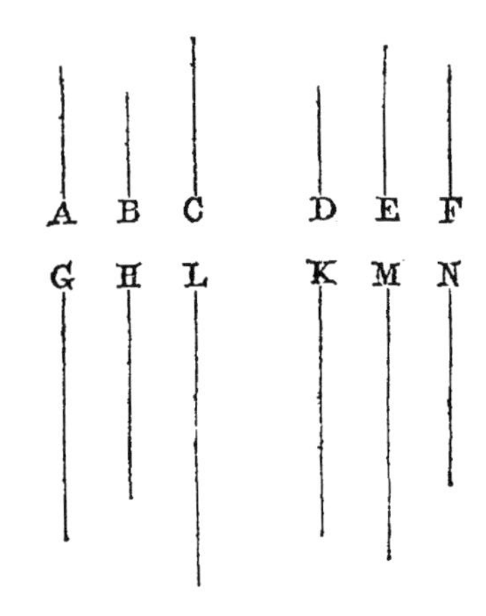

Then since there are three magnitudes G, H, L, and other three K, M, N, which have the same ratio, taken two and two in a cross order;

therefore if G be greater than L, K is greater than N; and if equal, equal; and if less, less. [V. 21.

But G and K are any equimultiples whatever of A and D, and L and N are any equimultiples whatever of C and F; therefore A is to C as D is to F. [V. *Definition* 5.

Next, let there be four magnitudes A, B, C, D, and other four E, F, G, H, which have the same ratio, taken two and two in a cross order; namely, let A be to B as G is to H, and B to C as F is to G, and C to D as E is to F: A shall be to D as E is to H.

A. B. C. D.
E. F. G. H.

For, because A, B, C are three magnitudes, and F, G, H other three, which have the same ratio, taken two and two in a cross order; [*Hypothesis.*

therefore, by the first case, A is to C as F is to H.

But C is to D as E is to F; [*Hypothesis.*

therefore also, by the first case, A is to D as E is to H.

And so on, whatever be the number of magnitudes.

Wherefore, *if there be any number* &c. Q.E.D.

PROPOSITION 24. *THEOREM.*

If the first have to the second the same ratio which the third has to the fourth, and the fifth have to the second the same ratio which the sixth has to the fourth, then the first and fifth together shall have to the second the same ratio which the third and sixth together have to the fourth.

Let AB the first have to C the second the same ratio which DE the third has to F the fourth; and let BG the fifth have to C the second the same ratio which EH the sixth has to F the fourth: AG, the first and fifth together, shall have to C the second the same ratio which DH, the third and sixth together, has to F the fourth.

For, because BG is to C as EH is to F, [*Hypothesis.*

therefore, by inversion, C is to BG as F is to EH. [V. *B.*

And because AB is to C as DE is to F, [*Hypothesis.*

and C is to BG as F is to EH;

therefore, ex æquali, AB is to BG as DE is to EH. [V. 22.

And, because these magnitudes are proportionals, they are also proportionals when taken jointly; [V. 18.

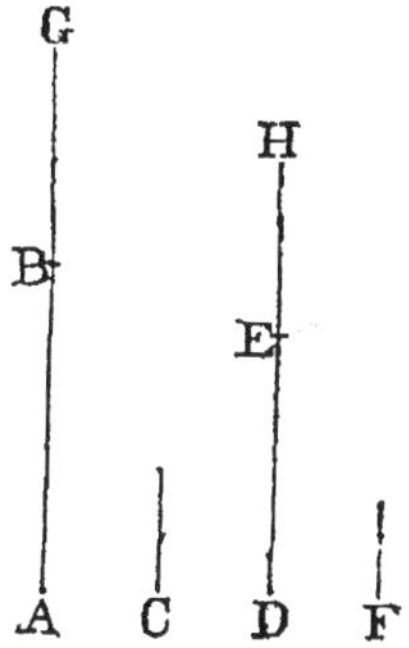

therefore AG is to BG as DH is to EH.

But BG is to C as EH is to F; [*Hypothesis.*

therefore, ex æquali, AG is to C as DH is to F. [V. 22.

Wherefore, *if the first* &c. Q.E.D.

COROLLARY 1. If the same hypothesis be made as in the proposition, the excess of the first and fifth shall be to the second as the excess of the third and sixth is to the fourth. The demonstration of this is the same as that of the proposition, if division be used instead of composition.

COROLLARY 2. The proposition holds true of two ranks of magnitudes, whatever be their number, of which each of the first rank has to the second magnitude the same ratio that the corresponding one of the second rank has to the fourth magnitude; as is manifest.

PROPOSITION 25. *THEOREM.*

If four magnitudes of the same kind be proportionals, the greatest and least of them together shall be greater than the other two together.

Let the four magnitudes AB, CD, E, F be proportionals; namely, let AB be to CD as E is to F; and let AB be the greatest of them, and consequently F the least: [V. *A*, V. 14.

AB and F together shall be greater than CD and E together.

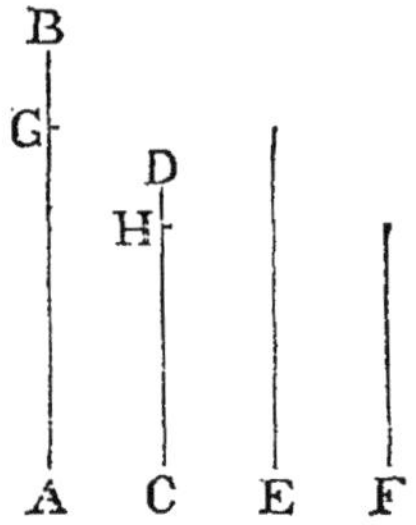

Take AG equal to E, and CH equal to F.

Then, because AB is to CD as E is to F, [*Hypothesis.*

and that AG is equal to E, and CH equal to F; [*Construction.*

therefore AB is to CD as AG is to CH. [V. 7, V. 11.

And because the whole AB is to the whole CD as AG is to CH;

therefore the remainder GB is to the remainder HD as the whole AB is to the whole CD. [V. 19.

But AB is greater than CD; [*Hypothesis.*

therefore BG is greater than DH. [V. *A.*

And because AG is equal to E and CH equal to F, [*Constr.*

therefore AG and F together are equal to CH and E together.

And if to the unequal magnitudes BG, DH, of which BG is the greater, there be added equal magnitudes, namely, AG and F to BG, and CH and E to DH, then AB and F together are greater than CD and E together.

Wherefore, *if four magnitudes* &c. Q.E.D.

BOOK VI.

DEFINITIONS.

1. SIMILAR rectineal figures are those which have their several angles equal, each to each, and the sides about the equal angles proportionals.

2. Reciprocal figures, namely, triangles and parallelograms, are such as have their sides about two of their angles proportionals in such a manner, that a side of the first figure is to a side of the other, as the remaining side of this other is to the remaining side of the first.

3. A straight line is said to be cut in extreme and mean ratio, when the whole is to the greater segment as the greater segment is to the less.

4. The altitude of any figure is the straight line drawn from its vertex perpendicular to the base.

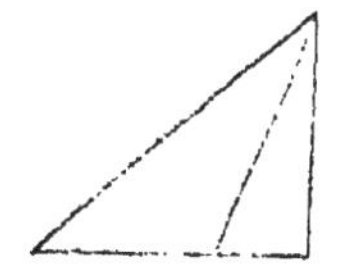

PROPOSITION 1. *THEOREM.*

Triangles and parallelograms of the same altitude are to one another as their bases.

Let the triangles ABC, ACD, and the parallelograms EC, CF have the same altitude, namely, the perpendicular drawn from the point A to BD: as the base BC is to the base CD, so shall the triangle ABC be to the triangle ACD, and the parallelogram EC to the parallelogram CF.

Produce BD both ways;
take any number of straight lines BG, GH, each equal to BC, and any number of straight lines DK, KL, each equal to CD; [I. 3.
and join AG, AH, AK, AL.

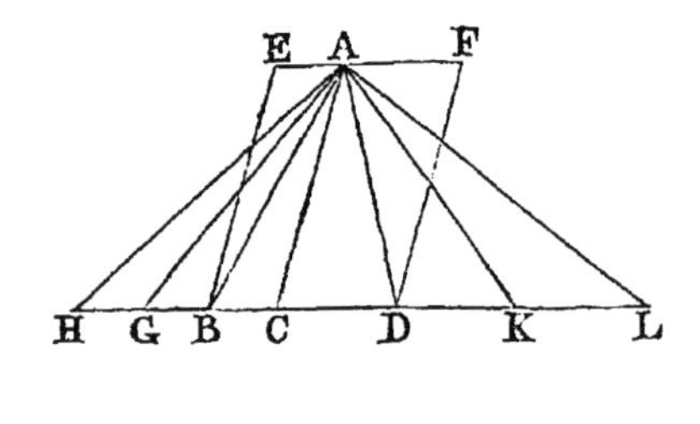

Then, because CB, BG, GH are all equal, [*Construction.*
the triangles ABC, AGB, AHG are all equal. [I. 38.

Therefore whatever multiple the base HC is of the base BC, the same multiple is the triangle AHC of the triangle ABC.

For the same reason, whatever multiple the base CL is of the base CD, the same multiple is the triangle ACL of the triangle ACD.

And if the base HC be equal to the base CL, the triangle AHC is equal to the triangle ACL; and if the base HC be greater than the base CL, the triangle AHC is greater than the triangle ACL; and if less, less. [I. 38.

Therefore, since there are four magnitudes, namely, the two bases BC, CD, and the two triangles ABC, ACD; and of the base BC, and the triangle ABC, the first and the third, any equimultiples whatever have been taken, namely, the base HC and the triangle AHC; and of the base CD and the triangle ACD, the second and the fourth, any equimultiples whatever have been taken, namely, the base CL and the triangle ACL;

and since it has been shewn that if the base *HC* be greater than the base *CL*, the triangle *AHC* is greater than the triangle *ACL*; and if equal, equal; and if less, less;

therefore as the base *BC* is to the base *CD*, so is the triangle *ABC* to the triangle *ACD*. [V. *Definition* 5.

And, because the parallelogram *CE* is double of the triangle *ABC*, and the parallelogram *CF* is double of the triangle *ACD*; [I. 41.

and that magnitudes have the same ratio which their equimultiples have; [V. 15.

therefore the parallelogram *EC* is to the parallelogram *CF* as the triangle *ABC* is to the triangle *ACD*.

But it has been shewn that the triangle *ABC* is to the triangle *ACD* as the base *BC* is to the base *CD*;

therefore the parallelogram *EC* is to the parallelogram *CF* as the base *BC* is to the base *CD*. [V. 11.

Wherefore, *triangles* &c. Q.E.D.

COROLLARY. From this it is plain that triangles and parallelograms which have equal altitudes, are to one another as their bases.

For, let the figures be placed so as to have their bases in the same straight line, and to be on the same side of it; and having drawn perpendiculars from the vertices of the triangles to the bases, the straight line which joins the vertices is parallel to that in which their bases are; [I. 33.

because the perpendiculars are both equal and parallel to one another. [I. 28.

Then, if the same construction be made as in the proposition, the demonstration will be the same.

PROPOSITION 2. *THEOREM.*

If a straight line be drawn parallel to one of the sides of a triangle, it shall cut the other sides, or those sides produced, proportionally; and if the sides, or the sides produced, be cut proportionally, the straight line which joins the points of section, shall be parallel to the remaining side of the triangle.

Let DE be drawn parallel to BC, one of the sides of the triangle ABC: BD shall be to DA as CE is to EA.

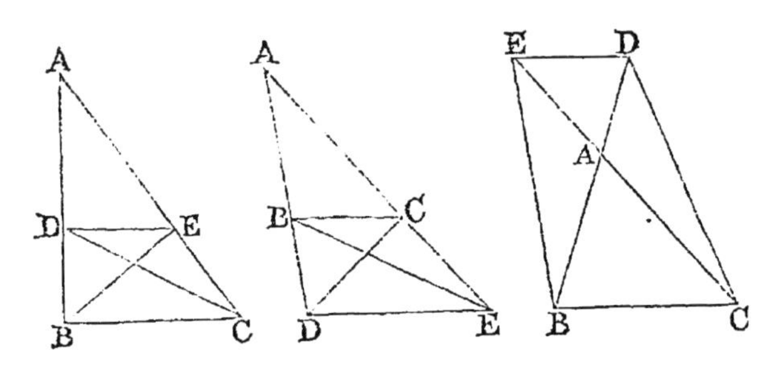

Join BE, CD.

Then the triangle BDE is equal to the triangle CDE, because they are on the same base DE and between the same parallels DE, BC. [I. 37.

And ADE is another triangle;

and equal magnitudes have the same ratio to the same magnitude; [V. 7.

therefore the triangle BDE is to the triangle ADE as the triangle CDE is to the triangle ADE.

But the triangle BDE is to the triangle ADE as BD is to DA;

because the triangles have the same altitude, namely, the perpendicular drawn from E to AB, and therefore they are to one another as their bases. [VI. 1.

For the same reason the triangle CDE is to the triangle ADE as CE is to EA.

Therefore BD is to DA as CE is to EA. [V. 11.

Next, let BD be to DA as CE is to EA, and join DE: DE shall be parallel to BC.

For, the same construction being made,

because BD is to DA as CE is to EA, [*Hypothesis.*

and as BD is to DA, so is the triangle BDE to the triangle ADE, [VI. 1.

and as CE is to EA so is the triangle CDE to the triangle ADE; [VI. 1.

therefore the triangle BDE is to the triangle ADE as the triangle CDE is to the triangle ADE; [V. 11.

that is, the triangles BDE and CDE have the same ratio to the triangle ADE.

Therefore the triangle BDE is equal to the triangle CDE. [V. 9.

And these triangles are on the same base DE and on the same side of it;

but equal triangles on the same base, and on the same side of it, are between the same parallels; [I. 39.

therefore DE is parallel to BC.

Wherefore, *if a straight line* &c. Q.E.D.

PROPOSITION 3. *THEOREM.*

If the vertical angle of a triangle be bisected by a straight line which also cuts the base, the segments of the base shall have the same ratio which the other sides of the triangle have to one another; and if the segments of the base have the same ratio which the other sides of the triangle have to one another, the straight line drawn from the vertex to the point of section shall bisect the vertical angle.

Let ABC be a triangle, and let the angle BAC be bisected by the straight line AD, which meets the base at D: BD shall be to DC as BA is to AC.

Through C draw CE parallel to DA, [I. 31.
and let BA produced meet CE at E.

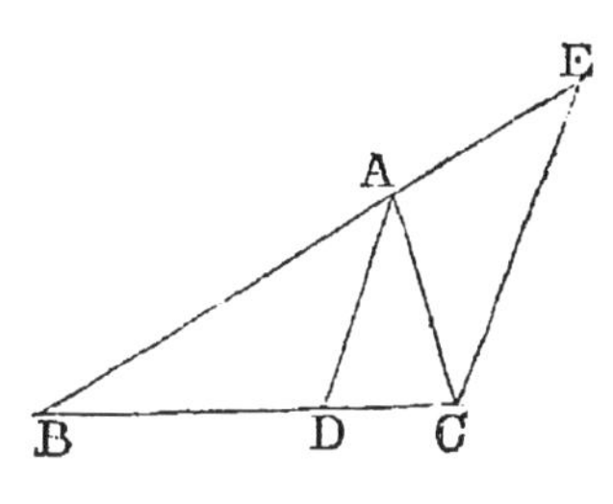

Then, because the straight line AC meets the parallels AD, EC, the angle ACE is equal to the alternate angle CAD; [I. 29.

but the angle CAD is, by hypothesis, equal to the angle BAD;

therefore the angle BAD is equal to the angle ACE. [*Ax.* 1.

Again, because the straight line BAE meets the parallels AD, EC, the exterior angle BAD is equal to the interior and opposite angle AEC; [I. 29.

but the angle BAD has been shewn equal to the angle ACE;

therefore the angle ACE is equal to the angle AEC; [*Axiom* 1.

and therefore AC is equal to AE. [I. 6.

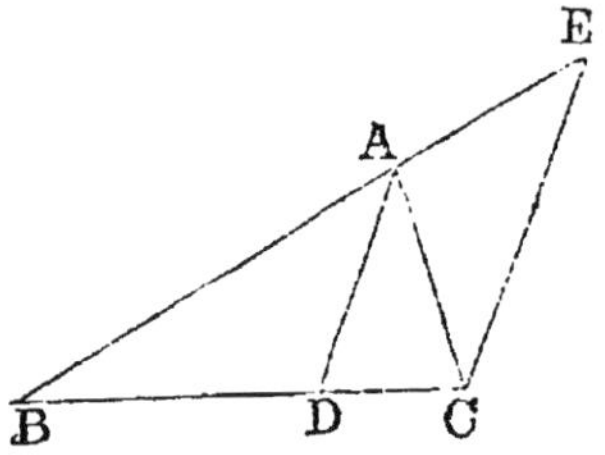

And, because AD is parallel to EC, [*Constr.* one of the sides of the triangle BCE,

therefore BD is to DC as BA is to AE; [VI. 2.

but AE is equal to AC;

therefore BD is to DC as BA is to AC. [V. 7.

Next, let BD be to DC as BA is to AC, and join AD: the angle BAC shall be bisected by the straight line AD.

For, let the same construction be made.

Then BD is to DC as BA is to AC; [*Hypothesis.*

and BD is to DC as BA is to AE, [VI. 2.

because AD is parallel to EC; [*Construction.*

therefore BA is to AC as BA is to AE; [V. 11.

therefore AC is equal to AE; [V. 9.

and therefore the angle AEC is equal to the angle ACE. [I. 5.

But the angle AEC is equal to the exterior angle BAD; [I. 29.

and the angle ACE is equal to the alternate angle CAD; [I. 29.

therefore the angle BAD is equal to the angle CAD; [*Ax.* 1.

that is, the angle BAC is bisected by the straight line AD.

Wherefore, *if the vertical angle* &c. Q.E.D.

PROPOSITION *A*. *THEOREM*.

If the exterior angle of a triangle, made by producing one of its sides, be bisected by a straight line which also cuts the base produced, the segments between the dividing straight line and the extremities of the base shall have the same ratio which the other sides of the triangle have to one another; and if the segments of the base produced have the same ratio which the other sides of the triangle have to one another, the straight line drawn from the vertex to the point of section shall bisect the exterior angle of the triangle.

Let ABC be a triangle, and let one of its sides BA be produced to E; and let the exterior angle CAE be bisected by the straight line AD which meets the base produced at D: BD shall be to DC as BA is to AC.

Through C draw CF parallel to AD, [I. 31. meeting AB at F.

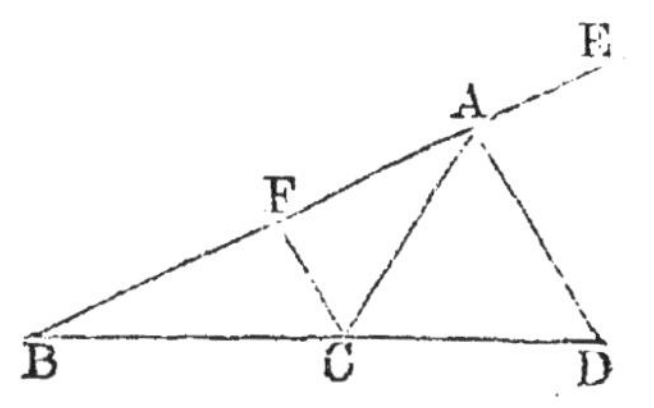

Then, because the straight line AC meets the parallels AD, FC, the angle ACF is equal to the alternate angle CAD; [I. 29.
but the angle CAD is, by hypothesis, equal to the angle DAE;

therefore the angle DAE is equal to the angle ACF. [*Ax.* 1.

Again, because the straight line FAE meets the parallels AD, FC, the exterior angle DAE is equal to the interior and opposite angle AFC; [I. 29.

but the angle DAE has been shewn equal to the angle ACF;

therefore the angle ACF is equal to the angle AFC; [*Ax.* 1.

and therefore AC is equal to AF. [I. 6.

And, because AD is paralled to FC, [*Construction.*

one of the sides of the triangle BCF;

therefore BD is to DC as BA is to AF; [VI. 2.

but AF is equal to AC;

therefore BD is to DC as BA is to AC. [V. 7.

Next, let BD be to DC as BA is to AC; and join AD: the exterior angle CAE shall be bisected by the straight line AD.

For, let the same construction be made.

Then BD is to DC as BA is to AC; [*Hypothesis.*

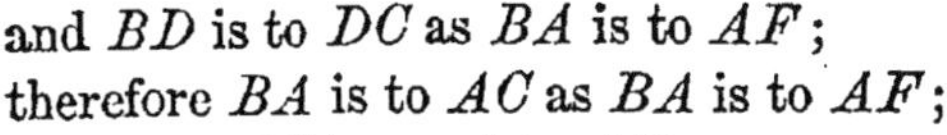

and BD is to DC as BA is to AF; [VI. 2.

therefore BA is to AC as BA is to AF; [V. 11.

therefore AC is equal to AF, [V. 9.

and therefore the angle ACF is equal to the angle AFC. [I. 5.

But the angle AFC is equal to the exterior angle DAE; [I. 29.

and the angle ACF is equal to the alternate angle CAD; [I. 29.

therefore the angle CAD is equal to the angle DAE; [*Ax.* 1.

that is, the angle CAE is bisected by the straight line AD.

Wherefore, *if the exterior angle* &c. Q.E.D.

PROPOSITION 4. *THEOREM.*

The sides about the equal angles of triangles which are equiangular to one another are proportionals; and those which are opposite to the equal angles are homologous sides, that is, are the antecedents or consequents of the ratios.

Let the triangle ABC be equiangular to the triangle DCE, having the angle ABC equal to the angle DCE, and the angle ACB equal to the angle DEC, and consequently the angle BAC equal to the angle CDE: the sides about the equal angles of the triangles ABC, DCE, shall be proportionals; and those shall be the homologous sides, which are opposite to the equal angles.

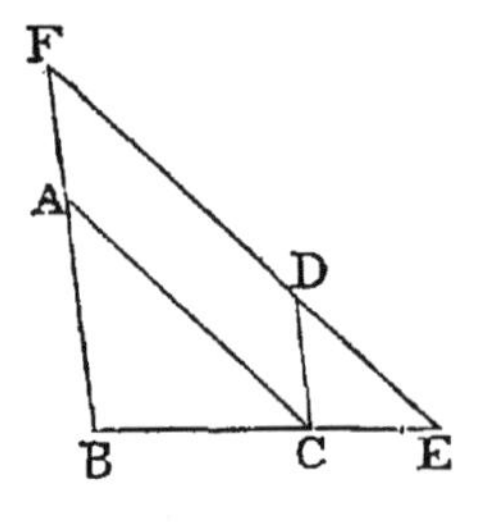

Let the triangle DCE be placed so that its side CE may be contiguous to BC, and in the same straight line with it. [I. 22.

Then the angle *BCA* is equal to the angle *CED*; [*Hyp.*
add to each the angle *ABC*;
therefore the two angles *ABC, BCA* are equal to the two angles *ABC, CED*; [*Axiom* 2.
but the angles *ABC, BCA* are together less than two right angles; [I. 17.
therefore the angles *ABC, CED* are together less than two right angles;
therefore *BA* and *ED*, if produced, will meet. [*Axiom* 12.
Let them be produced and meet at the point *F*.

Then, because the angle *ABC* is equal to the angle *DCE*, [*Hypothesis.*
BF is parallel to *CD*; [I. 28.
and because the angle *ACB* is equal to the angle *DEC*, [*Hyp.*
AC is parallel to *FE*. [I. 28.
Therefore *FACD* is a parallelogram;
and therefore *AF* is equal to *CD*, and *AC* is equal to *FD*. [I. 34.

And, because *AC* is parallel to *FE*, one of the sides of the triangle *FBE*,
therefore *BA* is to *AF* as *BC* is to *CE*; [VI. 2.
but *AF* is equal to *CD*;
therefore *BA* is to *CD* as *BC* is to *CE*; [V. 7.
and, alternately, *AB* is to *BC* as *DC* is to *CE*. [V. 16.
Again, because *CD* is parallel to *BF*,
therefore *BC* is to *CE* as *FD* is to *DE*; [VI. 2.
but *FD* is equal to *AC*;
therefore *BC* is to *CE* as *AC* is to *DE*; [V. 7.
and, alternately, *BC* is to *CA* as *CE* is to *ED*. [V. 16.
Then, because it has been shewn that *AB* is to *BC* as *DC* is to *CE*, and that *BC* is to *CA* as *CE* is to *ED*;
therefore, ex æquali, *BA* is to *AC* as *CD* is to *DE*. [V. 22.

Wherefore, *the sides* &c. Q.E.D.

PROPOSITION 5. *THEOREM.*

If the sides of two triangles, about each of their angles, be proportionals, the triangles shall be equiangular to one another, and shall have those angles equal which are opposite to the homologous sides.

Let the triangles *ABC, DEF* have their sides proportional, so that *AB* is to *BC* as *DE* is to *EF*; and *BC* to *CA* as *EF* is to *FD*; and, consequently, ex æquali, *BA* to *AC* as *ED* is to *DF*: the triangle *ABC* shall be equiangular to the triangle *DEF*, and they shall have those angles equal which are opposite to the homologous sides, namely, the angle *ABC* equal to the angle *DEF*, and the angle *BCA* equal to the angle *EFD*, and the angle *BAC* equal to the angle *EDF*.

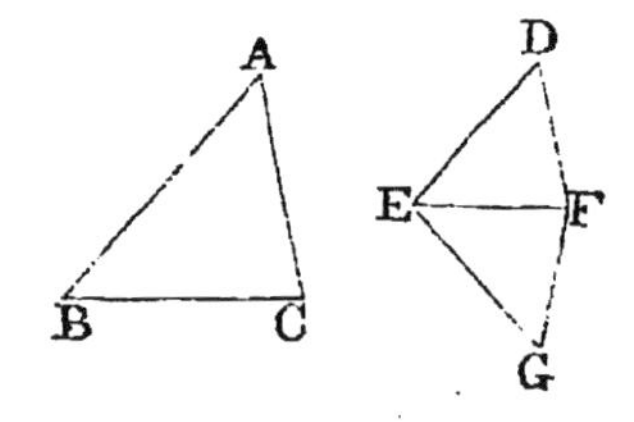

At the point *E*, in the straight line *EF*, make the angle *FEG* equal to the angle *ABC*; and at the point *F*, in the straight line *EF*, make the angle *EFG* equal to the angle *BCA*; [I. 23.
therefore the remaining angle *EGF* is equal to the remaining angle *BAC*.

Therefore the triangle *ABC* is equiangular to the triangle *GEF*;

and therefore they have their sides opposite to the equal angles proportionals; [VI. 4.

therefore *AB* is to *BC* as *GE* is to *EF*.

But *AB* is to *BC* as *DE* is to *EF*: [*Hypothesis.*

therefore *DE* is to *EF* as *GE* is to *EF*; [V. 11.

therefore *DE* is equal to *GE*. [V. 9.

For the same reason, *DF* is equal to *GF*.

Then, because in the two triangles *DEF, GEF*, *DE* is equal to *GE*, and *EF* is common;

the two sides *DE, EF* are equal to the two sides *GE, EF*, each to each;

and the base *DF* is equal to the base *GF*;

therefore the angle *DEF* is equal to the angle *GEF*, [I. 8.

and the other angles to the other angles, each to each, to which the equal sides are opposite. [I. 4.

therefore the angle *DFE* is equal to the angle *GFE*, and the angle *EDF* is equal to the angle *EGF*.

And, because the angle DEF is equal to the angle GEF, and the angle GEF is equal to the angle ABC, [*Constr.*
therefore the angle ABC is equal to the angle DEF. [*Ax.* 1.
For the same reason, the angle ACB is equal to the angle DFE, and the angle at A is equal to the angle at D.
Therefore the triangle ABC is equiangular to the triangle DEF.

Wherefore, *if the sides* &c. Q.E.D.

PROPOSITION 6. *THEOREM.*

If two triangles have one angle of the one equal to one angle of the other, and the sides about the equal angles proportionals, the triangles shall be equiangular to one another, and shall have those angles equal which are opposite to the homologous sides.

Let the triangles ABC, DEF have the angle BAC in the one, equal to the angle EDF in the other, and the sides about those angles proportionals, namely, BA to AC as ED is to DF: the triangle ABC shall be equiangular to the triangle DEF, and shall have the angle ABC equal to the angle DEF, and the angle ACB equal to the angle DFE.

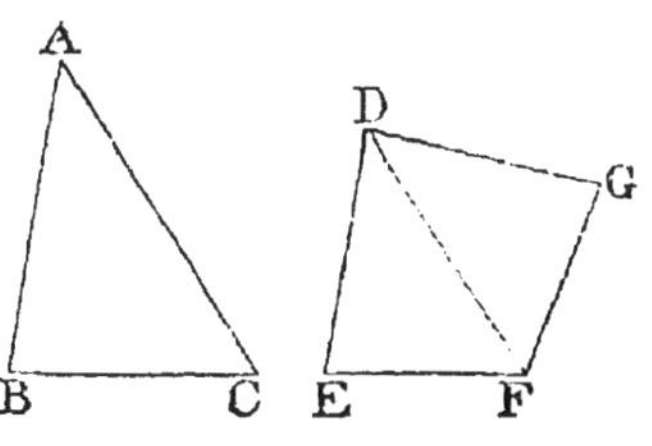

At the point D, in the straight line DF, make the angle FDG equal to either of the angles BAC, EDF; and at the point F, in the straight line DF, make the angle DFG equal to the angle ACB; [I. 23.
therefore the remaining angle at G is equal to the remaining angle at B.

Therefore the triangle ABC is equiangular to the triangle DGF;

therefore BA is to AC as GD is to DF. [VI. 4.
But BA is to AC as ED is to DF; [*Hypothesis.*
therefore ED is to DF as GD is to DF; [V. 11.
therefore ED is equal to GD. [V. 9.

And DF is common to the two triangles EDF, GDF; therefore the two sides ED, DF are equal to the two sides GD, DF, each to each;

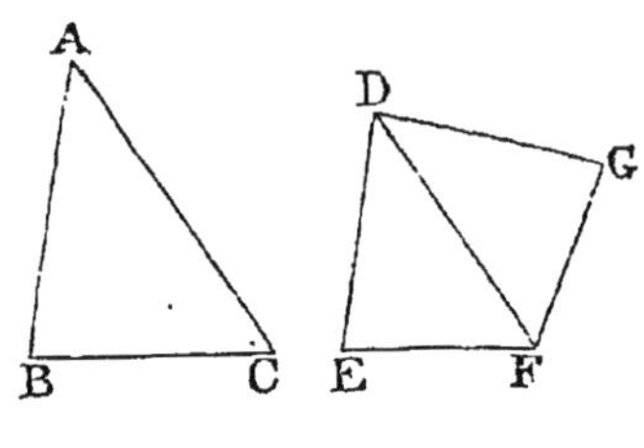

and the angle EDF is equal to the angle GDF; [*Constr.*
therefore the base EF is equal to the base GF, and the triangle EDF to the triangle GDF, and the remaining angles to the remaining angles, each to each, to which the equal sides are opposite; [I. 4.
therefore the angle DFG is equal to the angle DFE, and the angle at G is equal to the angle at E.

But the angle DFG is equal to the angle ACB; [*Constr.*
therefore the angle ACB is equal to the angle DFE. [*Ax.* 1.
And the angle BAC is equal to the angle EDF; [*Hypothesis.*
therefore the remaining angle at B is equal to the remaining angle at E.
Therefore the triangle ABC is equiangular to the triangle DEF.

Wherefore, *if two triangles* &c. Q.E.D.

PROPOSITION 7. *THEOREM.*

If two triangles have one angle of the one equal to one angle of the other, and the sides about two other angles proportionals; then, if each of the remaining angles be either less, or not less, than a right angle, or if one of them be a right angle, the triangles shall be equiangular to one another, and shall have those angles equal about which the sides are proportionals.

Let the triangles ABC, DEF have one angle of the one equal to one angle of the other, namely, the angle BAC equal to the angle EDF, and the sides about two other angles ABC, DEF, proportionals, so that AB is to BC as DE is to EF; and, first, let each of the remaining angles at C and F be less than a right angle: the triangle ABC shall be equiangular to the triangle DEF, and shall have the angle ABC equal to the angle DEF, and the angle at C equal to the angle at F.

For, if the angles ABC, DEF be not equal, one of them must be greater than the other.

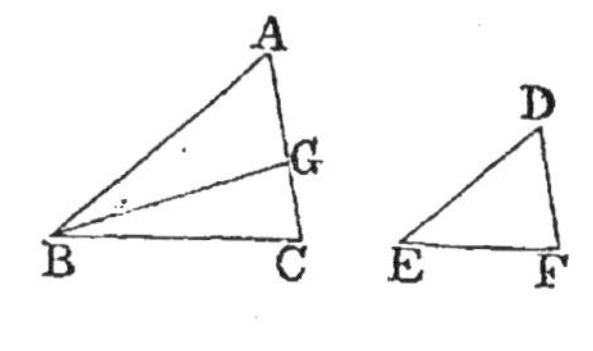

Let ABC be the greater, and at the point B, in the straight line AB, make the angle ABG equal to the angle DEF. [I. 23.

Then, because the angle at A is equal to the angle at D, [*Hyp.*

and the angle ABG is equal to the angle DEF, [*Constr.*

therefore the remaining angle AGB is equal to the remaining angle DFE;

therefore the triangle ABG is equiangular to the triangle DEF.

Therefore AB is to BG as DE is to EF. [VI. 4.

But AB is to BC as DE is to EF; [*Hypothesis.*

therefore AB is to BC as AB is to BG; [V. 11.

therefore BC is equal to BG; [V. 9.

and therefore the angle BCG is equal to the angle BGC. [I. 5.

But the angle BCG is less than a right angle; [*Hyp.*

therefore the angle BGC is less than a right angle;

and therefore the adjacent angle AGB must be greater than a right angle. [I. 13.

But the angle AGB was shewn to be equal to the angle at F;

therefore the angle at F is greater than a right angle.

But the angle at F is less than a right angle; [*Hypothesis.*

which is absurd.

Therefore the angles ABC and DEF are not unequal; that is, they are equal.

And the angle at A is equal to the angle at D; [*Hypothesis.*

therefore the remaining angle at C is equal to the remaining angle at F;

therefore the triangle ABC is equiangular to the triangle DEF.

Next, let each of the angles at C and F be not less than a right angle: the triangle ABC shall be equiangular to the triangle DEF.

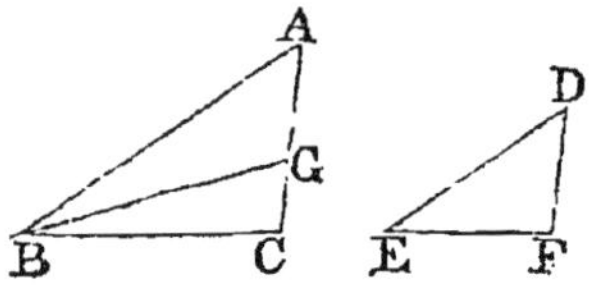

For, the same construction being made, it may be shewn in the same manner, that BC is equal to BG; therefore the angle BCG is equal to the angle BGC. [I. 5.
But the angle BCG is not less than a right angle; [*Hyp.*
therefore the angle BGC is not less than a right angle;
that is, two angles of the triangle BCG are together not less than two right angles; which is impossible. [I. 17.
Therefore the triangle ABC may be shewn to be equiangular to the triangle DEF, as in the first case.

Lastly, let one of the angles at C and F be a right angle, namely, the angle at C: the triangle ABC shall be equiangular to the triangle DEF.

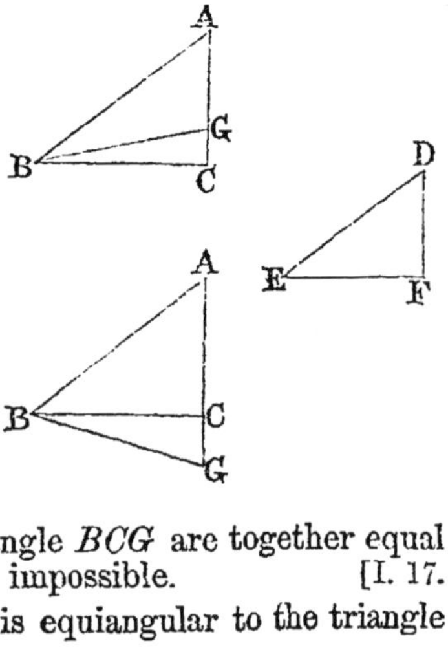

For, if the triangle ABC be not equiangular to the triangle DEF, at the point B, in the straight line AB, make the angle ABG equal to the angle DEF. [I. 23.
Then it may be shewn, as in the first case, that BC is equal to BG;
therefore the angle BCG is equal to the angle BGC. [I. 5.
But the angle BCG is a right angle: [*Hypothesis.*
therefore the angle BGC is a right angle;
that is, two angles of the triangle BCG are together equal to two right angles; which is impossible. [I. 17.
Therefore the triangle ABC is equiangular to the triangle DEF.

Wherefore, *if two triangles* &c. Q.E.D.

PROPOSITION 8. *THEOREM.*

In a right-angled triangle, if a perpendicular be drawn from the right angle to the base, the triangles on each side of it are similar to the whole triangle, and to one another.

Let ABC be a right-angled triangle, having the right angle BAC; and from the point A, let AD be drawn perpendicular to the base BC: the triangles DBA, DAC shall be similar to the whole triangle ABC, and to one another.

For, the angle BAC is equal to the angle BDA, each of them being a right angle, [*Axiom* 11.
and the angle at B is common to the two triangles ABC, DBA;
therefore the remaining angle ACB is equal to the remaining angle DAB.

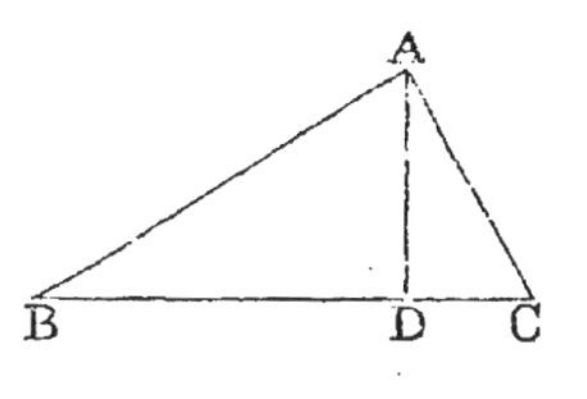

Therefore the triangle ABC is equiangular to the triangle DBA, and the sides about their equal angles are proportionals; [VI. 4.
therefore the triangles are similar. [VI. *Definition* 1.

In the same manner it may be shewn that the triangle DAC is similar to the triangle ABC.
And the triangles DBA, DAC being both similar to the triangle ABC, are similar to each other.

Wherefore, *in a right-angled triangle* &c. Q.E.D.

COROLLARY. From this it is manifest, that the perpendicular drawn from the right angle of a right-angled triangle to the base, is a mean proportional between the segments of the base, and also that each of the sides is a mean proportional between the base and the segment of the base adjacent to that side.

For, in the triangles DBA, DAC,
BD is to DA as DA is to DC; [VI. 4.
and in the triangles ABC, DBA,
BC is to BA as BA is to BD; [VI. 4.
and in the triangles ABC, DAC,
BC is to CA as CA is to CD. [VI. 4.

PROPOSITION 9. *PROBLEM.*

From a given straight line to cut off any part required.

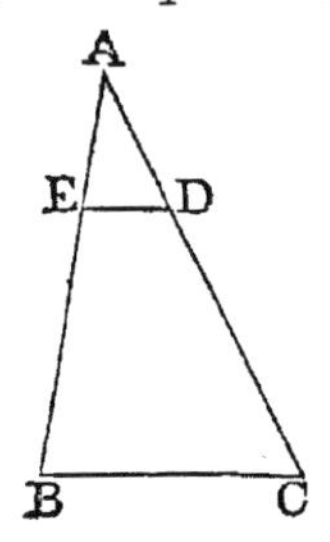

Let AB be the given straight line: it is required to cut off any part from it.

From the point A draw a straight line AC, making any angle with AB; in AC take any point D, and take AC the same multiple of AD, that AB is of the part which is to be cut off from it; join BC, and draw DE parallel to it. AE shall be the part required to be cut off.

For, because ED is parallel to BC, [Construction.
one of the sides of the triangle ABC,
therefore CD is to DA as BE is to EA; [VI. 2.
and, by composition, CA is to AD as BA is to AE. [V. 18.
But CA is a multiple of AD; [Construction.
therefore BA is the same multiple of AE; [V. *D.*
that is, whatever part AD is of AC, AE is the same part of AB.

Wherefore, *from the given straight line AB, the part required has been cut off.* Q.E.F.

PROPOSITION 10. *PROBLEM.*

To divide a given straight line similarly to a given divided straight line, that is, into parts which shall have the same ratios to one another, that the parts of the given divided straight line have.

Let AB be the straight line given to be divided, and AC the given divided straight line: it is required to divide AB similarly to AC.

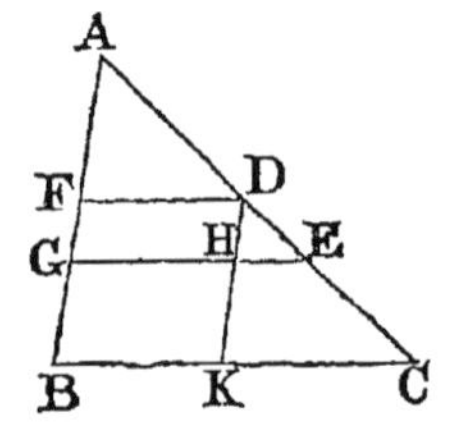

Let AC be divided at the points D, E; and let AB, AC be placed so as to contain any angle, and join BC; through the point D, draw DF parallel to BC, and through the point E draw EG parallel to BC. [I. 31.
AB shall be divided at the points F, G, similarly to AC.

Through D draw DHK parallel to AB. [I. 31.
Then each of the figures FH, HB is a parallelogram;
therefore DH is equal to FG, and HK is equal to GB. [I. 34.
Then, because HE is parallel to KC, [*Construction.*
one of the sides of the triangle DKC,
therefore KH is to HD as CE is to ED. [VI. 2.
But KH is equal to BG, and HD is equal to GF;
therefore BG is to GF as CE is to ED. [V. 7.

Again, because FD is parallel to GE, [*Construction.*
one of the sides of the triangle AGE,
therefore GF is to FA as ED is to DA. [VI. 2.
And it has been shewn that BG is to GF as CE is to ED.
Therefore BG is to GF as CE is to ED, and GF is to FA as ED is to DA.

Wherefore *the given straight line AB is divided similarly to the given divided straight line AC.* Q.E.F.

PROPOSITION 11. *PROBLEM.*

To find a third proportional to two given straight lines.

Let AB, AC be the two given straight lines: it is required to find a third proportional to AB, AC.

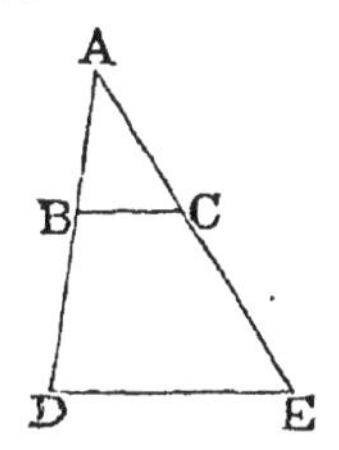

Let AB, AC be placed so as to contain any angle; produce AB, AC, to the points D, E; and make BD equal to AC; [I. 3.
join BC, and through D draw DE parallel to BC. [I. 31.

CE shall be a third proportional to AB, AC.

For, because BC is parallel to DE, [*Construction.*
one of the sides of the triangle ADE,
therefore AB is to BD as AC is to CE; [VI. 2.
but BD is equal to AC; [*Construction.*
therefore AB is to AC as AC is to CE. [V. 7.

Wherefore *to the two given straight lines AB, AC, a third proportional CE is found.* Q.E.F.

PROPOSITION 12. *THEOREM.*

To find a fourth proportional to three given straight lines.

Let A, B, C be the three given straight lines: it is required to find a fourth proportional to A, B, C.

Take two straight lines, DE, DF, containing any angle EDF; and in these make DG equal to A, GE equal to B, and DH equal to C; [I. 3. join GH, and through E draw EF parallel to GH. [I. 31.

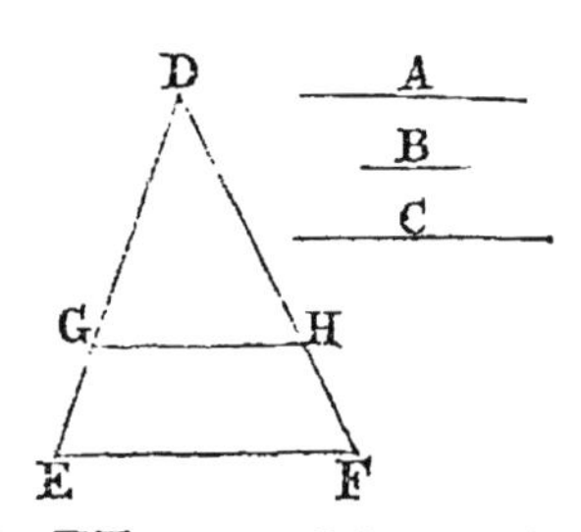

HF shall be a fourth proportional to A, B, C.

For, because GH is parallel to EF, [*Construction.*
one of the sides of the triangle DEF,
therefore DG is to GE as DH is to HF. [VI. 2.
But DG is equal to A, GE is equal to B, and DH is equal to C; [*Construction.*
therefore A is to B as C is to HF. [V. 7.

Wherefore *to the three given straight lines A, B, C, a fourth proportional HF is found.* Q.E.F.

PROPOSITION 13. *PROBLEM.*

To find a mean proportional between two given straight lines.

Let AB, BC be the two given straight lines: it is required to find a mean proportional between them.

Place AB, BC in a straight line, and on AC describe the semicircle ADC; from the point B draw BD at right angles to AC. [I. 11.

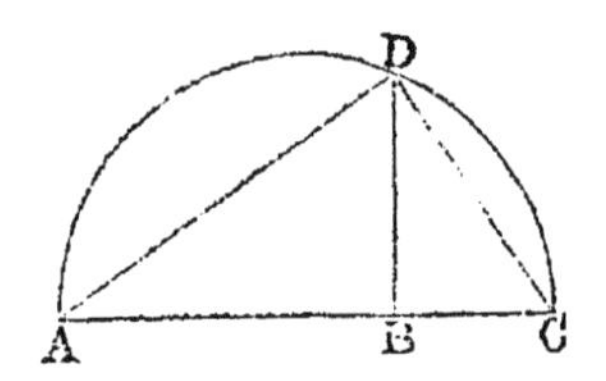

BD shall be a mean proportional between AB and BC.

Join AD, DC.

Then, the angle ADC, being in a semicircle, is a right angle; [III. 31.

and because in the right-angled triangle ADC, DB is drawn from the right angle perpendicular to the base,

therefore DB is a mean proportional between AB, BC, the segments of the base. [VI. 8, *Corollary.*

Wherefore, *between the two given straight lines AB, BC, a mean proportional DB is found.* Q.E.F.

PROPOSITION 14. *THEOREM.*

Equal parallelograms which have one angle of the one equal to one angle of the other, have their sides about the equal angles reciprocally proportional; and parallelograms which have one angle of the one equal to one angle of the other, and their sides about the equal angles reciprocally proportional, are equal to one another.

Let AB, BC be equal parallelograms, which have the angle FBD equal to the angle EBG: the sides of the parallelograms about the equal angles shall be reciprocally proportional, that is, DB shall be to BE as GB is to BF.

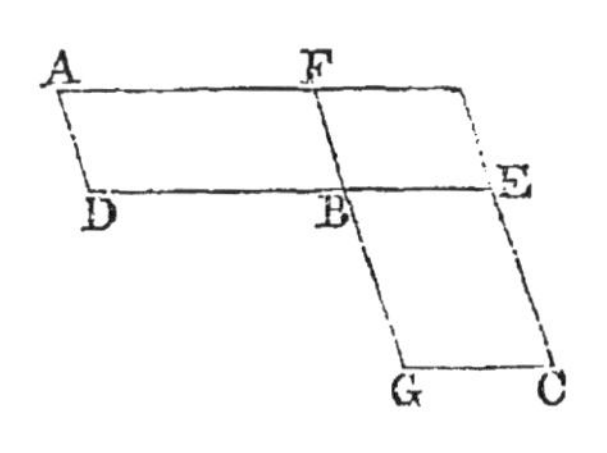

Let the parallelograms be placed, so that the sides DB, BE may be in the same straight line;

therefore also FB, BG are in one straight line. [I. 14.

Complete the parallelogram FE.

Then, because the parallelogram AB is equal to the parallelogram BC, [*Hypothesis.*

and that FE is another parallelogram,

therefore AB is to FE as BC is to FE. [V. 7.

But AB is to FE as the base DB is to the base BE, [VI. 1.

and BC is to FE as the base GB is to the base BF; [VI. 1.

therefore DB is to BE as GB is to BF. [V. 11.

Next, let the angle FBD be equal to the angle EBG, and let the sides about the equal angles be reciprocally proportional, namely, DB to BE as GB is to BF: the parallelogram AB shall be equal to the parallelogram BC.

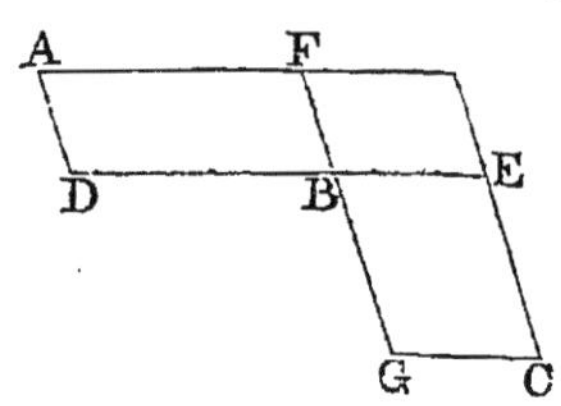

For, let the same construction be made.

Then, because DB is to BE as GB is to BF, [*Hypothesis.*

and that DB is to BE as the parallelogram AB is to the parallelogram FE, [VI. 1.

and that GB is to BF as the parallelogram BC is to the parallelogram FE; [VI. 1.

therefore the parallelogram AB is to the parallelogram FE as the parallelogram BC is to the parallelogram FE; [V. 11.

therefore the parallelogram AB is equal to the parallelogram BC. [V. 9.

Wherefore, *equal parallelograms* &c. Q.E.D.

PROPOSITION 15. *THEOREM.*

Equal triangles which have one angle of the one equal to one angle of the other, have their sides about the equal angles reciprocally proportional; and triangles which have one angle of the one equal to one angle of the other, and their sides about the equal angles reciprocally proportional, are equal to one another.

Let ABC, ADE be equal triangles, which have the angle BAC equal to the angle DAE: the sides of the triangles about the equal angles shall be reciprocally proportional; that is, CA shall be to AD as EA is to AB.

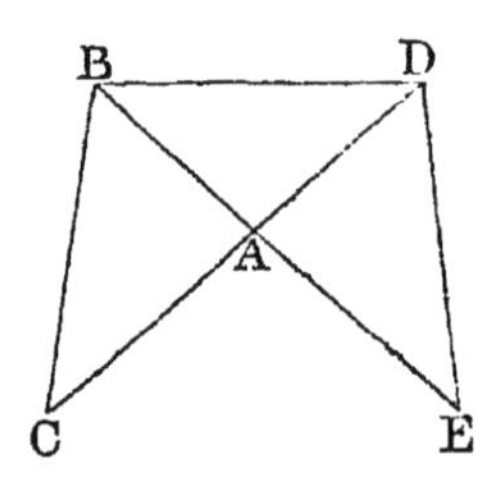

Let the triangles be placed so that the sides CA, AD may be in the same straight line,

therefore also EA, AB are in one straight line; [I. 14.
join BD.

Then, because the triangle ABC is equal to the triangle ADE, [*Hypothesis.*
and that ABD is another triangle,
therefore the triangle ABC is to the triangle ABD as the triangle ADE is to the triangle ABD. [V. 7.
But the triangle ABC is to the triangle ABD as the base CA is to the base AD, [VI. 1.
and the triangle ADE is to the triangle ABD as the base EA is to the base AB; [VI. 1.
therefore CA is to AD as EA is to AB. [V. 11.

Next, let the angle BAC be equal to the angle DAE, and let the sides about the equal angles be reciprocally proportional, namely, CA to AD as EA is to AB: the triangle ABC shall be equal to the triangle ADE.

For, let the same construction be made.
Then, because CA is to AD as EA is to AB, [*Hypothesis.*
and that CA is to AD as the triangle ABC is to the triangle ABD, [VI. 1.
and that EA is to AB as the triangle ADE is to the triangle ABD, [VI. 1.
therefore the triangle ABC is to the triangle ABD as the triangle ADE is to the triangle ABD; [V. 11.
therefore the triangle ABC is equal to the triangle ADE. [V. 9.

Wherefore, *equal triangles* &c. Q.E.D.

PROPOSITION 16. *THEOREM.*

If four straight lines be proportionals, the rectangle contained by the extremes is equal to the rectangle contained by the means; and if the rectangle contained by the extremes be equal to the rectangle contained by the means, the four straight lines are proportionals.

Let the four straight lines AB, CD, E, F, be proportionals, namely, let AB be to CD as E is to F: the rectangle contained by AB and F shall be equal to the rectangle contained by CD and E.

From the points A, C, draw AG, CH at right angles to AB, CD; [I. 11.
make AG equal to F, and CH equal to E; [I. 3.
and complete the parallelograms BG, DH. [I. 31.

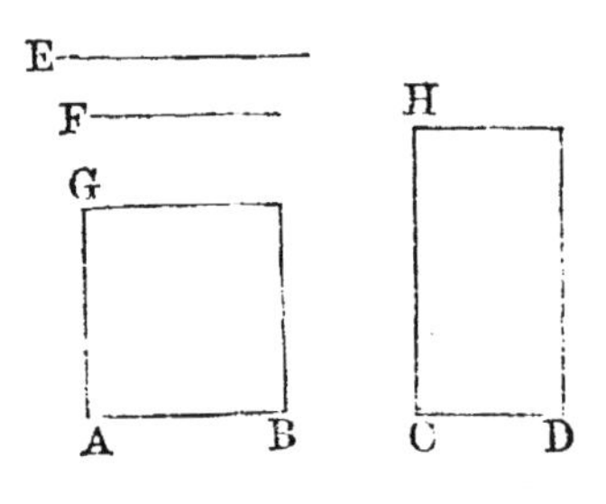

Then, because AB is to CD as E is to F, [*Hyp.*
and that E is equal to CH, and F is equal to AG, [*Construction.*
therefore AB is to CD as CH is to AG; [V. 7.
that is, the sides of the parallelograms BG, DH about the equal angles are reciprocally proportional;
therefore the parallelogram BG is equal to the parallelogram DH. [VI. 14.

But the parallelogram BG is contained by the straight lines AB and F, because AG is equal to F; [*Construction.*
and the parallelogram DH is contained by the straight lines CD and E, because CH is equal to E;
therefore the rectangle contained by AB and F is equal to the rectangle contained by CD and E.

Next, let the rectangle contained by AB and F be equal to the rectangle contained by CD and E: these four straight lines shall be proportional, namely, AB shall be to CD as E is to F.

For, let the same construction be made.

Then, because the rectangle contained by AB and F is equal to the rectangle contained by CD and E, [*Hypothesis.*
and that the rectangle BG is contained by AB and F, because AG is equal to F, [*Construction.*
and that the rectangle DH is contained by CD and E, because CH is equal to E, [*Construction.*

therefore the parallelogram BG is equal to the parallelogram DH. [*Axiom* 1.

And these parallelograms are equiangular to one another; therefore the sides about the equal angles are reciprocally proportional; [VI. 14.

therefore AB is to CD as CH is to AG.

But CH is equal to E, and AG is equal to F; [*Constr.*

therefore AB is to CD as E is to F. [V. 7.

Wherefore, *if four straight lines* &c. Q.E.D.

PROPOSITION 17. *THEOREM.*

If three straight lines be proportionals, the rectangle contained by the extremes is equal to the square on the mean; and if the rectangle contained by the extremes be equal to the square on the mean, the three straight lines are proportionals.

Let the three straight lines A, B, C be proportionals, namely, let A be to B as B is to C: the rectangle contained by A and C shall be equal to the square on B.

Take D equal to B.

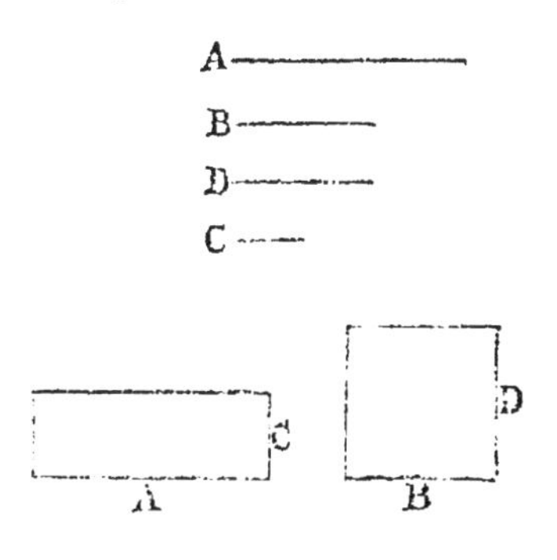

Then, because A is to B as B is to C, [*Hyp.*

and that B is equal to D, therefore A is to B as D is to C. [V. 7.

But if four straight lines be proportionals, the rectangle contained by the extremes is equal to the rectangle contained by the means; [VI. 16.

therefore the rectangle contained by A and C is equal to the rectangle contained by B and D.

But the rectangle contained by B and D is the square on B, because B is equal to D; [*Construction.*

therefore the rectangle contained by A and C is equal to the square on B.

Next, let the rectangle contained by A and C be equal to the square on B: A shall be to B as B is to C.

For, let the same construction be made.

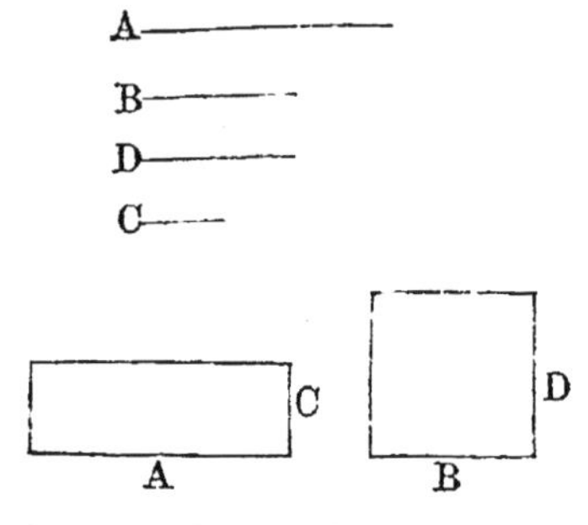

Then, because the rectangle contained by A and C is equal to the square on B, [*Hypothesis.*

and that the square on B is equal to the rectangle contained by B and D, because B is equal to D, [*Construction.*

therefore the rectangle contained by A and C is equal to the rectangle contained by B and D.

But if the rectangle contained by the extremes be equal to the rectangle contained by the means, the four straight lines are proportionals; [VI. 16.

therefore A is to B as D is to C.

But B is equal to D; [*Construction.*

Therefore A is to B as B is to C. [V. 7.

Wherefore, *if three straight lines* &c. Q.E.D.

PROPOSITION 18. *PROBLEM.*

On a given straight line to describe a rectilineal figure similar and similarly situated to a given rectilineal figure.

Let AB be the given straight line, and $CDEF$ the given rectilineal figure of four sides: it is required to describe on the given straight line AB. a rectilineal figure, similar and similarly situated to $CDEF$.

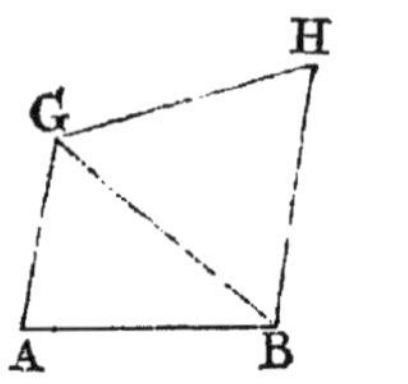

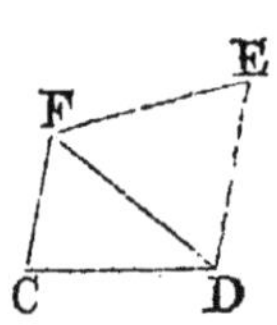

Join DF; at the point A, in the straight line AB, make the angle BAG equal to the angle DCF; and at the point B, in the straight line AB, make the angle ABG equal to the angle CDF; [I. 23.

therefore the remaining angle AGB is equal to the remaining angle CFD,

and the triangle AGB is equiangular to the triangle CFD.

Again, at the point B, in the straight line BG, make the angle GBH equal to the angle FDE; and at the point G, in the straight line BG, make the angle BGH equal to the angle DFE; [I. 23.

therefore the remaining angle BHG is equal to the remaining angle DEF,

and the triangle BHG is equiangular to the triangle DEF.

Then, because the angle AGB is equal to the angle CFD, and the angle BGH equal to the angle DFE; [*Construction.*

therefore the whole angle AGH is equal to the whole angle CFE. [*Axiom* 2.

For the same reason the angle ABH is equal to the angle CDE.

And the angle BAG is equal to the angle DCF, and the angle BHG is equal to the angle DEF.

Therefore the rectilineal figure $ABHG$ is equiangular to the rectilineal figure $CDEF$.

Also these figures have their sides about the equal angles proportionals.

For, because the triangle BAG is equiangular to the triangle DCF, therefore BA is to AG as DC is to CF. [VI. 4.

And, for the same reason, AG is to GB as CF is to FD, and BG is to GH as DF is to FE;

therefore, ex æquali, AG is to GH as CF is to FE. [V. 22.

In the same manner it may be shewn that AB is to BH as CD is to DE.

And GH is to HB as FE is to ED. [VI. 4.

Therefore, the rectilineal figures $ABHG$ and $CDEF$ are equiangular to one another, and have their sides about the equal angles proportionals;

therefore they are similar to one another. [VI. *Definition* 1.

Next, let it be required to describe on the given straight line AB, a rectilineal figure, similar, and similarly situated, to the rectilineal figure $CDKEF$ of five sides.

Join DE, and on the given straight line AB describe, as in the former case, the rectilineal figure $ABHG$, similar, and similarly situated to the rectilineal figure $CDEF$ of four sides. At the point B, in the straight line BH, make the angle HBL equal to the angle EDK; and at the point H, in the straight line BH, make the angle BHL equal to the angle DEK; [I. 23.

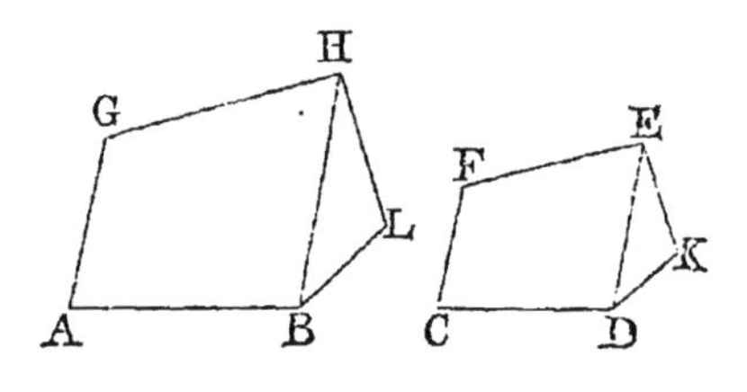

therefore the remaining angle at L is equal to the remaining angle at K.

Then, because the figures $ABHG$, $CDEF$ are similar, the angle ABH is equal to the angle CDE; [VI. *Def.* 1.
and the angle HBL is equal to the angle EDK; [*Constr.*
therefore the whole angle ABL is equal to the whole angle CDK. [*Axiom* 2.
For the same reason the whole angle GHL is equal to the whole angle FEK.
Therefore the five-sided figures $ABLHG$ and $CDKEF$ are equiangular to one another.

And, because the figures $ABHG$ and $CDEF$ are similar, therefore AB is to BH as CD is to DE; [VI. *Definition* 1.
but BH is to BL as DE is to DK; [VI. 4.
therefore, ex æquali, AB is to BL as CD is to DK. [V. 22.
For the same reason, GH is to HL as FE is to EK.
And BL is to LH as DK is to KE. [VI. 4.

Therefore, the five-sided figures $ABLHG$ and $CDKEF$ are equiangular to one another, and have their sides about the equal angles proportionals;
therefore they are similar to one another. [VI. *Definition* 1.

In the same manner a rectilineal figure of six sides may be described on a given straight line, similar and similarly situated to a given rectilineal figure of six sides; and so on. Q.E.F.

PROPOSITION 19. *THEOREM.*

Similar triangles are to one another in the duplicate ratio of their homologous sides.

Let ABC and DEF be similar triangles, having the angle B equal to the angle E, and let AB be to BC as DE is to EF, so that the side BC is homologous to the side EF: the triangle ABC shall be to the triangle DEF in the duplicate ratio of BC to EF.

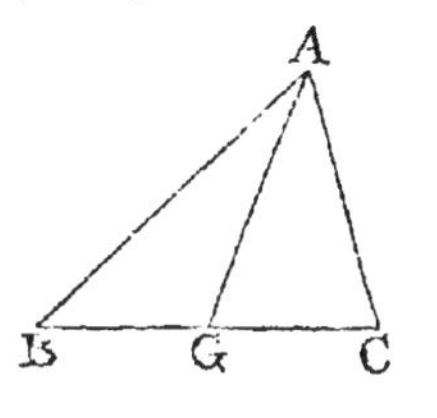

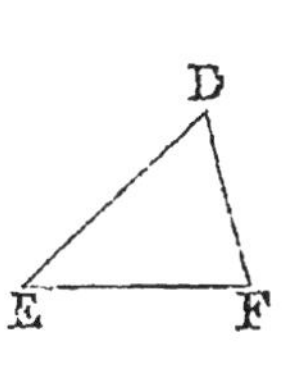

Take BG a third proportional to BC and EF, so that BC may be to EF as EF is to BG; [VI. 11.
and join AG.

Then, because AB is to BC as DE is to EF, [*Hypothesis.*
therefore, alternately, AB is to DE as BC is to EF; [V. 16.
but BC is to EF as EF is to BG; [*Construction.*
therefore AB is to DE as EF is to BG; [V. 11.
that is, the sides of the triangles ABG and DEF, about their equal angles, are reciprocally proportional;
but triangles which have their sides about two equal angles reciprocally proportional are equal to one another, [VI. 15.
therefore the triangle ABG is equal to the triangle DEF.

And, because BC is to EF as EF is to BG,
therefore BC has to BG the duplicate ratio of that which BC has to EF. [V. *Definition* 10.
But the triangle ABC is to the triangle ABG as BC is to BG; [VI. 1.
therefore the triangle ABC has to the triangle ABG the duplicate ratio of that which BC has to EF.
But the triangle ABG was shewn equal to the triangle DEF;
therefore the triangle ABC has to the triangle DEF the duplicate ratio of that which BC has to EF. [V. 7.

Wherefore, *similar triangles* &c. Q.E.D.

COROLLARY. From this it is manifest, that if three

straight lines be proportionals, as the first is to the third, so is any triangle described on the first to a similar and similarly described triangle on the second.

PROPOSITION 20. *THEOREM.*

Similar polygons may be divided into the same number of similar triangles, having the same ratio to one another that the polygons have; and the polygons are to one another in the duplicate ratio of their homologous sides.

Let *ABCDE, FGHKL* be similar polygons, and let *AB* be the side homologous to the side *FG*: the polygons *ABCDE, FGHKL* may be divided into the same number of similar triangles, of which each shall have to each the same ratio which the polygons have; and the polygon *ABCDE* shall be to the polygon *FGHKL* in the duplicate ratio of *AB* to *FG*.

Join *BE, EC, GL, LH.*

Then, because the polygon *ABCDE* is similar to the polygon *FGHKL*, [*Hypothesis.*
the angle *BAE* is equal to the angle *GFL*, and *BA* is to *AE* as *GF* is to *FL*. [VI. *Definition* 1.
And, because the triangles *ABE* and *FGL* have one angle of the one equal to one angle of the other, and the sides

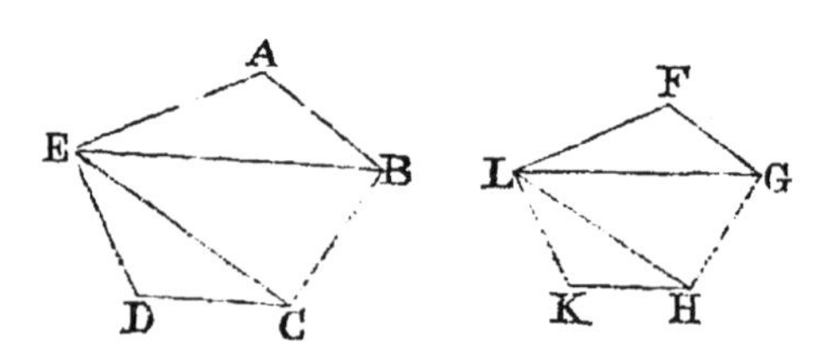

about these equal angles proportionals,
therefore the triangle *ABE* is equiangular to the triangle *FGL*, [VI. 6.
and therefore these triangles are similar; [VI. 4.
therefore the angle *ABE* is equal to the angle *FGL*.

But, because the polygons are similar, [*Hypothesis.*
therefore the whole angle ABC is equal to the whole angle FGH; [VI. *Definition* 1.
therefore the remaining angle EBC is equal to the remaining angle LGH. [*Axiom* 3.

And, because the triangles ABE and FGL are similar,
therefore EB is to BA as LG is to GF;
and also, because the polygons are similar, [*Hypothesis.*
therefore AB is to BC as FG is to GH; [VI. *Definition* 1.
therefore, ex æquali, EB is to BC as LG is to GH; [V. 22.
that is, the sides about the equal angles EBC and LGH are proportionals;
therefore the triangle EBC is equiangular to the triangle LGH; [VI. 6.
and therefore these triangles are similar. [VI. 4.

For the same reason the triangle ECD is similar to the triangle LHK.

Therefore the similar polygons $ABCDE$, $FGHKL$ may be divided into the same number of similar triangles.

Also these triangles shall have, each to each, the same ratio which the polygons have, the antecedents being ABE, EBC, ECD, and the consequents FGL, LGH, LHK; and the polygon $ABCDE$ shall be to the polygon $FGHKL$ in the duplicate ratio of AB to FG.

For, because the triangle ABE is similar to the triangle FGL,
therefore ABE is to FGL in the duplicate ratio of EB to LG. [VI. 19.

For the same reason the triangle EBC is to the triangle LGH in the duplicate ratio of EB to LG.

Therefore the triangle ABE is to the triangle FGL as the triangle EBC is to the triangle LGH. [V. 11.

Again, because the triangle EBC is similar to the triangle LGH,
therefore EBC is to LGH in the duplicate ratio of EC to LH. [VI. 19.

For the same reason the triangle *ECD* is to the triangle *LHK* in the duplicate ratio of *EC* to *LH*.

Therefore the triangle *EBC* is to the triangle *LGH* as the triangle *ECD* is to the triangle *LHK*. [V. 11.

But it has been shewn that the triangle *EBC* is to the triangle *LGH* as the triangle *ABE* is to the triangle *FGL*.

Therefore as the triangle *ABE* is to the triangle *FGL*, so is the triangle *EBC* to the triangle *LGH*, and the triangle *ECD* to the triangle *LHK*; [V. 11.

and therefore as one of the antecedents is to its consequent so are all the antecedents to all the consequents; [V. 12.

that is, as the triangle *ABE* is to the triangle *FGL* so is the polygon *ABCDE* to the polygon *FGHKL*.

But the triangle *ABE* is to the triangle *FGL* in the duplicate ratio of the side *AB* to the homologous side *FG*; [VI. 19.

therefore the polygon *ABCDE* is to the polygon *FGHKL* in the duplicate ratio of the side *AB* to the homologous side *FG*.

Wherefore, *similar polygons* &c. Q.E.D.

COROLLARY 1. In like manner it may be shewn that similar four-sided figures, or figures of any number of sides, are to one another in the duplicate ratio of their homologous sides; and it has already been shewn for triangles; therefore universally, similar rectilineal figures are to one another in the duplicate ratio of their homologous sides.

COROLLARY 2. If to *AB* and *FG*, two of the homologous sides, a third proportional *M* be taken, [VI. 11.

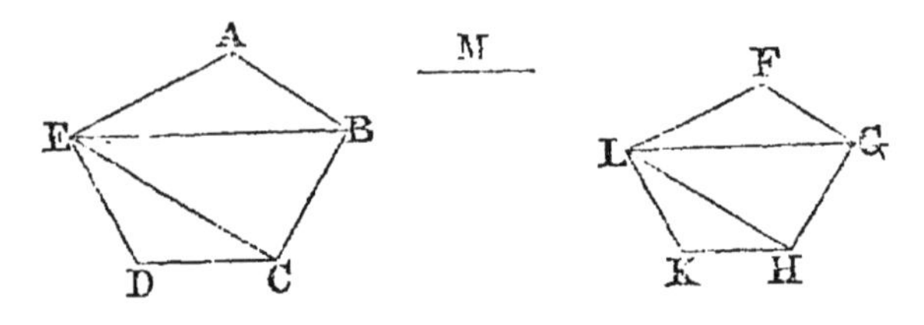

then *AB* has to *M* the duplicate ratio of that which *AB* has to *FG*. [V. *Definition* 10.

But any rectilineal figure described on AB is to the similar and similarly described rectilineal figure on FG in the duplicate ratio of AB to FG, [*Corollary* 1.

Therefore as AB is to M, so is the figure on AB to the figure on FG; [V. 11.

and this was shewn before for triangles. [VI. 19, *Corollary*.

Wherefore, universally, if three straight lines be proportionals, as the first is to the third, so is any rectilineal figure described on the first to a similar and similarly described rectilineal figure on the second.

PROPOSITION 21. *THEOREM.*

Rectilineal figures which are similar to the same rectilineal figure, are also similar to each other.

Let each of the rectilineal figures A and B be similar to the rectilineal figure C: the figure A shall be similar to the figure B.

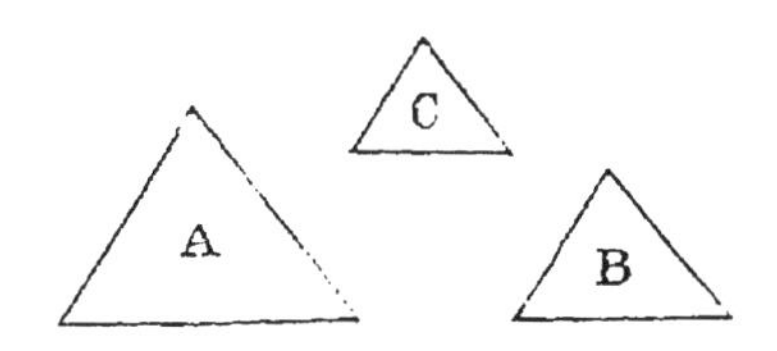

For, because A is similar to C, [*Hyp.*

A is equiangular to C, and A and C have their sides about the equal angles proportionals. [VI. *Def.* 1.

Again, because B is similar to C, [*Hyp.*

B is equiangular to C, and B and C have their sides about the equal angles proportionals. [VI. *Definition* 1.

Therefore the figures A and B are each of them equiangular to C, and have the sides about the equal angles of each of them and of C proportionals.

Therefore A is equiangular to B, [*Axiom* 1.

and A and B have their sides about the equal angles proportionals; [V. 11.

therefore the figure A is similar to the figure B. [VI. *Def.* 1.

Wherefore, *rectilineal figures* &c. Q.E.D.

PROPOSITION 22. *THEOREM.*

If four straight lines be proportionals, the similar rectilineal figures similarly described on them shall also be proportionals; and if the similar rectilineal figures similarly described on four straight lines be proportionals, those straight lines shall be proportionals.

Let the four straight lines *AB*, *CD*, *EF*, *GH* be proportionals, namely, *AB* to *CD* as *EF* is to *GH*; and on *AB*, *CD* let the similar rectilineal figures *KAB*, *LCD* be similarly described; and on *EF*, *GH* let the similar rectilineal figures *MF*, *NH* be similarly described: the figure *KAB* shall be to the figure *LCD* as the figure *MF* is to the figure *NH*.

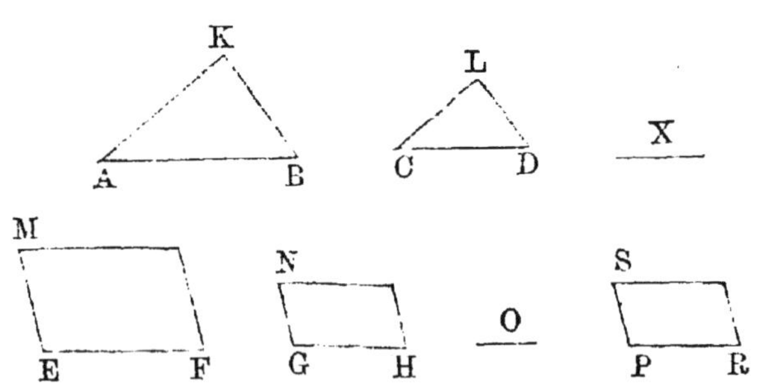

To *AB* and *CD* take a third proportional *X*, and to *EF* and *GH* a third proportional *O*. [VI. 11.

Then, because *AB* is to *CD* as *EF* is to *GH*, [*Hypothesis.*
and *AB* is to *CD* as *CD* is to *X*; [*Construction.*
and *EF* is to *GH* as *GH* is to *O*; [*Construction.*
therefore *CD* is to *X* as *GH* is to *O*. [V. 11.
And *AB* is to *CD* as *EF* is to *GH*;
therefore, ex æquali, *AB* is to *X* as *EF* is to *O*. [V. 22.
But as *AB* is to *X*, so is the rectilineal figure *KAB* to the rectilineal figure *LCD*; [VI. 20, *Corollary* 2.
and as *EF* is to *O*, so is the rectilineal figure *MF* to the rectilineal figure *NH*; [VI. 20, *Corollary* 2.

therefore the figure KAB is to the figure LCD as the figure MF is to the figure NH. [V. 11.

Next, let the figure KAB be to the similar figure LCD as the figure MF is to the similar figure NH: AB shall be to CD as EF is to GH.

Make as AB is to CD so EF to PR: [VI. 12.
and on PR describe the rectilineal figure SR, similar and similarly situated to either of the figures MF, NH. [VI. 18.
Then, because AB is to CD as EF is to PR,
and that on AB, CD are described the similar and similarly situated rectilineal figures KAB, LCD,
and on EF, PR the similar and similarly situated rectilineal figures MF, SR;
therefore, by the former part of this proposition, KAB is to LCD as MF is to SR.
But, by hypothesis, KAB is to LCD as MF is to NH;
therefore MF is to SR as MF is to NH; [V. 11.
therefore SR is equal to NH. [V. 9.
But the figures SR and NH are similar and similarly situated, [*Construction.*
therefore PR is equal to GH.
And because AB is to CD as EF is to PR,
and that PR is equal to GH;
therefore AB is to CD as EF is to GH. [V. 7.

Wherefore, *if four straight lines* &c. Q.E.D.

PROPOSITION 23. *THEOREM.*

Parallelograms which are equiangular to one another have to one another the ratio which is compounded of the ratios of their sides.

Let the parallelogram AC be equiangular to the parallelogram CF, having the angle BCD equal to the angle ECG: the parallelogram AC shall have to the parallelogram CF the ratio which is compounded of the ratios of their sides.

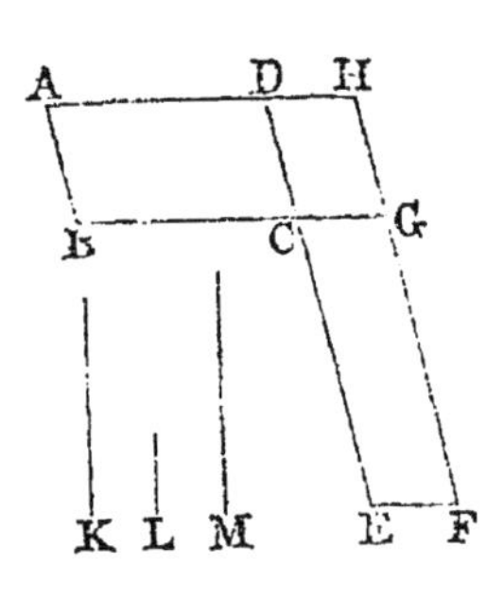

Let BC and CG be placed in a straight line;
therefore DC and CE are also in a straight line; [I. 14.
complete the parallelogram DG;
take any straight line K, and make K to L as BC is to CG, and L to M as DC is to CE; [VI. 12.
then the ratios of K to L and of L to M are the same with the ratios of the sides, namely, of BC to CG and of DC to CE.
But the ratio of K to M is that which is said to be compounded of the ratios of K to L and of L to M; [V. *Def. A.*
therefore K has to M the ratio which is compounded of the ratios of the sides.

Now the parallelogram AC is to the parallelogram CH as BC is to CG; [VI. 1.
but BC is to CG as K is to L; [*Construction.*
therefore the parallelogram AC is to the parallelogram CH as K is to L. [V. 11.
Again, the parallelogram CH is to the parallelogram CF as DC is to CE; [VI. 1.
but DC is to CE as L is to M; [*Construction.*
therefore the parallelogram CH is to the parallelogram CF as L is to M. [V. 11.
Then, since it has been shewn that the parallelogram AC is to the parallelogram CH as K is to L,
and that the parallelogram CH is to the parallelogram CF as L is to M,
therefore, ex æquali, the parallelogram AC is to the parallelogram CF as K is to M. [V. 22.
But K has to M the ratio which is compounded of the ratios of the sides;

therefore also the parallelogram AC has to the parallelogram CF the ratio which is compounded of the ratios of the sides.

Wherefore, *parallelograms* &c. Q.E.D.

PROPOSITION 24. *THEOREM.*

Parallelograms about the diameter of any parallelogram are similar to the whole parallelogram and to one another.

Let $ABCD$ be a parallelogram, of which AC is a diameter; and let EG and HK be parallelograms about the diameter: the parallelograms EG and HK shall be similar both to the whole parallelogram and to one another.

A E B G F H D K C

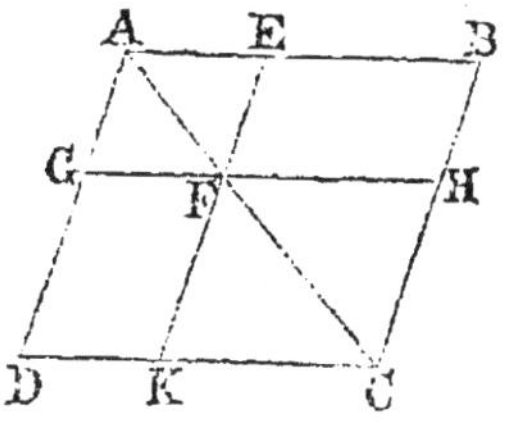

For, because DC and GF are parallels, the angle ADC is equal to the angle AGF. [I. 29.

And because BC and EF are parallels, the angle ABC is equal to the angle AEF. [I. 29.

And each of the angles BCD and EFG is equal to the opposite angle BAD, [I. 34.
and therefore they are equal to one another.

Therefore the parallelograms $ABCD$ and $AEFG$ are equiangular to one another.

And because the angle ABC is equal to the angle AEF, and the angle BAC is common to the two triangles BAC and EAF,

therefore these triangles are equiangular to one another;
and therefore AB is to BC as AE is to EF. [VI. 4.

And the opposite sides of parallelograms are equal to one another; [I. 34.

therefore AB is to AD as AE is to AG,
and DC is to CB as GF is to FE,
and CD is to DA as FG is to GA. [V. 7.

Therefore the sides of the parallelograms *ABCD* and *AEFG* about their equal angles are proportional,

and the parallelograms are therefore similar to one another. [VI. *Definition* 1.

For the same reason the parallelogram *ABCD* is similar to the parallelogram *FHCK*.

Therefore each of the parallelograms *EG* and *HK* is similar to *BD*;

therefore the parallelogram *EG* is similar to the parallelogram *HK*. [VI. 21.

Wherefore, *parallelograms* &c. Q.E.D.

PROPOSITION 25. *PROBLEM.*

To describe a rectilineal figure which shall be similar to one given rectilineal figure and equal to another given rectilineal figure.

Let *ABC* be the given rectilineal figure to which the figure to be described is to be similar, and *D* that to which it is to be equal: it is required to describe a rectilineal figure similar to *ABC* and equal to *D*.

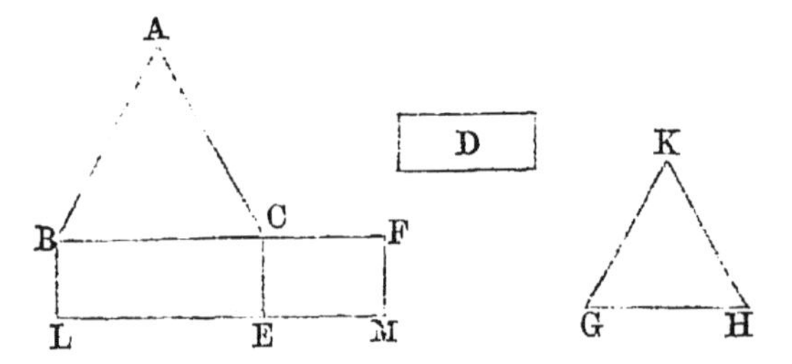

On the straight line *BC* describe the parallelogram *BE* equal to the figure *ABC*.

On the straight line *CE* describe the parallelogram *CM* equal to *D*, and having the angle *FCE* equal to the angle *CBL*; [I. 45, *Corollary*.

therefore BC and CF will be in one straight line, and LE and EM will be in one straight line.

Between BC and CF find a mean proportional GH, [VI. 13.
and on GH describe the rectilineal figure KGH, similar and similarly situated to the rectilineal figure ABC. [VI. 18.
KGH shall be the rectilineal figure required.

For, because BC is to GH as GH is to CF, [*Construction.*
and that if three straight lines be proportionals, as the first is to the third so is any figure on the first to a similar and similarly described figure on the second, [VI. 20, *Cor.* 2.
therefore as BC is to CF so is the figure ABC to the figure KGH.

But as BC is to CF so is the parallelogram BE to the parallelogram CM; [VI. 1.
therefore the figure ABC is to the figure KGH as the parallelogram BE is to the parallelogram CM. [V. 11.
And the figure ABC is equal to the parallelogram BE;
therefore the rectilineal figure KGH is equal to the parallelogram CM. [V. 14.
But the parallelogram CM is equal to the figure D; [*Constr.*
therefore the figure KGH is equal to the figure D, [*Axiom* 1.
and it is similar to the figure ABC. [*Construction.*

Wherefore *the rectilineal figure KGH has been described similar to the figure ABC, and equal to D.* Q.E.F.

PROPOSITION 26. *THEOREM.*

If two similar parallelograms have a common angle, and be similarly situated, they are about the same diameter.

Let the parallelograms $ABCD$, $AEFG$ be similar and similarly situated, and have the common angle BAD: $ABCD$ and $AEFG$ shall be about the same diameter.

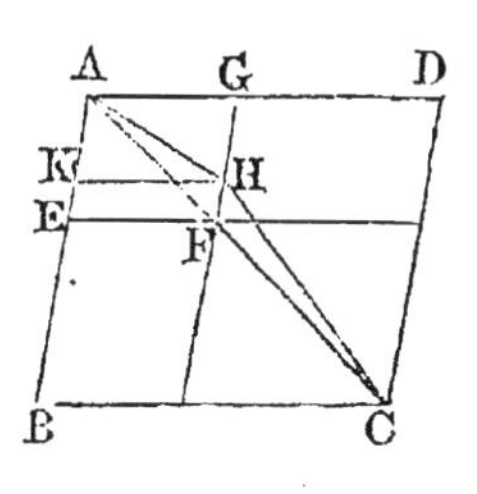

For, if not, let, if possible, the parallelogram BD have its diameter AHC in a different straight line from AF, the diameter of the parallelogram EG; let GF meet AHC at H, and through H draw HK parallel to AD or BC. [I. 31.

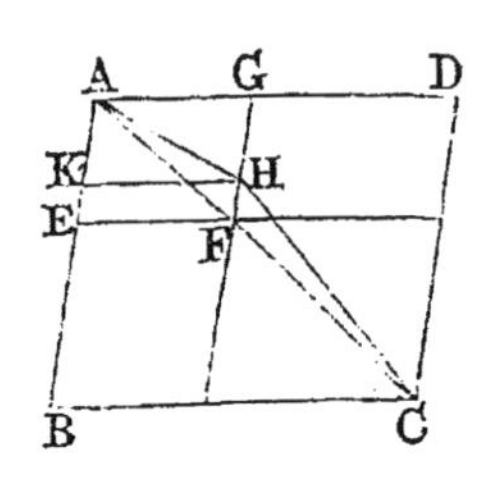

Then the parallelograms $ABCD$ and $AKHG$ are about the same diameter, and are therefore similar to one another; [VI. 24.

therefore DA is to AB as GA is to AK.

But because $ABCD$ and $AEFG$ are similar parallelograms, [*Hypothesis.*

therefore DA is to AB as GA is to AE. [VI. *Definition* 1.

Therefore GA is to AK as GA is to AE, [V. 11.

that is, GA has the same ratio to each of the straight lines AK and AE,

and therefore AK is equal to AE, [V. 9.

the less to the greater; which is impossible.

Therefore the parallelograms $ABCD$ and $AEFG$ must have their diameters in the same straight line, that is, they are about the same diameter.

Wherefore, *if two similar parallelograms* &c. Q.E.D.

PROPOSITION 30. *PROBLEM.*

To cut a given straight line in extreme and mean ratio.

Let AB be the given straight line: it is required to cut it in extreme and mean ratio.

Divide AB at the point C, so that the rectangle contained by AB, BC may be equal to the square on AC. [II. 11.

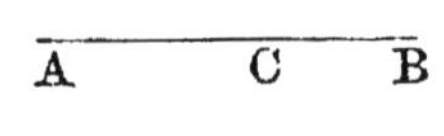

Then, because the rectangle AB, BC is equal to the square on AC, [*Construction.*

therefore AB is to AC as AC is to CB. [VI. 17.

Therefore AB is cut in extreme and mean ratio at the point C. Q.E.F. [VI. *Definition* 3.

PROPOSITION 31. *THEOREM.*

In any right-angled triangle, any rectilineal figure described on the side subtending the right angle is equal to the similar and similarly described figures on the sides containing the right angle.

Let ABC be a right-angled triangle, having the right angle BAC: the rectilineal figure described on BC shall be equal to the similar and similarly described figures on BA and CA.

Draw the perpendicular AD. [I. 12.

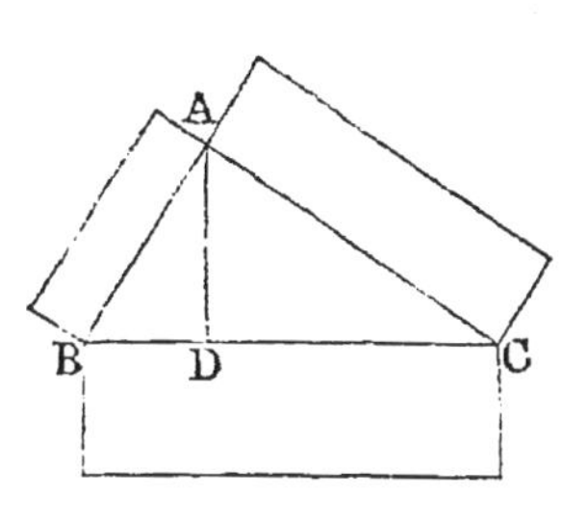

Then, because in the right-angled triangle ABC, AD is drawn from the right angle at A, perpendicular to the base BC, the triangles ABD, CAD are similar to the whole triangle CBA, and to one another. [VI. 8.

And because the triangle CBA is similar to the triangle ABD,

therefore CB is to BA as BA is to BD. [VI. *Def.* 1.

And when three straight lines are proportionals, as the first is to the third so is the figure described on the first to the similar and similarly described figure on the second; [VI. 20, *Corollary* 2.

therefore as CB is to BD so is the figure described on CB to the similar and similarly described figure on BA;

and inversely, as BD is to BC so is the figure described on BA to that described on CB. [V. *B*.

In the same manner, as CD is to CB so is the figure described on CA to the similar figure described on CB.

Therefore as BD and CD together are to CB so are the figures described on BA and CA together to the figure described on CB. [V. 24.

But BD and CD together are equal to CB;

therefore the figure described on BC is equal to the similar and similarly described figures on BA and CA. [V. *A*.

Wherefore, *in any right-angled triangle* &c. Q.E.D.

PROPOSITION 32. *THEOREM.*

If two triangles, which have two sides of the one proportional to two sides of the other, be joined at one angle so as to have their homologous sides parallel to one another, the remaining sides shall be in a straight line.

Let *ABC* and *DCE* be two triangles, which have the two sides *BA*, *AC* proportional to the two sides *CD*, *DE*, namely, *BA* to *AC* as *CD* is to *DE*; and let *AB* be parallel to *DC* and *AC* parallel to *DE*: *BC* and *CE* shall be in one straight line.

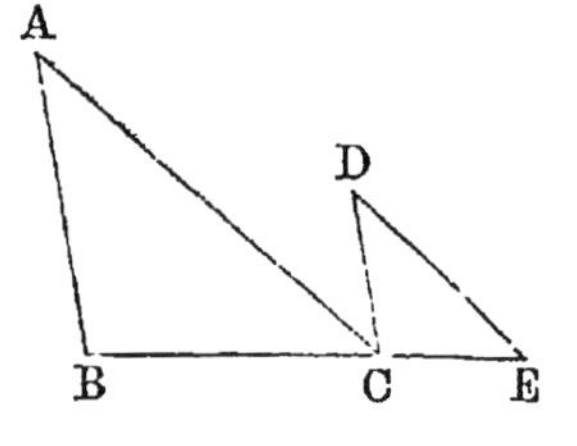

For, because *AB* is parallel to *DC*, [*Hypothesis.*
and *AC* meets them,
the alternate angles *BAC*, *ACD* are equal; [I. 29.
for the same reason the angles *ACD*, *CDE* are equal;
therefore the angle *BAC* is equal to the angle *CDE*. [*Ax.* 1.
And because the triangles *ABC*, *DCE* have the angle at *A* equal to the angle at *D*, and the sides about these angles proportionals, namely, *BA* to *AC* as *CD* is to *DE*, [*Hyp.*
therefore the triangle *ABC* is equiangular to the triangle *DCE*; [VI. 6.
therefore the angle *ABC* is equal to the angle *DCE*.
And the angle *BAC* was shewn equal to the angle *ACD*;
therefore the whole angle *ACE* is equal to the two angles *ABC* and *BAC*. [*Axiom* 2.
Add the angle *ACB* to each of these equals;
then the angles *ACE* and *ACB* are together equal to the angles *ABC*, *BAC*, *ACB*.
But the angles *ABC*, *BAC*, *ACB* are together equal to two right angles; [I. 32.
therefore the angles *ACE* and *ACB* are together equal to two right angles.

And since at the point *C*, in the straight line *AC*, the two straight lines *BC*, *CE* which are on the opposite sides of it, make the adjacent angles *ACE*, *ACB* together equal to two right angles,

therefore *BC* and *CE* are in one straight line. [I. 14.

Wherefore, *if two triangles* &c. Q.E.D.

PROPOSITION 33. *THEOREM.*

In equal circles, angles, whether at the centres or at the circumferences, have the same ratio which the arcs on which they stand have to one another; so also have the sectors.

Let *ABC* and *DEF* be equal circles, and let *BGC* and *EHF* be angles at their centres, and *BAC* and *EDF* angles at their circumferences: as the arc *BC* is to the arc *EF* so shall the angle *BGC* be to the angle *EHF*, and the angle *BAC* to the angle *EDF*; and so also shall the sector *BGC* be to the sector *EHF*.

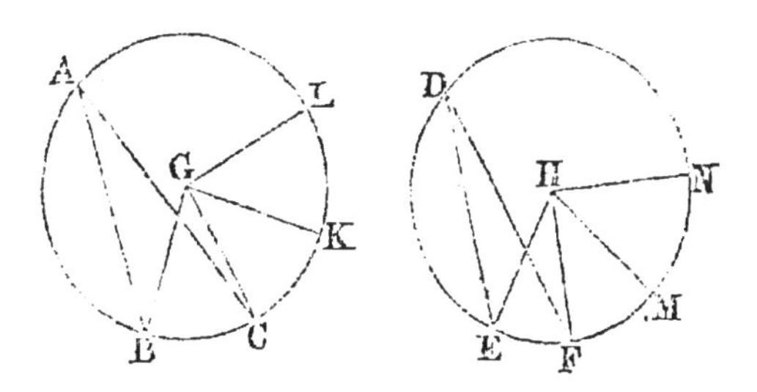

Take any number of arcs *CK*, *KL*, each equal to *BC*, and also any number of arcs *FM*, *MN* each equal to *EF*; and join *GK*, *GL*, *HM*, *HN*.

Then, because the arcs *BC*, *CK*, *KL*, are all equal, [*Constr.*
the angles *BGC*, *CGK*, *KGL* are also all equal; [III. 27.
and therefore whatever multiple the arc *BL* is of the arc *BC*, the same multiple is the angle *BGL* of the angle *BGC*.

For the same reason, whatever multiple the arc *EN* is of the arc *EF*, the same multiple is the angle *EHN* of the angle *EHF*.

And if the arc BL be equal to the arc EN, the angle BGL is equal to the angle EHN; [III. 27.

and if the arc BL be greater than the arc EN, the angle BGL is greater than the angle EHN; and if less, less.

Therefore since there are four magnitudes, the two arcs BC, EF, and the two angles BGC, EHF;

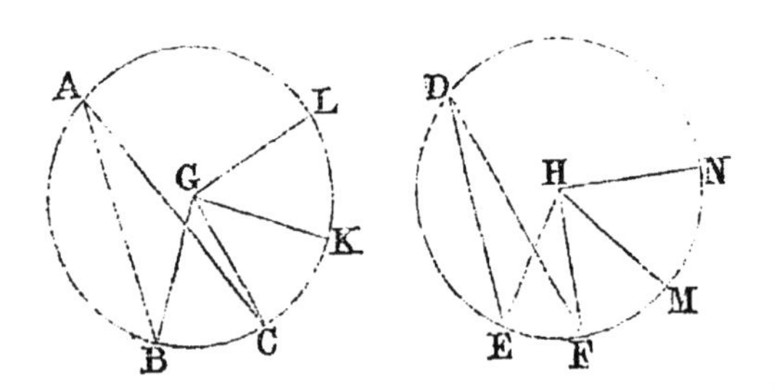

and that of the arc BC and of the angle BGC have been taken any equimultiples whatever, namely, the arc BL and the angle BGL;

and of the arc EF and of the angle EHF have been taken any equimultiples whatever, namely, the arc EN and the angle EHN;

and since it has been shewn that if the arc BL be greater than the arc EN, the angle BGL is greater than the angle EHN; and if equal, equal; and if less, less;

therefore as the arc BC is to the arc EF, so is the angle BGC to the angle EHF. [V. *Definition* 5.

But as the angle BGC is to the angle EHF, so is the angle BAC to the angle EDF, [V. 15.

for each is double of each; [III. 20.

therefore, as the arc BC is to the arc EF so is the angle BGC to the angle EHF, and the angle BAC to the angle EDF.

Also as the arc BC is to the arc EF, so shall the sector BGC be to the sector EHF.

Join BC, CK, and in the arcs BC, CK take any points X, O, and join BX, XC, CO, OK.

Then, because in the triangles BGC, CGK, the two sides BG, GC are equal to the two sides CG, GK, each to each;
and that they contain equal angles; [III. 27.
therefore the base BC is equal to the base CK, and the triangle BGC is equal to the triangle CGK. [I. 4.

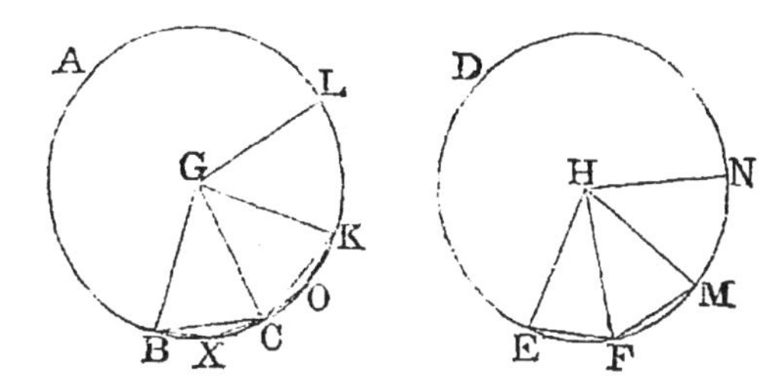

And because the arc BC is equal to the arc CK, [*Constr.*
the remaining part when BC is taken from the circumference is equal to the remaining part when CK is taken from the circumference;
therefore the angle BXC is equal to the angle COK. [III. 27.
Therefore the segment BXC is similar to the segment COK; [III. *Definition* 11.
and they are on equal straight lines BC, CK.
But similar segments of circles on equal straight lines are equal to one another; [III. 24.
therefore the segment BXC is equal to the segment COK.

And the triangle BGC was shewn to be equal to the triangle CGK;
therefore the whole, the sector BGC, is equal to the whole, the sector CGK. [*Axiom* 2.
For the same reason the sector KGL is equal to each of the sectors BGC, CGK.
In the same manner the sectors EHF, FHM, MHN may be shewn to be equal to one another.

Therefore whatever multiple the arc BL is of the arc BC, the same multiple is the sector BGL of the sector BGC;

and for the same reason whatever multiple the arc *EN* is of the arc *EF*, the same multiple is the sector *EHN* of the sector *EHF*.

And if the arc *BL* be equal to the arc *EN*, the sector *BGL* is equal to the sector *EHN*;

and if the arc *BL* be greater than the arc *EN*, the sector *BGL* is greater than the sector *EHN*; and if less, less.

Therefore, since there are four magnitudes, the two arcs *BC*, *EF*, and the two sectors *BGC*, *EHF*;

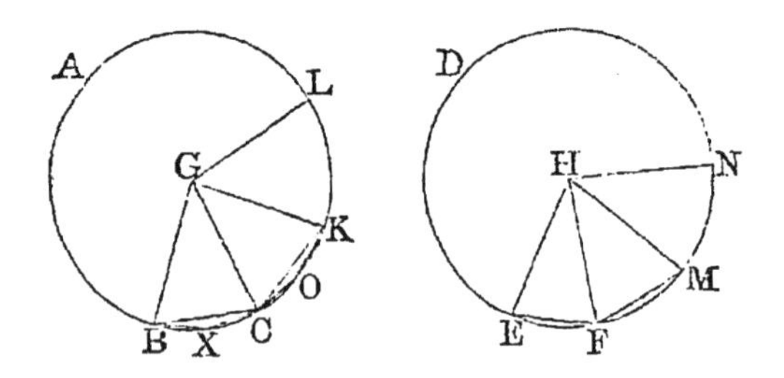

and that of the arc *BC* and of the sector *BGC* have been taken any equimultiples whatever, namely, the arc *BL* and the sector *BGL*;

and of the arc *EF* and of the sector *EHF* have been taken any equimultiples whatever, namely, the arc *EN* and the sector *EHN*;

and since it has been shewn that if the arc *BL* be greater than the arc *EN*, the sector *BGL* is greater than the sector *EHN*; and if equal, equal; and if less, less;

therefore as the arc *BC* is to the arc *EF*, so is the sector *BGC* to the sector *EHF*. [V. *Definition* 5.

Wherefore, *in equal circles* &c. Q.E.D.

PROPOSITION *B.* *THEOREM.*

If the vertical angle of a triangle be bisected by a straight line which likewise cuts the base, the rectangle contained by the sides of the triangle is equal to the rectangle contained by the segments of the base, together with the square on the straight line which bisects the angle.

Let ABC be a triangle, and let the angle BAC be bisected by the straight line AD: the rectangle BA, AC shall be equal to the rectangle BD, DC together with the square on AD.

Describe the circle ACB about the triangle, [IV. 5.
and produce AD to meet the circumference at E,
and join EC.

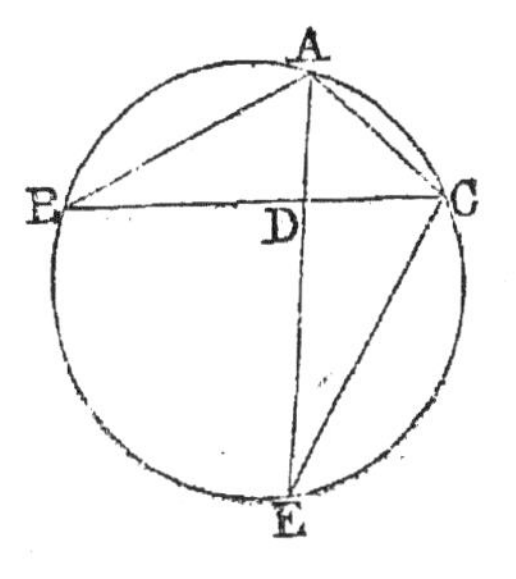

Then, because the angle BAD is equal to the angle EAC, [*Hypothesis.*
and the angle ABD is equal to the angle AEC, for they are in the same segment of the circle, [III. 21.
therefore the triangle BAD is equiangular to the triangle EAC.

Therefore BA is to AD as EA is to AC; [VI. 4.
therefore the rectangle BA, AC is equal to the rectangle EA, AD, [VI. 16.
that is, to the rectangle ED, DA, together with the square on AD. [II. 3.

But the rectangle ED, DA is equal to the rectangle BD, DC; [III. 35.
therefore the rectangle BA, AC is equal to the rectangle BD, DC, together with the square on AD.

Wherefore, *if the vertical angle* &c. Q.E.D.

PROPOSITION *C.* *THEOREM.*

If from the vertical angle of a triangle a straight line be drawn perpendicular to the base, the rectangle contained by the sides of the triangle is equal to the rectangle contained by the perpendicular and the diameter of the circle described about the triangle.

Let ABC be a triangle, and let AD be the perpendicular from the angle A to the base BC: the rectangle BA, AC shall be equal to the rectangle contained by AD and the diameter of the circle described about the triangle.

Describe the circle ACB about the triangle; [IV. 5.
draw the diameter AE, and join EC.

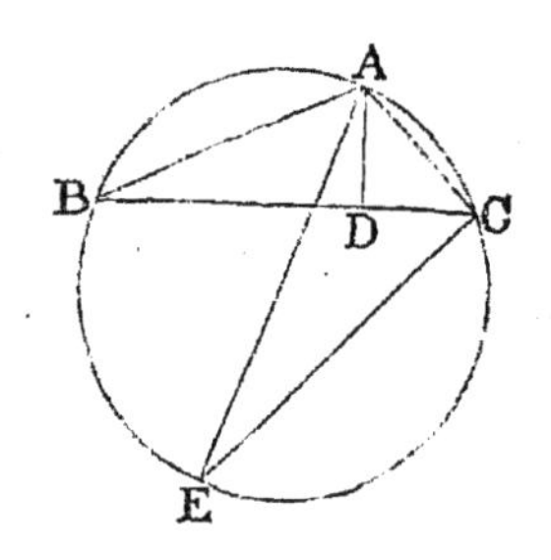

Then, because the right angle BDA is equal to the angle ECA in a semicircle; [III. 31.
and the angle ABD is equal to the angle AEC, for they are in the same segment of the circle; [III. 21.
therefore the triangle ABD is equiangular to the triangle AEC.
Therefore BA is to AD as EA is to AC; [VI. 4.
therefore the rectangle BA, AC is equal to the rectangle EA, AD. [VI. 16.

Wherefore, *if from the vertical angle* &c. Q.E.D.

PROPOSITION *D.* *THEOREM.*

The rectangle contained by the diagonals of a quadrilateral figure inscribed in a circle is equal to both the rectangles contained by its opposite sides.

Let $ABCD$ be any quadrilateral figure inscribed in a circle, and join AC, BD: the rectangle contained by AC, BD shall be equal to the two rectangles contained by AB, CD and by AD, BC.

Make the angle ABE equal to the angle DBC; [I. 23.

add to each of these equals the angle EBD,

then the angle ABD is equal to the angle EBC. [*Axiom* 2.

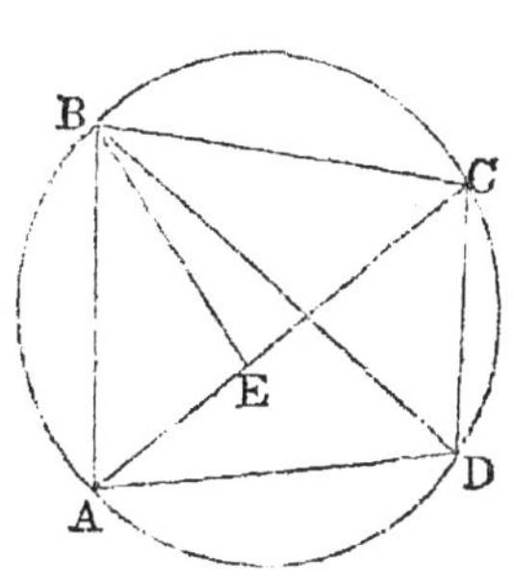

And the angle BDA is equal to the angle BCE, for they are in the same segment of the circle; [III. 21.

therefore the triangle ABD is equiangular to the triangle EBC.

Therefore AD is to DB as EC is to CB; [VI. 4.

therefore the rectangle AD, CB is equal to the rectangle DB, EC. [VI. 16.

Again, because the angle ABE is equal to the angle DBC, [*Construction.*

and the angle BAE is equal to the angle BDC, for they are in the same segment of the circle; [III. 21.

therefore the triangle ABE is equiangular to the triangle DBC.

Therefore BA is to AE as BD is to DC; [VI. 4.

therefore the rectangle BA, DC is equal to the rectangle AE, BD. [VI. 16.

But the rectangle AD, CB has been shewn equal to the rectangle DB, EC;

therefore the rectangles AD, CB and BA, DC are together equal to the rectangles BD, EC and BD, AE;

that is, to the rectangle BD, AC. [II. 1.

Wherefore, *the rectangle contained* &c. Q.E.D.

BOOK XI.

DEFINITIONS.

1. A SOLID is that which has length, breadth, and thickness.

2. That which bounds a solid is a superficies.

3. A straight line is perpendicular, or at right angles, to a plane, when it makes right angles with every straight line meeting it in that plane.

4. A plane is perpendicular to a plane, when the straight lines drawn in one of the planes perpendicular to the common section of the two planes, are perpendicular to the other plane.

5. The inclination of a straight line to a plane is the acute angle contained by that straight line, and another drawn from the point at which the first line meets the plane to the point at which a perpendicular to the plane drawn from any point of the first line above the plane, meets the same plane.

6. The inclination of a plane to a plane is the acute angle contained by two straight lines drawn from any the same point of their common section at right angles to it, one in one plane, and the other in the other plane.

7. Two planes are said to have the same or a like inclination to one another, which two other planes have, when the said angles of inclination are equal to one another.

8. Parallel planes are such as do not meet one another though produced.

9. A solid angle is that which is made by more than two plane angles, which are not in the same plane, meeting at one point.

10. Equal and similar solid figures are such as are contained by similar planes equal in number and magnitude. [*See the Notes.*]

11. Similar solid figures are such as have all their solid angles equal, each to each, and are contained by the same number of similar planes.

12. A pyramid is a solid figure contained by planes which are constructed between one plane and one point above it at which they meet.

13. A prism is a solid figure contained by plane figures, of which two that are opposite are equal, similar, and parallel to one another; and the others are parallelograms.

14. A sphere is a solid figure described by the revolution of a semicircle about its diameter, which remains fixed.

15. The axis of a sphere is the fixed straight line about which the semicircle revolves.

16. The centre of a sphere is the same with that of the semicircle.

17. The diameter of a sphere is any straight line which passes through the centre, and is terminated both ways by the superficies of the sphere.

18. A cone is a solid figure described by the revolution of a right-angled triangle about one of the sides containing the right angle, which side remains fixed.

If the fixed side be equal to the other side containing the right angle, the cone is called a right-angled cone; if it be less than the other side, an obtuse-angled cone; and if greater, an acute-angled cone.

19. The axis of a cone is the fixed straight line about which the triangle revolves.

20. The base of a cone is the circle described by that side containing the right angle which revolves.

21. A cylinder is a solid figure described by the revolution of a right-angled parallelogram about one of its sides which remains fixed.

22. The axis of a cylinder is the fixed straight line about which the parallelogram revolves.

23. The bases of a cylinder are the circles described by the two revolving opposite sides of the parallelogram.

24. Similar cones and cylinders are those which have their axes and the diameters of their bases proportionals.

25. A cube is a solid figure contained by six equal squares.

26. A tetrahedron is a solid figure contained by four equal and equilateral triangles.

27. An octahedron is a solid figure contained by eight equal and equilateral triangles.

28. A dodecahedron is a solid figure contained by twelve equal pentagons which are equilateral and equiangular.

29. An icosahedron is a solid figure contained by twenty equal and equilateral triangles.

A. A parallelepiped is a solid figure contained by six quadrilateral figures, of which every opposite two are parallel.

PROPOSITION 1. *THEOREM.*

One part of a straight line cannot be in a plane, and another part without it.

If it be possible, let AB, part of the straight line ABC, be in a plane, and the part BC without it.

Then since the straight line AB is in the plane, it can be produced in that plane; let it be produced to D; and let any plane pass through the straight line AD, and be turned about until it pass through the point C.

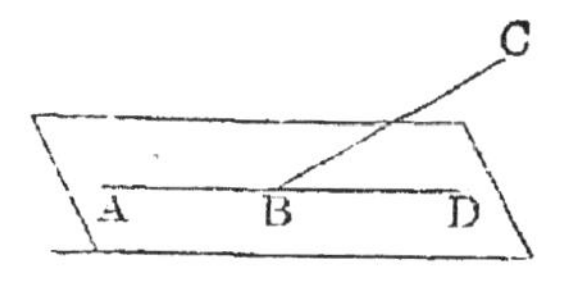

Then, because the points B and C are in this plane, the straight line BC is in it. [I. *Definition* 7.

Therefore there are two straight lines ABC, ABD in the same plane, that have a common segment AB;

but this is impossible. [I. 11, *Corollary*.

Wherefore, *one part of a straight line* &c. Q.E.D.

PROPOSITION 2. *THEOREM.*

Two straight lines which cut one another are in one plane; and three straight lines which meet one another are in one plane.

Let the two straight lines AB, CD cut one another at E: AB and CD shall be in one plane; and the three straight lines EC, CB, BE which meet one another, shall be in one plane.

Let any plane pass through the straight line EB, and let the plane be turned about EB, produced if necessary, until it pass through the point C.

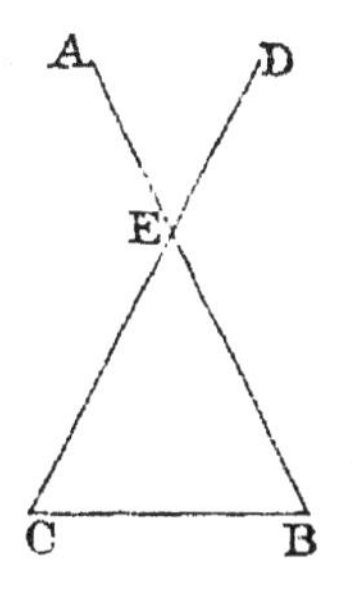

Then, because the points E and C are in this plane, the straight line EC is in it; [I. *Definition* 7.

for the same reason, the straight line BC is in the same plane;

and, by hypothesis, EB is in it.

Therefore the three straight lines EC, CB, BE are in one plane.

But AB and CD are in the plane in which EB and EC are; [XI. 1.

therefore AB and CD are in one plane.

Wherefore, *two straight lines* &c. Q.E.D.

PROPOSITION 3. *THEOREM.*

If two planes cut one another their common section is a straight line.

Let two planes *AB*, *BC* cut one another, and let *BD* be their common section: *BD* shall be a straight line.

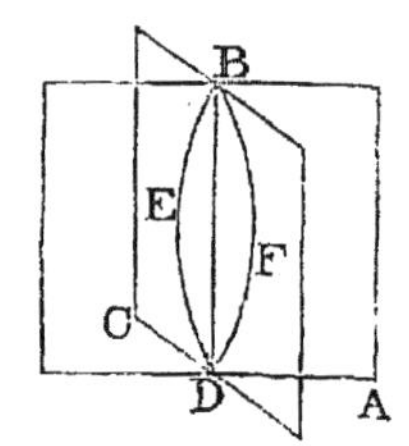

If it be not, from *B* to *D*, draw in the plane *AB* the straight line *BED*, and in the plane *BC* the straight line *BFD*. [*Postulate* 1.

Then the two straight lines *BED*, *BFD* have the same extremities, and therefore include a space between them;

but this is impossible. [*Axiom* 10.

Therefore *BD*, the common section of the planes *AB* and *BC* cannot but be a straight line.

Wherefore, *if two planes* &c. Q.E.D.

PROPOSITION 4. *THEOREM.*

If a straight line stand at right angles to each of two straight lines at the point of their intersection, it shall also be at right angles to the plane which passes through them, that is, to the plane in which they are.

Let the straight line *EF* stand at right angles to each of the straight lines *AB*, *CD*, at *E*, the point of their intersection: *EF* shall also be at right angles to the plane passing through *AB*, *CD*.

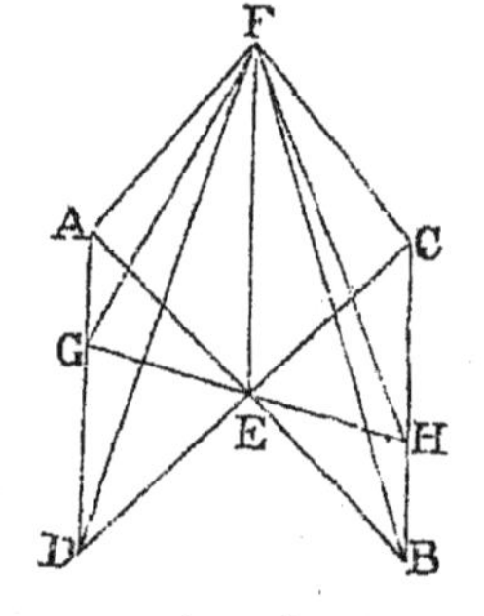

Take the straight lines *AE*, *EB*, *CE*, *ED*, all equal to one another; join *AD*, *CB*; through *E* draw in the plane in which are *AB*, *CD*, any straight line cutting *AD* at *G*, and *CB* at *H*; and from any point *F* in *EF* draw *FA*, *FG*, *FD*, *FC*, *FH*, *FB*.

Then, because the two sides AE, ED are equal to the two sides BE, EC, each to each, [*Construction.*
and that they contain equal angles AED, BEC; [I. 15.
therefore the base AD is equal to the base BC, and the angle DAE is equal to the angle EBC. [I. 4.
And the angle AEG is equal to the angle BEH; [I. 15.
therefore the triangles AEG, BEH have two angles of the one equal to two angles of the other, each to each;
and the sides EA, EB adjacent to the equal angles are equal to one another; [*Construction.*
therefore EG is equal to EH, and AG is equal to BH. [I. 26.

And because EA is equal to EB, [*Construction.*
and EF is common and at right angles to them, [*Hypothesis.*
therefore the base AF is equal to the base BF. [I. 4.
For the same reason CF is equal to DF.

And since it has been shewn that the two sides DA, AF are equal to the two sides CB, BF, each to each,
and that the base DF is equal to the base CF;
therefore the angle DAF is equal to the angle CBF. [I. 8.

Again, since it has been shewn that the two sides FA, AG are equal to the two sides FB, BH, each to each,
and that the angle FAG is equal to the angle FBH;
therefore the base FG is equal to the base FH. [I. 4.

Lastly, since it has been shewn that GE is equal to HE, and EF is common to the two triangles FEG, FEH;
and the base FG has been shewn equal to the base FH;
therefore the angle FEG is equal to the angle FEH. [I. 8.
Therefore each of these angles is a right angle. [I. *Defin.* 10.

In like manner it may be shewn that EF makes right angles with every straight line which meets it in the plane passing through AB, CD.
Therefore EF is at right angles to the plane in which are AB, CD. [XI. *Definition* 3.

Wherefore, *if a straight line* &c. Q.E.D.

PROPOSITION 5. *THEOREM.*

If three straight lines meet all at one point, and a straight line stand at right angles to each of them at that point, the three straight lines shall be in one and the same plane.

Let the straight line AB stand at right angles to each of the straight lines BC, BD, BE, at B the point where they meet: BC, BD, BE shall be in one and the same plane.

For, if not, let, if possible, BD and BE be in one plane, and BC without it; let a plane pass through AB and BC; the common section of this plane with the plane in which are BD and BE is a straight line; [XI. 3.

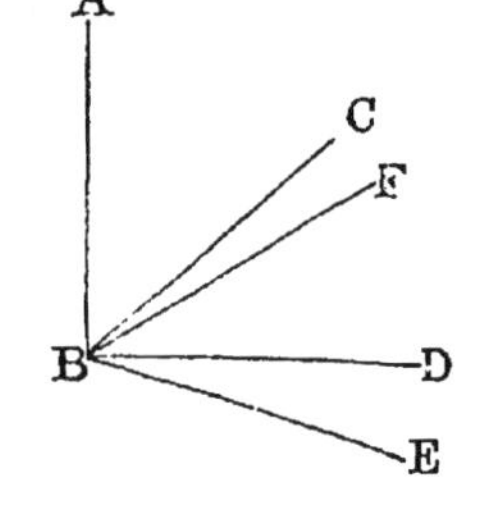

let this straight line be BF.

Then the three straight lines AB, BC, BF are all in one plane, namely, the plane which passes through AB and BC.

And because AB stands at right angles to each of the straight lines BD, BE, [*Hypothesis.*

therefore it is at right angles to the plane passing through them; [XI. 4.

therefore it makes right angles with every straight line meeting it in that plane. [XI. *Definition* 3.

But BF meets it, and is in that plane;

therefore the angle ABF is a right angle.

But the angle ABC is, by hypothesis, a right angle;

therefore the angle ABC is equal to the angle ABF; [*Ax.* 11.

and they are in one plane; which is impossible. [*Axiom* 9.

Therefore the straight line BC is not without the plane in which are BD and BE,

therefore the three straight lines BC, BD, BE are in one and the same plane.

Wherefore, *if three straight lines* &c. Q.E.D.

PROPOSITION 6. *THEOREM.*

If two straight lines be at right angles to the same plane, they shall be parallel to one another.

Let the straight lines AB, CD be at right angles to the same plane: AB shall be parallel to CD.

Let them meet the plane at the points B, D; join BD; and in the plane draw DE at right angles to BD; [I. 11.
make DE equal to AB; [I. 3.
and join BE, AE, AD.

Then, because AB is perpendicular to the plane, [*Hypothesis.*
it makes right angles with every straight line meeting it in that plane. [XI. *Def.* 3.

But BD and BE meet AB, and are in that plane,

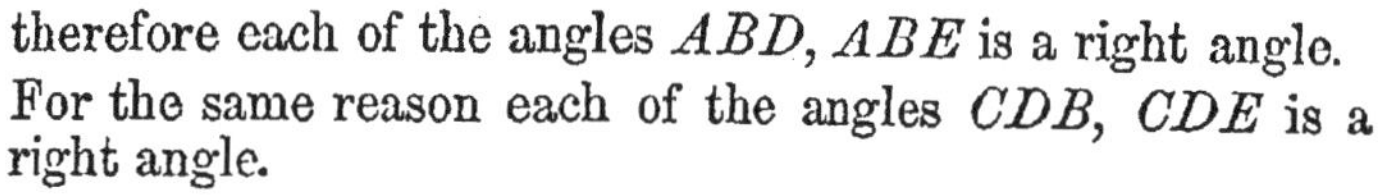

therefore each of the angles ABD, ABE is a right angle.
For the same reason each of the angles CDB, CDE is a right angle.

And because AB is equal to ED, [*Construction.*
and BD is common to the two triangles ABD, EDB,
the two sides AB, BD are equal to the two sides ED, DB, each to each;
and the angle ABD is equal to the angle EDB, each of them being a right angle; [*Axiom* 11.
therefore the base AD is equal to the base EB. [I. 4.

Again, because AB is equal to ED, [*Construction.*
and it has been shewn that BE is equal to DA;
therefore the two sides AB, BE are equal to the two sides ED, DA, each to each;
and the base AE is common to the two triangles ABE, EDA;
therefore the angle ABE is equal to the angle EDA. [I. 8.
But the angle ABE is a right angle,
therefore the angle EDA is a right angle,
that is, ED is at right angles to AD.

But ED is also at right angles to each of the two BD, CD;
therefore ED is at right angles to each of the three straight lines BD, AD, CD, at the point at which they meet;
therefore these three straight lines are all in the same plane. [XI. 5.

But AB is in the plane in which are BD, DA; [XI. 2.
therefore AB, BD, CD are in one plane.
And each of the angles ABD, CDB is a right angle;
therefore AB is parallel to CD. [I. 28.

Wherefore, *if two straight lines* &c. Q.E.D.

PROPOSITION 7. *THEOREM.*

If two straight lines be parallel, the straight line drawn from any point in one to any point in the other, is in the same plane with the parallels.

Let AB, CD be parallel straight lines, and take any point E in one and any point F in the other: the straight line which joins E and F shall be in the same plane with the parallels.

For, if not, let it be, if possible, without the plane, as EGF; and in the plane $ABCD$, in which the parallels are, draw the straight line EHF from E to F.

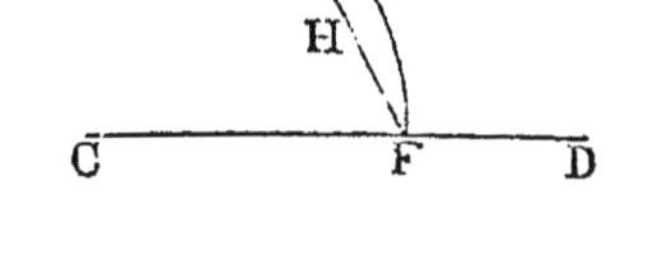

Then, since EGF is also a straight line, [*Hypothesis.*
the two straight lines EGF, EHF include a space between them; which is impossible. [*Axiom* 10.
Therefore the straight line joining the points E and F is not without the plane in which the parallels AB, CD are;
therefore it is in that plane.

Wherefore, *if two straight lines* &c. Q.E.D.

PROPOSITION 8. *THEOREM.*

If two straight lines be parallel, and one of them be at right angles to a plane, the other also shall be at right angles to the same plane.

Let AB, CD be two parallel straight lines; and let one of them AB be at right angles to a plane: the other CD shall be at right angles to the same plane.

Let AB, CD meet the plane at the points B, D; join BD; therefore AB, CD, BD are in one plane. [XI. 7.

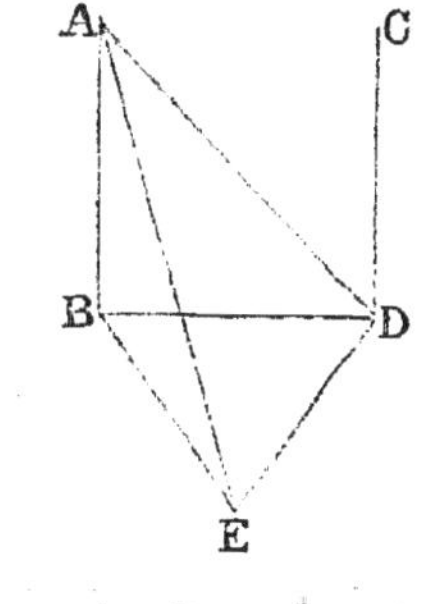

In the plane to which AB is at right angles, draw DE at right angles to BD; [I. 11.

make DE equal to AB; [I. 3.

and join BE, AE, AD.

Then, because AB is at right angles to the plane, [*Hypothesis.*

it makes right angles with every straight line meeting it in that plane; [XI. *Definition* 3.

therefore each of the angles ABD, ABE is a right angle.

And because the straight line BD meets the parallel straight lines AB, CD,

the angles ABD, CDB are together equal to two right angles. [I. 29.

But the angle ABD is a right angle, [*Hypothesis.*

therefore the angle CDB is a right angle;

that is, CD is at right angles to BD.

And because AB is equal to ED, [*Construction.*

and BD is common to the two triangles ABD, EDB;

the two sides AB, BD are equal to the two sides ED, DB, each to each;

and the angle ABD is equal to the angle EDB, each of them being a right angle; [*Axiom* 11.

therefore the base AD is equal to the base EB. [I. 4.

Again, because AB is equal to ED, [*Construction.*

and BE has been shewn equal to DA,

the two sides AB, BE are equal to the two sides ED, DA, each to each;

and the base AE is common to the two triangles ABE, EDA;

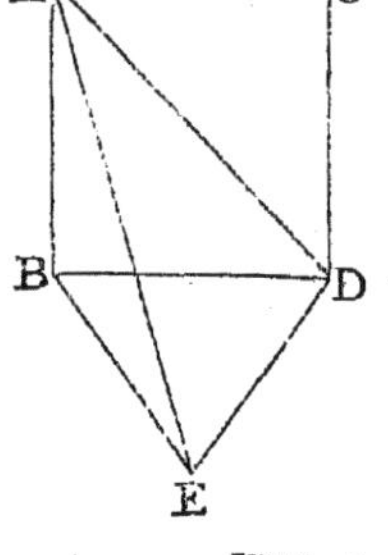

therefore the angle ABE is equal to the angle ADE. [I. 8.

But the angle ABE is a right angle;

therefore the angle ADE is a right angle;

that is, ED is at right angles to AD.

But ED is at right angles to BD, [*Const.*

therefore ED is at right angles to the plane which passes through BD, DA, [XI. 4.

and therefore makes right angles with every straight line meeting it in that plane. [XI. *Definition* 3.

But CD is in the plane passing through BD, DA, because all three are in the plane in which are the parallels AB, CD;

therefore ED is at right angles to CD,

and therefore CD is at right angles to ED.

But CD was shewn to be at right angles to BD;

therefore CD is at right angles to the two straight lines BD, ED, at the point of their intersection D,

and is therefore at right angles to the plane passing through BD, ED, [XI. 4.

that is, to the plane to which AB is at right angles.

Wherefore, *if two straight lines* &c. Q.E.D.

PROPOSITION 9. *THEOREM.*

Two straight lines which are each of them parallel to the same straight line, and not in the same plane with it, are parallel to one another.

Let AB and CD be each of them parallel to EF, and not in the same plane with it:

AB shall be parallel to CD.

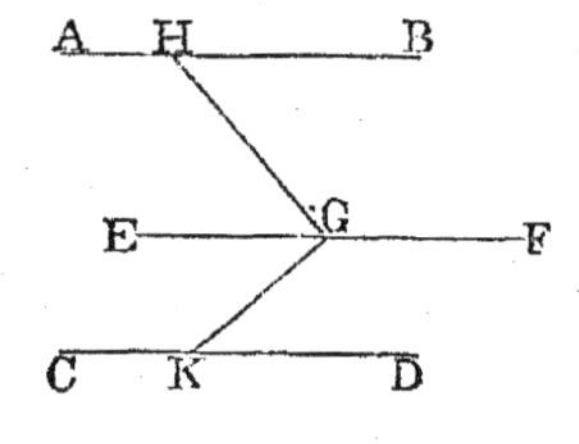

In EF take any point G; in the plane passing through EF and AB, draw from G the straight line GH at right angles to EF; and in the plane passing through EF and CD, draw from G the straight line GK at right angles to EF. [I. 11.

Then, because EF is at right angles to GH and GK, [*Construction.*

EF is at right angles to the plane HGK passing through them. [XI. 4.

And EF is parallel to AB; [*Hypothesis.*

therefore AB is at right angles to the plane HGK. [XI. 8.

For the same reason CD is at right angles to the plane HGK.

Therefore AB and CD are both at right angles to the plane HGK.

Therefore AB is parallel to CD. [XI. 6.

Wherefore, *if two straight lines* &c. Q.E.D.

PROPOSITION 10. *THEOREM.*

If two straight lines meeting one another be parallel to two others that meet one another, and are not in the same plane with the first two, the first two and the other two shall contain equal angles.

Let the two straight lines AB, BC, which meet one another, be parallel to the two straight lines DE, EF, which meet one another, and are not in the same plane with AB, BC: the angle ABC shall be equal to the angle DEF.

Take BA, BC, ED, EF all equal to one another, and join AD, BE, CF, AC, DF.

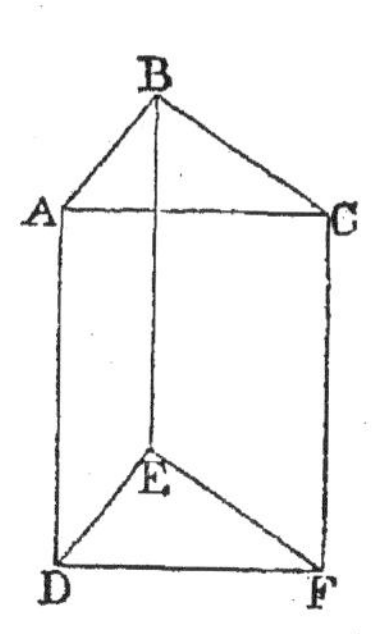

Then, because AB is equal and parallel to DE,

therefore AD is equal and parallel to BE. [I. 33.

For the same reason, CF is equal and parallel to BE.

Therefore AD and CF are each of them equal and parallel to BE.

Therefore AD is parallel to CF, [XI. 9.

and AD is also equal to CF. [*Axiom* 1.

Therefore AC is equal and parallel to DF. [I. 33.

And because AB, BC are equal to DE, EF, each to each,

and the base AC is equal to the base DF,

therefore the angle ABC is equal to the angle DEF. [I. 8.

Wherefore, *if two straight lines* &c. Q.E.D.

PROPOSITION 11. *PROBLEM.*

To draw a straight line perpendicular to a given plane, from a given point without it.

Let A be the given point without the plane BH: it is required to draw from the point A a straight line perpendicular to the plane BH.

Draw any straight line BC in the plane BH, and from the point A draw AD perpendicular to BC. [I. 12.

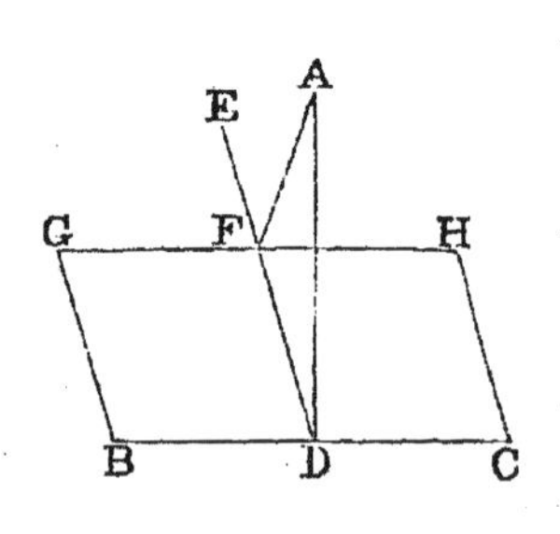

Then if AD be also perpendicular to the plane BH, the thing required is done. But, if not, from the point D draw, in the plane BH, the straight line DE at right angles to BC, [I. 11.
and from the point A draw AF perpendicular to DE. [I. 12.
AF shall be perpendicular to the plane BH.

Through F draw GH parallel to BC. [I. 31.

Then, because BC is at right angles to ED and DA, [*Constr.*
BC is at right angles to the plane passing through ED and DA. [XI. 4.

And GH is parallel to BC; [*Construction.*

therefore GH is at right angles to the plane passing through ED and DA; [XI. 8.

therefore GH makes right angles with every straight line meeting it in that plane. [XI. *Definition* 3.

But AF meets it, and is in the plane passing through ED and DA;

therefore GH is at right angles to AF,

and therefore AF is at right angles to GH.

But AF is also at right angles to DE; [*Construction.*

therefore AF is at right angles to each of the straight lines GH and DE at the point of their intersection;

therefore AF is perpendicular to the plane passing through GH and DE, that is, to the plane BH. [XI. 4.

Wherefore, *from the given point A, without the plane BH, the straight line AF has been drawn perpendicular to the plane.* Q.E.F.

PROPOSITION 12. *PROBLEM.*

To erect a straight line at right angles to a given plane, from a given point in the plane.

Let A be the given point in the given plane: it is required to erect a straight line from the point A, at right angles to the plane.

From any point B without the plane, draw BC perpendicular to the plane; [XI. 11.

and from the point A draw AD parallel to BC, [I. 31.

AD shall be the straight line required.

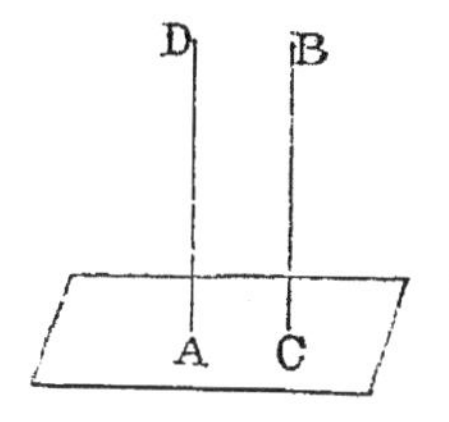

For, because AD and BC are two parallel straight lines, [*Constr.*

and that one of them BC is at right angles to the given plane, [*Construction.*

the other AD is also at right angles to the given plane. [XI. 8.

Wherefore *a straight line has been erected at right angles to a given plane, from a given point in it.* Q.E.F.

PROPOSITION 13. *THEOREM.*

From the same point in a given plane, there cannot be two straight lines at right angles to the plane, on the same side of it; and there can be but one perpendicular to a plane from a point without the plane.

For, if it be possible, let the two straight lines *AB*, *AC* be at right angles to a given plane, from the same point *A* in the plane, and on the same side of it.

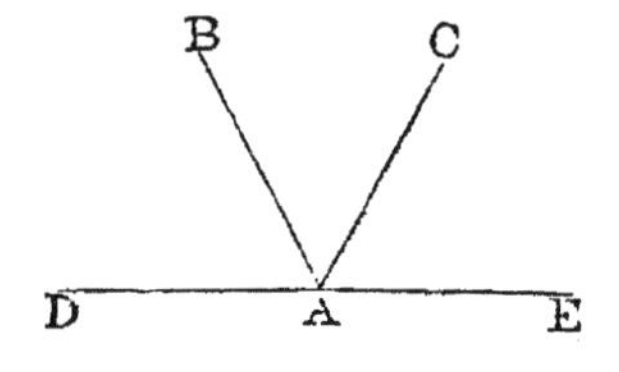

Let a plane pass through *BA*, *AC*;
the common section of this with the given plane is a straight line; [XI. 3.
let this straight line be *DAE*.

Then the three straight lines *AB*, *AC*, *DAE* are all in one plane.
And because *CA* is at right angles to the plane, [*Hypothesis.*
it makes right angles with every straight line meeting it in the plane. [XI. *Definition* 3.
But *DAE* meets *CA*, and is in that plane;
therefore the angle *CAE* is a right angle.
For the same reason the angle *BAE* is a right angle.
Therefore the angle *CAE* is equal to the angle *BAE*; [*Ax.* 11.
and they are in one plane;
which is impossible. [*Axiom* 9.

Also, from a point without the plane, there can be but one perpendicular to the plane.
For if there could be two, they would be parallel to one another, [XI. 6.
which is absurd.

Wherefore, *from the same point* &c. Q.E.D.

PROPOSITION 14. *THEOREM.*

Planes to which the same straight line is perpendicular are parallel to one another.

Let the straight line AB be perpendicular to each of the planes CD and EF: these planes shall be parallel to one another.

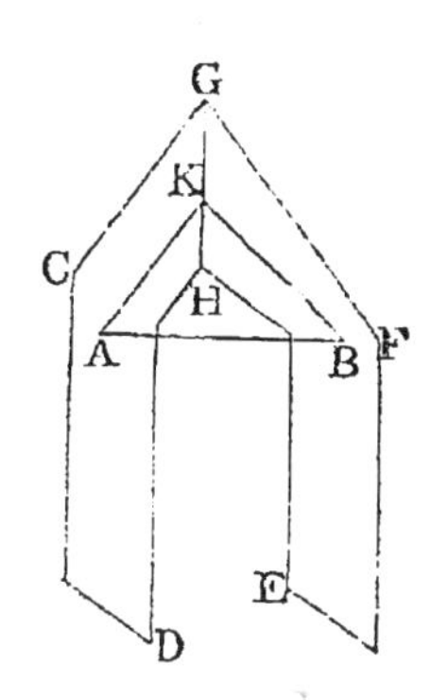

For, if not, they will meet one another when produced;

let them meet, then their common section will be a straight line;

let GH be this straight line; in it take any point K, and join AK, BK.

Then, because AB is perpendicular to the plane EF, [*Hyp.*

it is perpendicular to the straight line BK which is in that plane; [XI. *Definition* 3.

therefore the angle ABK is a right angle.

For the same reason the angle BAK is a right angle.

Therefore the two angles ABK, BAK of the triangle ABK are equal to two right angles;

which is impossible. [I. 17.

Therefore the planes CD and EF, though produced, do not meet one another;

that is, they are parallel. [XI. *Definition* 8.

Wherefore, *planes* &c. Q.E.D.

PROPOSITION 15. *THEOREM.*

If two straight lines which meet one another, be parallel to two other straight lines which meet one another, but are not in the same plane with the first two, the plane passing through these is parallel to the plane passing through the others.

Let AB, BC, two straight lines which meet one another, be parallel to two other straight lines DE, EF, which meet one another, but are not in the same plane with AB, BC: the plane passing through AB, BC, shall be parallel to the plane passing through DE, EF.

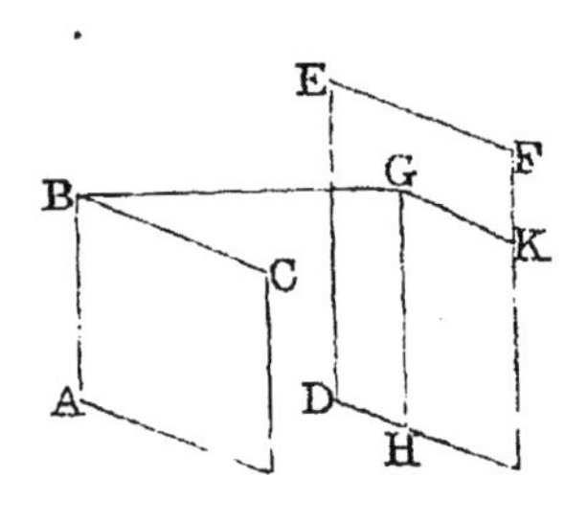

From the point B draw BG perpendicular to the plane passing through DE, EF, [XI. 11.
and let it meet that plane at G;
through G draw GH parallel to ED, and GK parallel to EF. [I. 31.

Then, because BG is perpendicular to the plane passing through DE, EF, [*Construction.*
it makes right angles with every straight line meeting it in that plane; [XI. *Definition* 3.

but the straight lines GH and GK meet it, and are in that plane;

therefore each of the angles BGH and BGK is a right angle.

Now because BA is parallel to ED, [*Hypothesis.*
and GH is parallel to ED, [*Construction.*

therefore BA is parallel to GH; [XI. 9.

therefore the angles ABG and BGH are together equal to two right angles. [I. 29.

And the angle BGH has been shewn to be a right angle;

therefore the angle ABG is a right angle.

For the same reason the angle CBG is a right angle.

Then, because the straight line GB stands at right angles to the two straight lines BA, BC, at their point of intersection B,

therefore GB is perpendicular to the plane passing through BA, BC. [XI. 4.

And GB is also perpendicular to the plane passing through DE, EF. [*Construction.*

But planes to which the same straight line is perpendicular are parallel to one another; [XI. 14.

therefore the plane passing through AB, BC is parallel to the plane passing through DE, EF.

Wherefore, *if two straight lines* &c. Q.E.D.

PROPOSITION 16. *THEOREM.*

If two parallel planes be cut by another plane, their common sections with it are parallel.

Let the parallel planes AB, CD be cut by the plane $EFHG$, and let their common sections with it be EF, GH: EF shall be parallel to GH.

For if not, EF and GH, being produced, will meet either towards F, H, or towards E, G. Let them be produced and meet towards F, H at the point K.

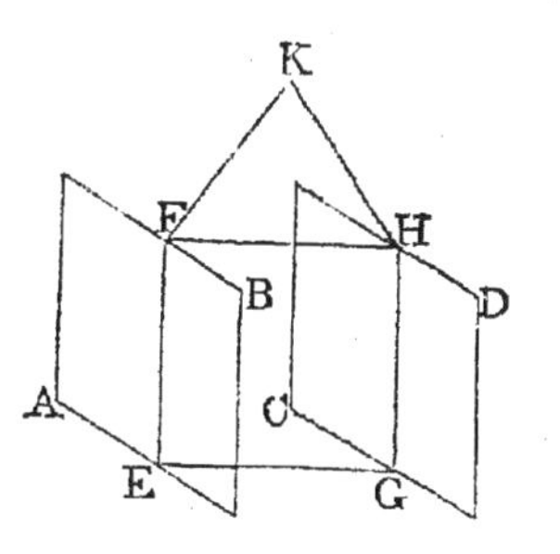

Then, since EFK is in the plane AB, every point in EFK is in that plane; [XI. 1.

therefore K is in the plane AB.

For the same reason K is in the plane CD.

Therefore the planes AB, CD, being produced, meet one another.

But they do not meet, since they are parallel by hypothesis.

Therefore EF and GH, being produced, do not meet towards F, H.

In the same manner it may be shewn that they do not meet towards E, G.

But straight lines which are in the same plane, and which being produced ever so far both ways do not meet are parallel;

therefore EF is parallel to GH.

Wherefore, *if two parallel planes* &c. Q.E.D.

PROPOSITION 17. *THEOREM.*

If two straight lines be cut by parallel planes, they shall be cut in the same ratio.

Let the straight lines *AB* and *CD* be cut by the parallel planes *GH, KL, MN*, at the points *A, E, B*, and *C, F, D*: *AE* shall be to *EB* as *CF* is to *FD*.

Join *AC, BD, AD*; let *AD* meet the plane *KL* at the point *X*; and join *EX, XF*.

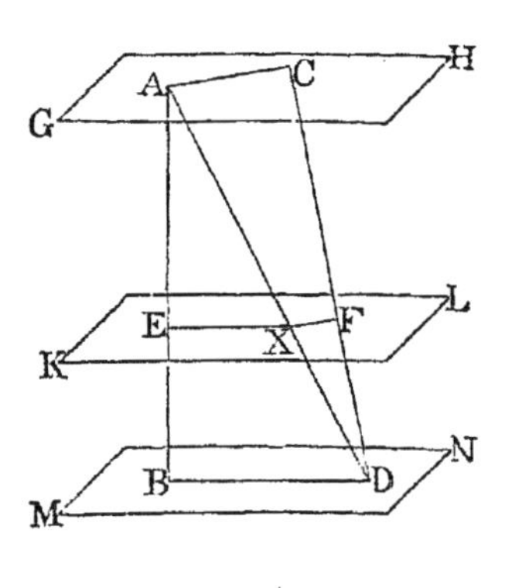

Then, because the two parallel planes *KL, MN* are cut by the plane *EBDX*, the common sections *EX*, *BD* are parallel; [XI. 16.

and because the two parallel planes *GH, KL* are cut by the plane *AXFC*, the common sections *AC*, *XF* are parallel. [XI. 16.

And, because *EX* is parallel to *BD*, a side of the triangle *ABD*,

therefore *AE* is to *EB* as *AX* is to *XD*. [VI. 2.

Again, because *XF* is parallel to *AC*, a side of the triangle *ADC*,

therefore *AX* is to *XD* as *CF* is to *FD*. [VI. 2.

And it was shewn that *AX* is to *XD* as *AE* is to *EB*;

therefore *AE* is to *EB* as *CF* is to *FD*. [V. 11.

Wherefore, *if two straight lines* &c. Q.E.D.

PROPOSITION 18. *THEOREM.*

If a straight line be at right angles to a plane, every plane which passes through it shall be at right angles to that plane.

Let the straight line *AB* be at right angles to the plane *CK*: every plane which passes through *AB* shall be at right angles to the plane *CK*.

Let any plane DE pass through AB, and let CE be the common section of the planes DE, CK; [XI. 3.
take any point F in CE, from which draw FG, in the plane DE, at right angles to CE. [I. 11.

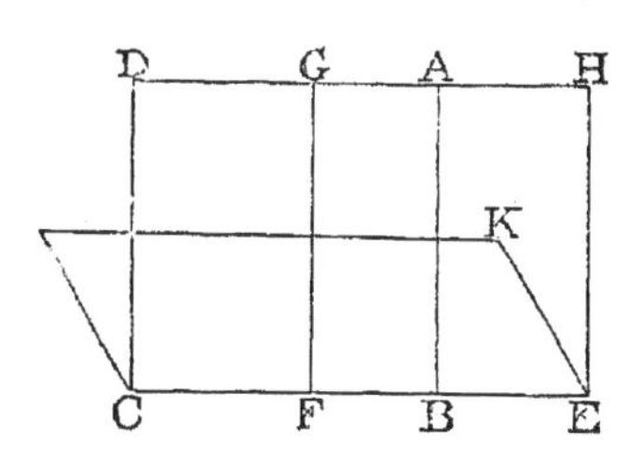

Then, because AB is at right angles to the plane CK, [*Hypothesis.*
therefore it makes right angles with every straight line meeting it in that plane; [XI. *Definition* 3.
but CB meets it, and is in that plane;
therefore the angle ABF is a right angle.
But the angle GFB is also a right angle; [*Construction.*
therefore FG is parallel to AB. [I. 28.
And AB is at right angles to the plane CK; [*Hypothesis.*
therefore FG is also at right angles to the same plane. [XI. 8.

But one plane is at right angles to another plane, when the straight lines drawn in one of the planes at right angles to their common section, are also at right angles to the other plane; [XI. *Definition* 4.
and it has been shewn that any straight line FG drawn in the plane DE, at right angles to CE, the common section of the planes, is at right angles to the other plane CK;
therefore the plane DE is at right angles to the plane CK.

In the same manner it may be shewn that any other plane which passes through AB is at right angles to the plane CK.

Wherefore, *if a straight line* &c. Q.E.D.

PROPOSITION 19. *THEOREM.*

If two planes which cut one another be each of them perpendicular to a third plane, their common section shall be perpendicular to the same plane.

Let the two planes BA, BC be each of them perpendicular to a third plane, and let BD be the common section of the planes BA, BC: BD shall be perpendicular to the third plane.

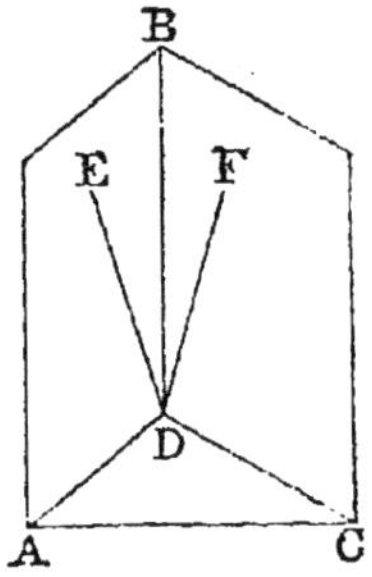

For, if not, from the point D, draw in the plane BA, the straight line DE at right angles to AD, the common section of the plane BA with the third plane; [I. 11.
and from the point D, draw in the plane BC, the straight line DF at right angles to CD, the common section of the plane BC with the third plane. [I. 11.

Then, because the plane BA is perpendicular to the third plane, [*Hypothesis.*
and DE is drawn in the plane BA at right angles to AD their common section; [*Construction.*
therefore DE is perpendicular to the third plane. [XI. *Def.* 4.

In the same manner it may be shewn that DF is perpendicular to the third plane.
Therefore from the point D two straight lines are at right angles to the third plane, on the same side of it; which is impossible. [XI. 13.
Therefore from the point D, there cannot be any straight line at right angles to the third plane, except BD the common section of the planes BA, BC;
therefore BD is perpendicular to the third plane.

Wherefore, *if two planes* &c. Q.E.D.

PROPOSITION 20. *THEOREM.*

If a solid angle be contained by three plane angles, any two of them are together greater than the third.

Let the solid angle at A be contained by the three plane angles BAC, CAD, DAB: any two of them shall be together greater than the third.

If the angles BAC, CAD, DAB be all equal, it is evident that any two of them are greater than the third.

If they are not all equal, let *BAC* be that angle which is not less than either of the other two, and is greater than one of them, *BAD*.

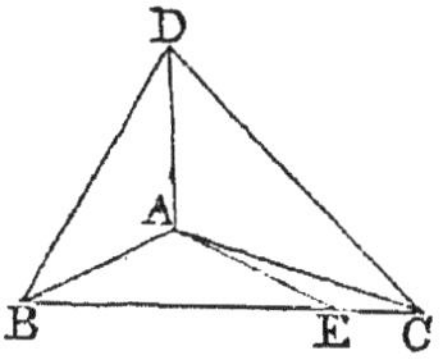

At the point *A*, in the straight line *BA*, make, in the plane which passes through *BA*, *AC*, the angle *BAE* equal to the angle *BAD*; [I. 23.

make *AE* equal to *AD*; [I. 3.

through *E* draw *BEC* cutting *AB*, *AC* at the points *B*, *C*; and join *DB*, *DC*.

Then, because *AD* is equal to *AE*, [*Construction.*

and *AB* is common to the two triangles *BAD*, *BAE*,

the two sides *BA*, *AD* are equal to the two sides *BA*, *AE*, each to each;

and the angle *BAD* is equal to the angle *BAE*; [*Constr.*

therefore the base *BD* is equal to the base *BE*. [I. 4.

And because *BD*, *DC* are together greater than *BC*, [I. 20.

and one of them *BD* has been shewn equal to *BE* a part of *BC*,

therefore the other *DC* is greater than the remaining part *EC*.

And because *AD* is equal to *AE*, [*Construction.*

and *AC* is common to the two triangles *DAC*, *EAC*,

but the base *DC* is greater than the base *EC*;

therefore the angle *DAC* is greater than the angle *EAC*. [I. 25.

And, by construction, the angle *BAD* is equal to the angle *BAE*;

therefore the angles *BAD*, *DAC* are together greater than the angles *BAE*, *EAC*, that is, than the angle *BAC*.

But the angle *BAC* is not less than either of the angles *BAD*, *DAC*;

therefore the angle *BAC* together with either of the other angles is greater than the third.

Wherefore, *if a solid angle* &c. Q.E.D.

PROPOSITION 21. *THEOREM.*

Every solid angle is contained by plane angles, which are together less than four right angles.

First let the solid angle at A be contained by three plane angles BAC, CAD, DAB: these three shall be together less than four right angles.

In the straight lines AB, AC, AD take any points B, C, D, and join BC, CD, DB.

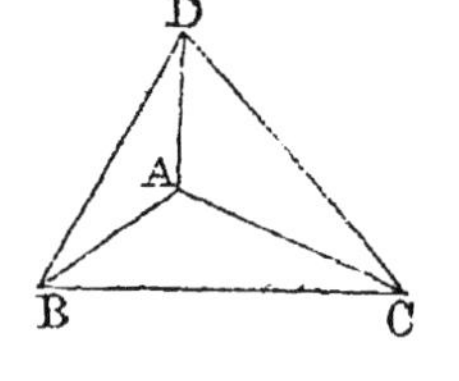

Then, because the solid angle at B is contained by the three plane angles CBA, ABD, DBC, any two of them are together greater than the third, [XI. 20.

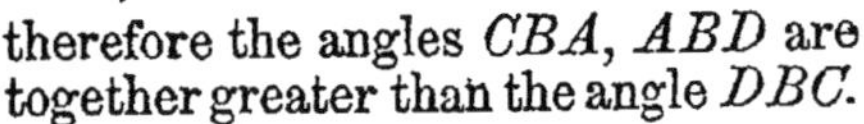
therefore the angles CBA, ABD are together greater than the angle DBC.

For the same reason, the angles BCA, ACD are together greater than the angle DCB,

and the angles CDA, ADB are together greater than the angle BDC.

Therefore the six angles CBA, ABD, BCA, ACD, CDA, ADB are together greater than the three angles DBC, DCB, BDC;

but the three angles DBC, DCB, BDC are together equal to two right angles. [I. 32.

Therefore the six angles CBA, ABD, BCA, ACD, CDA, ADB are together greater than two right angles.

And, because the three angles of each of the triangles ABC, ACD, ADB are together equal to two right angles, [I. 32.

therefore the nine angles of these triangles, namely, the angles CBA, BAC, ACB, ACD, CDA, CAD, ADB, DBA, DAB are equal to six right angles;

and of these, the six angles CBA, ACB, ACD, CDA, ADB, DBA are greater than two right angles,

therefore the remaining three angles BAC, CAD, DAB, which contain the solid angle at A, are together less than four right angles.

Next, let the solid angle at A be contained by any number of plane angles BAC, CAD, DAE, EAF, FAB: these shall be together less than four right angles.

Let the planes in which the angles are, be cut by a plane, and let the common sections of it with those planes be BC, CD, DE, EF, FB.

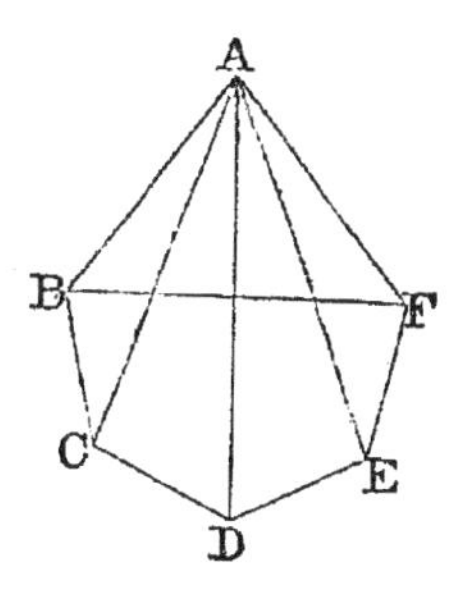

Then, because the solid angle at B is contained by the three plane angles CBA, ABF, FBC, any two of them are together greater than the third, [XI. 20.

therefore the angles CBA, ABF are together greater than the angle FBC.

For the same reason, at each of the points C, D, E, F, the two plane angles which are at the bases of the triangles having the common vertex A, are together greater than the third angle at the same point, which is one of the angles of the polygon $BCDEF$.

Therefore all the angles at the bases of the triangles are together greater than all the angles of the polygon.

Now all the angles of the triangles are together equal to twice as many right angles as there are triangles, that is, as there are sides in the polygon $BCDEF$; [I. 32.

and all the angles of the polygon, together with four right angles, are also equal to twice as many right angles as there are sides in the polygon; [I. 32, *Corollary* 1.

therefore all the angles of the triangles are equal to all the angles of the polygon, together with four right angles. [*Ax.* 1.

But it has been shewn that all the angles at the bases of the triangles are together greater than all the angles of the polygon;

therefore the remaining angles of the triangles, namely, those at the vertex, which contain the solid angle at A, are together less than four right angles.

Wherefore, *every solid angle* &c. Q.E.D.

BOOK XII.

LEMMA.

If from the greater of two unequal magnitudes there be taken more than its half, and from the remainder more than its half, and so on, there shall at length remain a magnitude less than the smaller of the proposed magnitudes.

Let AB and C be two unequal magnitudes, of which AB is the greater: if from AB there be taken more than its half, and from the remainder more than its half, and so on, there shall at length remain a magnitude less than C.

For C may be multiplied so as at length to become greater than AB.

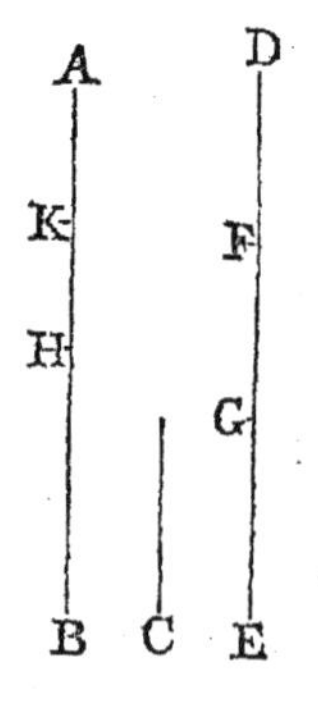

Let it be so multiplied, and let DE its multiple be greater than AB, and let DE be divided into DF, FG, GE, each equal to C.

From AB take BH, greater than its half, and from the remainder AH take HK greater than its half, and so on, until there be as many divisions in AB as in DE; and let the divisions in AB be AK, KH, HB, and the divisions in DE be DF, FG, GE.

Then, because DE is greater than AB;
and that EG taken from DE is not greater than its half;
but BH taken from AB is greater than its half;
therefore the remainder DG is greater than the remainder AH.

Again, because *DG* is greater than *AH*;
and that *GF* is not greater than the half of *DG*, but *HK* is greater than the half of *AH*;
therefore the remainder *DF* is greater than the remainder *AK*.

But *DF* is equal to *C*;
therefore *C* is greater than *AK*;
that is, *AK* is less than *C*. Q.E.D.

And if only the halves be taken away, the same thing may in the same way be demonstrated.

PROPOSITION 1. *THEOREM.*

Similar polygons inscribed in circles are to one another as the squares on their diameters.

Let *ABCDE*, *FGHKL* be two circles, and in them the similar polygons *ABCDE*, *FGHKL*; and let *BM*, *GN* be the diameters of the circles: the polygon *ABCDE* shall be to the polygon *FGHKL* as the square on *BM* is to the square on *GM*.

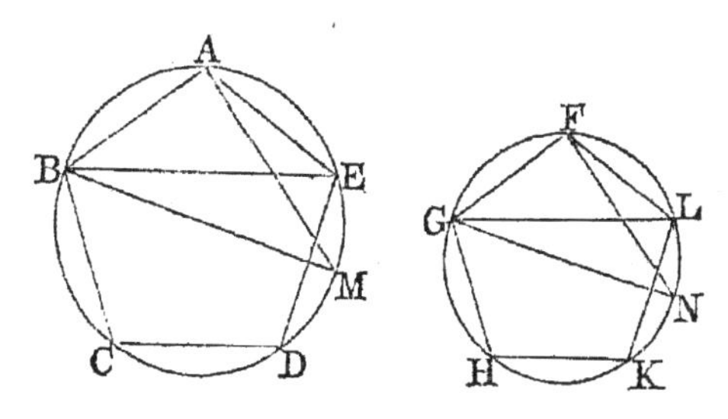

Join *AM*, *BE*, *FN*, *GL*.
Then, because the polygons are similar,
therefore the angle *BAE* is equal to the angle *GFL*,
and *BA* is to *AE* as *GF* is to *FL*. [VI. *Definition* 1.
Therefore the triangle *BAE* is equiangular to the triangle *GFL*; [VI. 6.
therefore the angle *AEB* is equal to the angle *FLG*.
But the angle *AEB* is equal to the angle *AMB*, and the angle *FLG* is equal to the angle *FNG*; [III. 21.
therefore the angle *AMB* is equal to the angle *FNG*.

And the angle BAM is equal to the angle GFN, for each of them is a right angle. [III. 31.

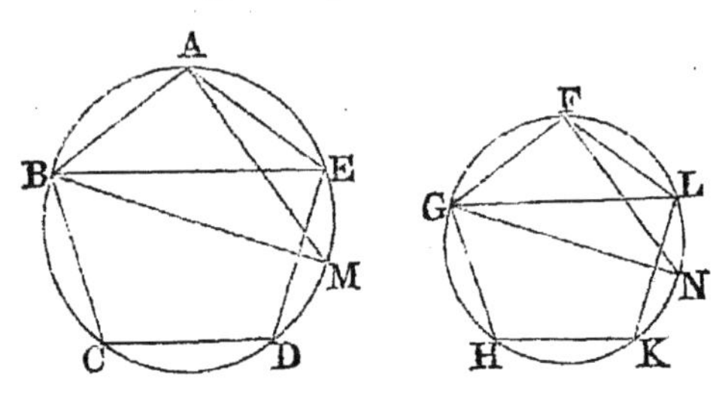

Therefore the remaining angles in the triangles AMB, FNG are equal, and the triangles are equiangular to one another;

therefore BA is to BM as GF is to GN, [VI. 4.

and, alternately, BA is to GF as BM is to GN; [V. 16.

therefore the duplicate ratio of BA to GF is the same as the duplicate ratio of BM to GN. [V. *Definition* 10, V. 22.

But the polygon $ABCDE$ is to the polygon $FGHKL$ in the duplicate ratio of BA to GF; [VI. 20.

and the square on BM is to the square on GN in the duplicate ratio of BM to GN; [VI. 20.

therefore the polygon $ABCDE$ is to the polygon $FGHKL$ as the square on BM is to the square on GN. [V. 11.

Wherefore, *similar polygons* &c. Q.E.D.

PROPOSITION 2. *THEOREM.*

Circles are to one another as the squares on their diameters.

Let $ABCD$, $EFGH$ be two circles, and BD, FH their diameters: the circle $ABCD$ shall be to the circle $EFGH$ as the square on BD is to the square on FH.

For, if not, the square on BD must be to the square on FH as the circle $ABCD$ is to some space either less than the circle $EFGH$, or greater than it.

First, if possible, let it be as the circle $ABCD$ is to a space S less than the circle $EFGH$.

In the circle $EFGH$ inscribe the square $EFGH$. [IV. 6.
This square shall be greater than half of the circle $EFGH$.

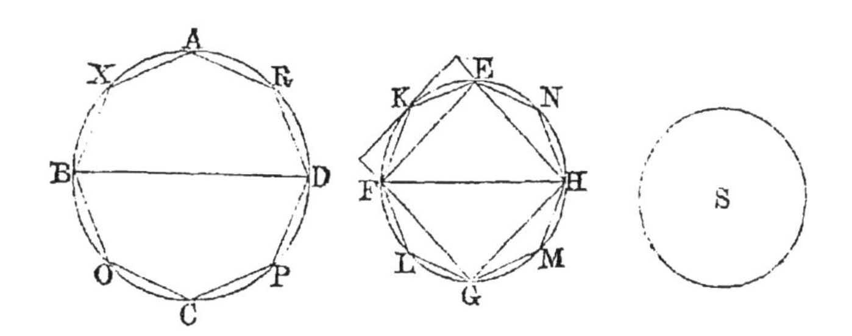

For the square $EFGH$ is half of the square which can be formed by drawing straight lines to touch the circle at the points E, F, G, H;

and the square thus formed is greater than the circle;

therefore the square $EFGH$ is greater than half of the circle.

Bisect the arcs EF, FG, GH, HE at the points K, L, M, N;

and join EK, KF, FL, LG, GM, MH, HN, NE. Then each of the triangles EKF, FLG, GMH, HNE shall be greater than half of the segment of the circle in which it stands.

For the triangle EKF is half of the parallelogram which can be formed by drawing a straight line to touch the circle at K, and parallel straight lines through E and F,

and the parallelogram thus formed is greater than the segment FEK; therefore the triangle EKF is greater than half of the segment.

And similarly for the other triangles.

Therefore the sum of all these triangles is together greater than half of the sum of the segments of the circle in which they stand.

Again, bisect EK, KF, &c. and form triangles as before;

then the sum of these triangles is greater than half of the sum of the segments of the circle in which they stand.

If this process be continued, and the triangles be supposed to be taken away, there will at length remain segments of circles which are together less than the excess of the circle *EFGH* above the space *S*, by the preceding Lemma.

Let then the segments *EK*, *KF*, *FL*, *LG*, *GM*, *MH*, *HN*, *NE* be those which remain, and which are together less than the excess of the circle above *S*;

therefore the rest of the circle, namely the polygon *EKFLGMHN*, is greater than the space *S*.

In the circle *ABCD* describe the polygon *AXBOCPDR* similar to the polygon *EKFLGMHN*.

Then the polygon *AXBOCPDR* is to the polygon *EKFLGMHN* as the square on *BD* is to the square on *FH*, [XII. 1.

that is, as the circle *ABCD* is to the space *S*. [*Hyp.*, V. 11.

But the polygon *AXBOCPDR* is less than the circle *ABCD* in which it is inscribed,

therefore the polygon *EKFLGMHN* is less than the space *S*; [V. 14.

but it is also greater, as has been shewn;

which is impossible.

Therefore the square on *BD* is not to the square on *FH* as the circle *ABCD* is to any space less than the circle *EFGH*.

In the same way it may be shewn that the square on *FH* is not to the square on *BD* as the circle *EFGH* is to any space less than the circle *ABCD*.

Nor is the square on *BD* to the square on *FH* as the circle *ABCD* is to any space greater than the circle *EFGH*.

For, if possible, let it be as the circle *ABCD* is to a space *T* greater than the circle *EFGH*.

Then, inversely, the square on *FH* is to the square on *BD* as the space *T* is to the circle *ABCD*.

But as the space *T* is to the circle *ABCD* so is the circle *EFGH* to some space, which must be less than the circle

ABCD, because, by hypothesis, the space *T* is greater than the circle *EFGH*. [V. 14.

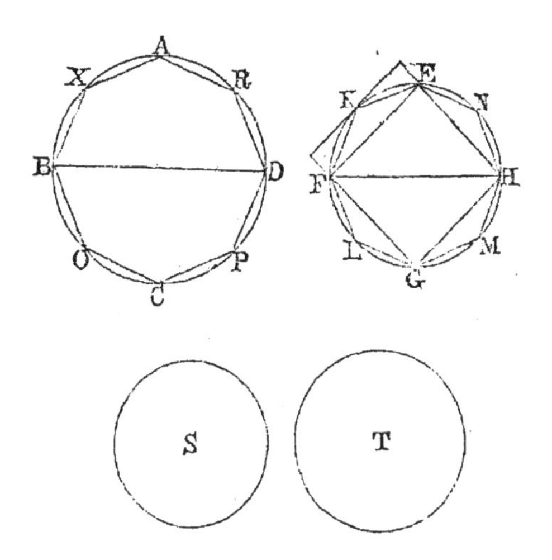

Therefore the square on *FH* is to the square on *BD* as the circle *EFGH* is to some space less than the circle *ABCD*;
which has been shewn to be impossible.

Therefore the square on *BD* is not to the square on *FH* as the circle *ABCD* is to any space greater than the circle *EFGH*.

And it has been shewn that the square on *BD* is not to the square on *FH* as the circle *ABCD* is to any space less than the circle *EFGH*.

Therefore the square on *BD* is to the square on *FH* as the circle *ABCD* is to the circle *EFGH*.

Wherefore, *circles* &c. Q.E.D.

NOTES ON EUCLID'S ELEMENTS.

THE article Eucleides in Dr Smith's *Dictionary of Greek and Roman Biography* was written by Professor De Morgan; it contains an account of the works of Euclid, and of the various editions of them which have been published. To that article we refer the student who desires full information on these subjects. Perhaps the only work of importance relating to Euclid which has been published since the date of that article is a work on the Porisms of Euclid by Chasles; Paris, 1860.

Euclid appears to have lived in the time of the first Ptolemy, B.C. 323—283, and to have been the founder of the Alexandrian mathematical school. The work on Geometry known as *The Elements of Euclid* consists of thirteen books; two other books have sometimes been added, of which it is supposed that Hypsicles was the author. Besides the *Elements*, Euclid was the author of other works, some of which have been preserved and some lost.

We will now mention the three editions which are the most valuable for those who wish to read the Elements of Euclid in the original Greek.

(1) The Oxford edition in folio, published in 1703 by David Gregory, under the title Εὐκλείδου τὰ σωζόμενα. "As an edition of the whole of Euclid's works, this stands alone, there being no other in Greek." *De Morgan.*

(2) *Euclidis Elementorum Libri sex priores...edidit Joannes Gulielmus Camerer.* This edition was published at Berlin in two volumes octavo, the first volume in 1824 and the second in 1825. It contains the first six books of the *Elements* in Greek with a Latin Translation, and very good notes which form a mathematical commentary on the subject.

(3) *Euclidis Elementa ex optimis libris in usum tironum Græce edita ab Ernesto Ferdinando August.* This edition was published at Berlin in two volumes octavo, the first volume in 1826 and the second in 1829. It contains the thirteen books of the *Elements* in Greek, with a collection of various readings.

A third volume, which was to have contained the remaining works of Euclid, never appeared. "To the scholar who wants one edition of the Elements we should decidedly recommend this, as bringing together all that has been done for the text of Euclid's greatest work." *De Morgan.*

An edition of the whole of Euclid's works in the original has long been promised by Teubner the well-known German publisher, as one of his series of compact editions of Greek and Latin authors; but we believe there is no hope of its early appearance.

Robert Simson's edition of the *Elements of Euclid,* which we have in substance adopted in the present work, differs considerably from the original. The English reader may ascertain the contents of the original by consulting the work entitled *The Elements of Euclid with dissertations...by James Williamson.* This work consists of two volumes quarto; the first volume was published at Oxford in 1781, and the second at London in 1788. Williamson gives a close translation of the thirteen books of the *Elements* into English, and he indicates by the use of Italics the words which are not in the original but which are required by our language.

Among the numerous works which contain notes on the *Elements of Euclid* we will mention four by which we have been aided in drawing up the selection given in this volume.

An Examination of the first six Books of Euclid's Elements by William Austin...Oxford, 1781.

Euclid's Elements of Plane Geometry with copious notes...by John Walker. London, 1827.

The first six books of the Elements of Euclid with a Commentary...by Dionysius Lardner, fourth edition. London, 1834.

Short supplementary remarks on the first six Books of Euclid's Elements, by Professor De Morgan, in the *Companion to the Almanac* for 1849.

We may also notice the following works:

Geometry, Plane, Solid, and Spherical,...London 1830; this forms part of the Library of Useful Knowledge.

Théorèmes et Problèmes de Géométrie Elementaire par Eugène Catalan...Troisième édition. Paris, 1858.

For the History of Geometry the student is referred to Montucla's *Histoire des Mathématiques,* and to Chasles's *Aperçu historique sur l'origine et le devéloppement des Méthodes en Géométrie...*

THE FIRST BOOK.

Definitions. The first seven definitions have given rise to considerable discussion, on which however we do not propose to enter. Such a discussion would consist mainly of two subjects, both of which are unsuitable to an elementary work, namely, an examination of the origin and nature of some of our elementary ideas, and a comparison of the original text of Euclid with the substitutions for it proposed by Simson and other editors. For the former subject the student may hereafter consult Whewell's *History of Scientific Ideas* and Mill's *Logic*, and for the latter the notes in Camerer's edition of the *Elements of Euclid.*

We will only observe that the ideas which correspond to the words *point, line*, and *surface*, do not admit of such definitions as will really supply the ideas to a person who is destitute of them. The so-called definitions may be regarded as cautions or restrictions. Thus a *point* is not to be supposed to have any *size*, but only *position;* a line is not to be supposed to have any *breadth* or *thickness*, but only length; a surface is not to be supposed to have any *thickness*, but only *length* and *breadth.*

The eighth definition seems intended to include the cases in which an angle is formed by the meeting of two *curved* lines, or of a *straight* line and a *curved* line; this definition however is of no importance, as the only angles ever considered are such as are formed by straight lines. The definition of a plane rectilineal angle is important; the beginner must carefully observe that no change is made in an angle by prolonging the lines which form it, away from the angular point.

Some writers object to such definitions as those of an equilateral triangle, or of a square, in which the existence of the object defined is *assumed* when it ought to be *demonstrated.* They would present them in such a form as the following: if there be a triangle having three equal sides, let it be called an equilateral triangle.

Moreover, some of the definitions are introduced prematurely. Thus, for example, take the definitions of a right-angled triangle and an obtuse-angled triangle; it is not shewn until I. 17, that a triangle cannot have both a right angle and an obtuse angle, and so cannot be at the same time right-angled and obtuse-angled. And before Axiom 11 has been given, it is conceivable that the same angle may be greater than one right angle, and less than another right angle, that is, obtuse and acute at the same time.

The definition of a square assumes more than is necessary. For if a four-sided figure have all its sides equal and *one* angle a right angle, it may be shewn that *all* its angles are right angles; or if a four-sided figure have all its angles *equal*, it may be shewn that they are all *right angles*.

Postulates. The postulates state what processes we assume that we can effect, namely, that we can draw a straight line between two given points, that we can produce a straight line to any length, and that we can describe a circle from a given centre with a given distance as radius. It is sometimes stated that the postulates amount to requiring the use of a *ruler* and *compasses.* It must however be observed that the ruler is not supposed to be a *graduated* ruler, so that we cannot use it to measure off assigned lengths. And we do not require the compasses for any other process than describing a circle from a given point with a given distance as radius; in other words, the compasses may be supposed to close of themselves, as soon as one of their points is removed from the paper.

Axioms. The axioms are called in the original *Common Notions.* It is supposed by some writers that Euclid intended his postulates to include all demands which are peculiarly geometrical, and his common notions to include only such notions as are applicable to all kinds of magnitude as well as to space magnitudes. Accordingly, these writers remove the last three axioms from their place and put them among the postulates; and this transposition is supported by some manuscripts and some versions of the *Elements.*

The fourth axiom is sometimes referred to in editions of Euclid when in reality more is required than this axiom expresses. Euclid says, that if A and B be unequal, and C and D equal, the sum of A and C is *unequal* to the sum of B and D. What Euclid often requires is something more, namely, that if A be greater than B, and C and D be equal, the sum of A and C is *greater* than the sum of B and D. Such an axiom as this is required, for example, in I. 17. A similar remark applies to the fifth axiom.

In the eighth axiom the words "that is, which exactly fill the same space," have been introduced without the authority of the original Greek. They are objectionable, because *lines* and *angles* are magnitudes to which the axiom may be applied, but they cannot be said to *fill space.*

On the *method of superposition* we may refer to a paper by Professor Kelland in the *Transactions of the Royal Society of Edinburgh,* Vol. XXI.

The eleventh axiom is not required before I. 14, and the twelfth axiom is not required before I. 28; we shall not consider these axioms until we arrive at the propositions in which they are respectively required for the first time.

The first book is chiefly devoted to the properties of triangles and parallelograms.

We may observe that Euclid himself does not distinguish between problems and theorems except by using at the end of the investigation phrases which correspond to Q.E.F. and Q.E.D. respectively.

I. 2. This problem admits of *eight* cases in its figure. For it will be found that the given point may be joined with *either* end of the given straight line, then the equilateral triangle may be described on *either* side of the straight line which is drawn, and the side of the equilateral triangle which is produced may be produced through *either* extremity. These various cases may be left for the exercise of the student, as they present no difficulty.

There will not however always be eight different straight lines obtained which solve the problem. For example, if the point *A* falls on *BC* produced, some of the solutions obtained coincide; this depends on the fact which follows from I. 32, that the angles of all equilateral triangles are equal.

I. 5. "Join *FC.*" Custom seems to allow this singular expression as an abbreviation for "draw the straight line *FC,*" or for "join *F* to *C* by the straight line *FC.*"

In comparing the triangles *BFC, CGB,* the words "and the base *BC* is common to the two triangles *BFC, CGB*" are usually inserted, with the authority of the original. As however these words are of no use, and tend to perplex a beginner, we have followed the example of some editors and omitted them.

A *corollary* to a proposition is an inference which may be deduced immediately from that proposition. Many of the corollaries in the *Elements* are not in the original text, but introduced by the editors.

It has been suggested to demonstrate I. 5 by *superposition*. Conceive the isosceles triangle ABC to be taken up, and then replaced so that AB falls on the old position of AC, and AC falls on the old position of AB. Thus, in the manner of I. 4, we can shew that the angle ABC is equal to the angle ACB.

I. 6 is the *converse* of part of I. 5. One proposition is said to be the converse of another when the conclusion of each is the hypothesis of the other. Thus in I. 5 the hypothesis is the equality of the sides, and one conclusion is the equality of the angles; in I. 6 the hypothesis is the equality of the angles and the conclusion is the equality of the sides. When there is more than one hypothesis or more than one conclusion to a proposition, we can form more than one converse proposition. For example, as another converse of I. 5 we have the following: if the angles formed by the base of a triangle and the sides produced be equal, the sides of the triangle are equal; this proposition is true and will serve as an exercise for the student.

The converse of a true proposition is not necessarily true; the student however will see, as he proceeds, that Euclid shews that the converses of many geometrical propositions are true.

I. 6 is an example of the *indirect* mode of demonstration, in which a result is established by shewing that some absurdity follows from supposing the required result to be untrue. Hence this mode of demonstration is called the *reductio ad absurdum*. Indirect demonstrations are often less esteemed than direct demonstrations; they are said to shew that a theorem *is* true rather than to shew *why* it is true. Euclid uses the *reductio ad absurdum* chiefly when he is demonstrating the converse of some former theorem; see I. 14, 19, 25, 40.

Some remarks on *indirect demonstration* by Professor Sylvester, Professor De Morgan, and Dr Adamson will be found in the volumes of the *Philosophical Magazine* for 1852 and 1853.

I. 6 is not required by Euclid before he reaches II. 4; so that I. 6 might be removed from its present place and demonstrated hereafter in other ways if we please. For example, I. 6 might be placed after I. 18 and demonstrated thus. Let the angle ABC be equal to the angle ACB: then the side AB shall be equal to the side AC. For if not, one of them must be greater than the other; suppose AB greater than AC. Then the angle ACB is greater than the angle ABC, by I. 18. But this is impossible, because the angle ACB is equal to the angle ABC, by hypothesis. Or I. 6 might be placed after I. 26 and demonstrated thus. Bisect the angle BAC by a straight line meeting the base at D. Then the triangles ABD and ACD are equal in all respects, by I. 26.

I. 7 is only required in order to lead to I. 8. The two might be superseded by another demonstration of I. 8, which has been recommended by many writers.

Let ABC, DEF be two triangles, having the sides AB, AC equal to the sides DE, DF, each to each, and the base BC equal to the base EF: the angle BAC shall be equal to the angle EDF.

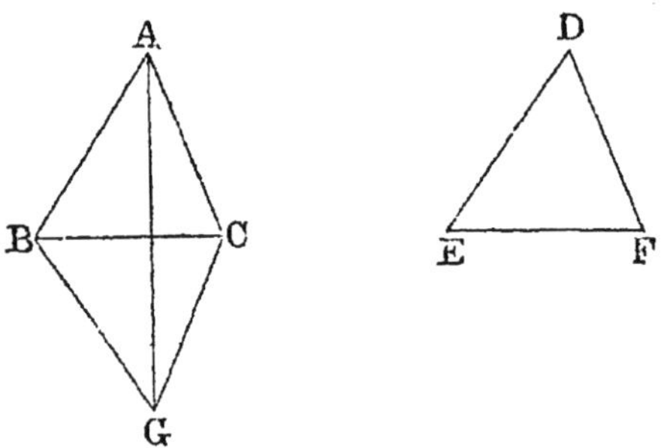

For, let the triangle DEF be applied to the triangle ABC, so that the bases may coincide, the equal sides be conterminous, and the vertices fall on opposite sides of the base. Let GBC represent the triangle DEF thus applied, so that G corresponds to D. Join AG. Since, by hypothesis, BA is equal to BG, the angle BAG is equal to the angle BGA, by I. 5. In the same manner the angle CAG is equal to the angle CGA. Therefore the whole angle BAC is equal to the whole angle BGC, that is, to the angle EDF.

There are two other cases; for the straight line AG may pass through B or C, or it may fall outside BC: these cases may be treated in the same manner as that which we have considered.

I. 8. It may be observed that the two triangles in I. 8 are equal in *all respects;* Euclid however does not assert more than the equality of the angles opposite to the bases, and when he requires more than this result he obtains it by using I. 4.

I. 9. Here the equilateral triangle DEF is to be described on the side *remote* from A, because if it were described on the *same* side, its vertex, F, might coincide with A, and then the construction would fail.

I. 11. The corollary was added by Simson. It is liable to serious objection. For we do not know how the perpendicular BE is to be drawn. If we are to use I. 11 we must produce AB, and then we must assume that there is only *one* way of producing AB, for otherwise we shall not know that there is only *one* perpendicular; and thus we assume what we have to demonstrate.

Simson's corollary might come after I. 13 and be demonstrated thus. If possible let the two straight lines ABC, ABD have the segment AB common to both. From the point B draw any straight line BE. Then the angles ABE and EBC are equal to two right angles, by I. 13, and the angles ABE and EBD are also equal to two right angles, by I. 13. Therefore the angles ABE and EBC are equal to the angles ABE and EBD. Therefore the angle EBC is equal to the angle EBD; which is absurd.

But if the question whether two straight lines can have a common segment is to be considered at all in the Elements, it might occur at an earlier place than Simson has assigned to it. For example, in the figure to I. 5, if two straight lines could have a common segment AB, and then separate at B, we should obtain two different angles formed on the other side of BC by these produced parts, and each of them would be equal to the angle BCG. The opinion has been maintained that even in I. 1, it is tacitly assumed that the straight lines AC and BC cannot have a common segment at C where they meet; see Camerer's *Euclid*, pages 30 and 36.

Simson never formally refers to his corollary until XI. 1. The corollary should be omitted, and the tenth axiom should be extended so as to amount to the following; if two straight lines coincide in two points they must coincide both beyond and between those points.

I. 12. Here the straight line is said to be of *unlimited* length, in order that we may ensure that it shall meet the circle.

Euclid distinguishes between the terms *at right angles* and *perpendicular*. He uses the term *at right angles* when the straight line is drawn from a point *in* another, as in I. 11; and he uses the term *perpendicular* when the straight line is drawn from a point *without* another, as in I. 12. This distinction however is often disregarded by modern writers.

I. 14. Here Euclid first requires his eleventh axiom. For

in the demonstration we have the angles ABC and ABE equal to two right angles, and also the angles ABC and ABD equal to two right angles; and then the former two right angles are equal to the latter two right angles by the aid of the eleventh axiom. Many modern editions of Euclid however refer *only* to the first axiom, as if that alone were sufficient; a similar remark applies to the demonstrations of I. 15, and I. 24. In these cases we have omitted the reference purposely, in order to avoid perplexing a beginner; but when his attention is thus drawn to the circumstance he will see that the first and eleventh axioms are both used.

We may observe that errors, in the references with respect to the eleventh axiom, occur in other places in many modern editions of Euclid. Thus for example in III. 1, at the step "therefore the angle FDB is equal to the angle GDB," a reference is given to the first axiom *instead* of to the eleventh.

There seems no objection on Euclid's principles to the following *demonstration* of his eleventh axiom.

Let AB be at right angles to DAC at the point A, and EF at right angles to HEG at the point E: then shall the angles BAC and FEG be equal.

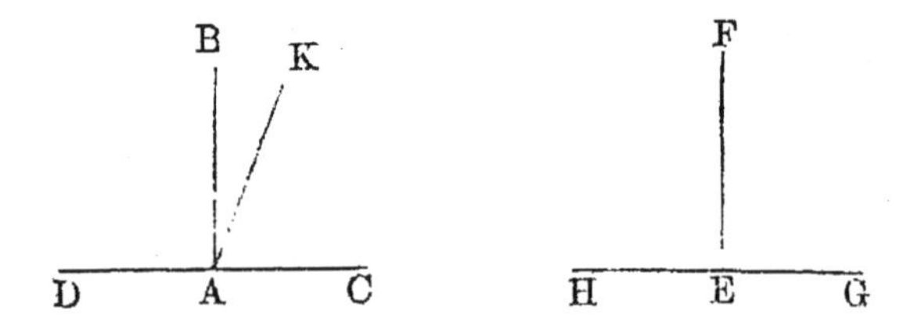

Take any length AC, and make AD, EH, EG all equal to AC. Now apply HEG to DAC, so that H may be on D, and HG on DC, and B and F on the same side of DC; then G will coincide with C, and E with A. Also EF shall coincide with AB; for if not, suppose, if possible, that it takes a different position as AK. Then the angle DAK is equal to the angle HEF, and the angle CAK to the angle GEF; but the angles HEF and GEF are equal, by hypothesis; therefore the angles DAK and CAK are equal. But the angles DAB and CAB are also equal, by hypothesis; and the angle CAB is greater than the angle CAK; therefore the angle DAB is greater than the angle CAK. Much more then is the angle DAK greater than the angle CAK. But the angle DAK was shewn to be equal to the angle CAK; which is absurd. Therefore EF must coincide with AB; and therefore the angle FEG coincides with the angle BAC, and is equal to it.

I. 18, I. 19. In order to assist the student in remembering which of these two propositions is demonstrated directly and which indirectly, it may be observed that the order is similar to that in I. 5 and I. 6.

I. 20. "Proclus, in his commentary, relates, that the Epicureans derided Prop. 20, as being manifest even to asses, and needing no demonstration; and his answer is, that though the truth of it be manifest to our senses, yet it is science which must give the reason why two sides of a triangle are greater than the third: but the right answer to this objection against this and the 21st, and some other plain propositions, is, that the number of axioms ought not to be increased without necessity, as it must be if these propositions be not demonstrated." *Simson.*

I. 21. Here it must be carefully observed that the two straight lines are to be drawn *from the ends of the side* of the triangle. If this condition be omitted the two straight lines will not necessarily be less than two sides of the triangle.

I. 22. "Some authors blame Euclid because he does not demonstrate that the two circles made use of in the construction of this problem must cut one another: but this is very plain from the determination he has given, namely, that any two of the straight lines DF, FG, GH, must be greater than the third. For who is so dull, though only beginning to learn the Elements, as not to perceive that the circle described from the centre F, at the distance FD, must meet FH betwixt F and H, because FD is less than FH; and that for the like reason, the circle described from the centre G, at the distance GH...must meet DG betwixt D and G; and that these circles must meet one another, because FD and GH are together greater than FG?" *Simson.*

The condition that B and C are greater than A, ensures that the circle described from the centre G shall not fall entirely within the circle described from the centre F; the condition that A and B are greater than C, ensures that the circle described from the centre F shall not fall entirely within the circle described from the centre G; the condition that A and C are greater than B, ensures that one of these circles shall not fall entirely without the other. Hence the circles must meet. It is easy to *see* this as Simson says, but there is something arbitrary in Euclid's selection of what is to be *demonstrated* and what is to be *seen*, and Simson's language suggests that he was really conscious of this.

I. 24. In the construction, the condition that DE is to be the side which is not greater than the other, was added by Simson; unless this condition be added there will be *three* cases to consider, for F may fall *on* EG, or *above* EG, or *below* EG. It may be objected that even if Simson's condition be added, it ought to be shewn that F will fall *below* EG. Simson accordingly says "...it is very easy to perceive, that DG being equal to DF, the point G is in the circumference of a circle described from the centre D at the distance DF, and must be in that part of it which is above the straight line EF, because DG falls above DF, the angle EDG being greater than the angle EDF." Or we may shew it in the following manner. Let H denote the point of intersection of DF and EG. Then, the angle DHG is greater than the angle DEG, by I. 16; the angle DEG is not less than the angle DGE, by I. 19; therefore the angle DHG is greater than the angle DGH. Therefore DH is less than DG, by I. 20. Therefore DH is less than DF.

If Simson's condition be omitted, we shall have two other cases to consider besides that in Euclid. If F falls *on* EG, it is obvious that EF is less than EG. If F falls *above* EG, the sum of DF and EF is less than the sum of DG and EG, by I. 21; and therefore EF is less than EG.

I. 26. It will appear after I. 32 that two triangles which have two angles of the one equal to two angles of the other, each to each, have also their third angles equal. Hence we are able to include the two cases of I. 26 in one enunciation thus; *if two triangles have all the angles of the one respectively equal to all the angles of the other, each to each, and have also a side of the one, opposite to any angle, equal to the side opposite to the equal angle in the other, the triangles shall be equal in all respects.*

The first twenty-six propositions constitute a distinct section of the first Book of the *Elements*. The principal results are those contained in Propositions 4, 8, and 26; in each of these Propositions it is shewn that two triangles which agree in three respects agree entirely. There are two other cases which will naturally occur to a student to consider besides those in Euclid; namely, (1) when two triangles have the three angles of the one respectively equal to the three angles of the other, (2) when two triangles have two sides of the one equal to two sides of the other, each to each, and an angle opposite to one side of one triangle equal to the angle opposite to the equal side of the other triangle. In the first of these two cases the student will easily see, after reading I. 29, that the two triangles are not necessarily equal. In the second case also the triangles are not necessarily equal, as may be shewn by an example; in the figure of I. 11, suppose the straight line *FB* drawn; then in the two triangles *FBE*, *FBD*, the side *FB* and the angle *FBC* are common, and the side *FE* is equal to the side *FD*, but the triangles are not equal in all respects. In certain cases, however, the triangles will be equal in all respects, as will be seen from a proposition which we shall now demonstrate.

If two triangles have two sides of the one equal to two sides of the other, each to each, and the angles opposite to a pair of equal sides equal; then if the angles opposite to the other pair of equal sides be both acute, or both obtuse, or if one of them be a right angle, the two triangles are equal in all respects.

Let *ABC* and *DEF* be two triangles; let *AB* be equal to *DE*, and *BC* equal to *EF*, and the angle *A* equal to the angle *D*.

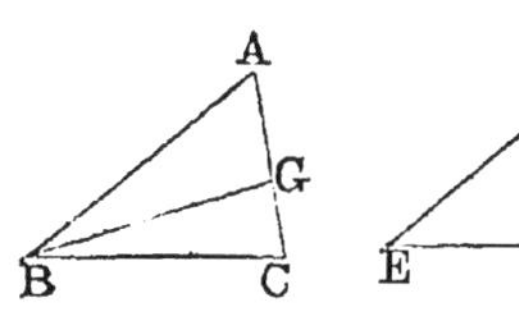

First, suppose the angles *C* and *F* acute angles.

If the angle *B* be equal to the angle *E*, the triangles *ABC*, *DEF* are equal in all respects, by I. 4. If the angle *B* be not equal to the angle *E*, one of them must be greater than the other; suppose the angle *B* greater than the angle *E*, and make the angle *ABG* equal to the angle *E*. Then the triangles *ABG*, *DEF* are equal in all respects, by I. 26; therefore *BG* is equal to *EF*, and the angle *BGA* is equal to the angle *EFD*. But the angle *EFD* is acute, by hypothesis; therefore the angle *BGA* is acute. Therefore the angle *BGC* is obtuse, by I. 13. But it has been shewn that BG is equal to EF; and EF is equal to BC, by hypothesis; therefore BG is equal to BC. Therefore the angle BGC is equal to the angle BCG, by I. 5; and the angle BCG is acute, by hypothesis; therefore the angle BGC is acute. But BGC was shewn to be obtuse; which is absurd. Therefore the angles ABC, DEF are not unequal; that is, they are equal. Therefore the triangles ABC, DEF are equal in all respects, by I. 4.

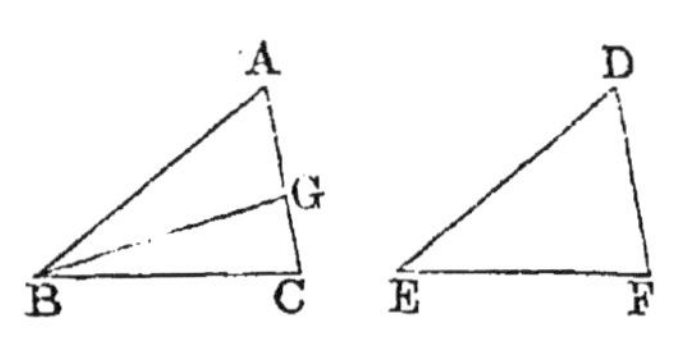

Next, suppose the angles at C and F obtuse angles. The demonstration is similar to the above.

Lastly, suppose one of the angles a right angle, namely, the angle C. If the angle B be not equal to the angle E, make the

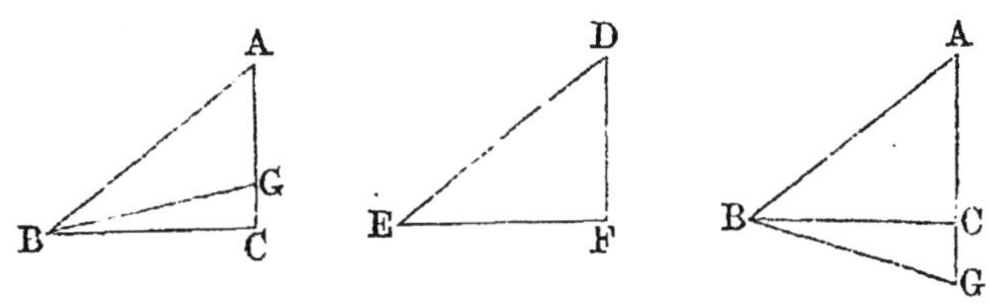

angle ABG equal to the angle E. Then it may be shewn, as before, that BG is equal to BC, and therefore the angle BGC is equal to the angle BCG, that is, equal to a right angle. Therefore two angles of the triangle BGC are equal to two right angles; which is impossible, by I. 17. Therefore the angles ABC and DEF are not unequal; that is, they are equal. Therefore the triangles ABC, DEF are equal in all respects, by I. 4.

If the angles A and D are both right angles, or both obtuse, the angles C and F must be both acute, by I. 17. If AB is less than BC, and DE less than EF, the angles at C and F must be both acute, by I. 18 and I. 17.

The propositions from I. 28 to I. 34 inclusive may be said to constitute the second section of the first Book of the *Elements*. They relate to the theory of parallel straight lines. In I. 29 Euclid uses for the first time his twelfth axiom. The theory of parallel straight lines has always been considered the great difficulty of elementary geometry, and many attempts have been made to overcome this difficulty in a better way than Euclid has done. We shall not give an account of these attempts. The student who wishes to examine them may consult Camerer's *Euclid*, Gergonne's *Annales de Mathématiques*, Volumes XV and XVI, the work by Colonel Perronet Thompson entitled *Geometry without Axioms*, the article *Parallels* in the *English Cyclopædia*, a memoir by Professor Baden Powell in the second volume of the *Memoirs of the Ashmolean Society*, an article by M. Bouniakofsky in the *Bulletin de l'Académie Impériale*, Volume V, St Pétersbourg, 1863, articles in the volumes of the *Philosophical Magazine* for 1856 and 1857, and a dissertation entitled *Sur un point de l'histoire de la Géométrie chez les Grecs......par A. J. H. Vincent. Paris*, 1857.

Speaking generally it may be said that the methods which differ substantially from Euclid's involve, in the first place an axiom as difficult as his, and then an intricate series of propositions; while in Euclid's method after the axiom is once admitted the remaining process is simple and clear.

One modification of Euclid's axiom has been proposed, which appears to diminish the difficulty of the subject. This consists in assuming instead of Euclid's axiom the following; *two intersecting straight lines cannot be both parallel to a third straight line.* The propositions in the *Elements* are then demonstrated as in Euclid up to I. 28, inclusive. Then, in I. 29, we proceed with Euclid up to the words, "therefore the angles BGH, GHD are less than two right angles." We then infer that BGH and GHD must meet: because if a straight line be drawn through G so as to make the interior angles together equal to two right angles this straight line will be parallel to CD, by I. 28; and, by our axiom, there cannot be two parallels to CD, both passing through G.

This form of making the necessary assumption has been recommended by various eminent mathematicians, among whom may be mentioned Playfair and De Morgan. By postponing the consideration of the axiom until it is wanted, that is, until after I. 28, and then presenting it in the form here given, the theory of parallel straight lines appears to be treated in the easiest manner that has hitherto been proposed.

I. 30. Here we may in the same way shew that if AB and EF are each of them parallel to CD, they are parallel to each other. It has been said that the case considered in the text is so obvious as to need no demonstration; for if AB and CD can never meet EF, which lies between them, they cannot meet one another.

I. 32. The corollaries to I. 32 were added by Simson. In the second corollary it ought to be stated what is meant by an *exterior* angle of a rectilineal figure. At each angular point let *one* of the sides meeting at that point be produced; then the exterior angle at that point is the angle contained between this produced part and the side which is not produced. *Either* of the sides may be produced, for the two angles which can thus be obtained are equal, by I. 15.

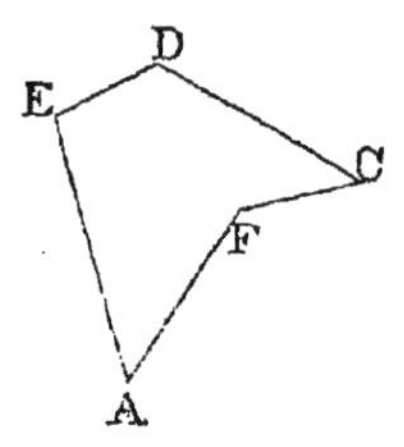

The rectilineal figures to which Euclid confines himself are those in which the angles all face inwards; we may here however notice another class of figures. In the accompanying diagram the angle AFC faces outwards, and it is an angle less than two right angles; this angle however is not one of the interior angles of the figure $AEDCF$. We may consider the corresponding interior angle to be the excess of four right angles above the angle AFC; such an angle, greater than two right angles, is called a *re-entrant* angle.

The *first* of the corollaries to I. 32 is true for a figure which has a re-entrant angle or re-entrant angles; but the *second* is not.

I. 32. If two triangles have two angles of the one equal to two angles of the other each to each they shall also have their third angles equal. This is a very important result, which is often required in the *Elements*. The student should notice how this result is established on Euclid's principles. By Axioms 11 and 2 one pair of right angles is equal to any other pair of right angles. Then, by I. 32, the three angles of one triangle are together equal to the three angles of any other triangle. Then, by Axiom 2, the sum of the two angles of one triangle is equal to the sum of the two equal angles of the other; and then, by Axiom 3, the third angles are equal.

After I. 32 we can draw a straight line at right angles to a given straight line from its extremity, without producing the given straight line.

Let AB be the given straight line. It is required to draw from A a straight line at right angles to AB.

On AB describe the equilateral triangle ABC. Produce BC to D, so that CD may be equal to CB. Join AD. Then AD shall be at right angles to AB. For, the angle CAD is equal to the angle CDA, and the angle CAB is equal to the angle CBA, by I. 5. Therefore the angle BAD is equal to the two angles ABD, BDA, by Axiom 2. Therefore the angle BAD is a right angle, by I. 32.

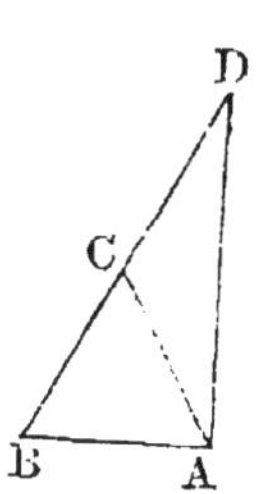

The propositions from I. 35 to I. 48 inclusive may be said to constitute the third section of the first Book of the *Elements.* They relate to equality of area in figures which are not necessarily identical in form.

I. 35. Here Simson has altered the demonstration given by Euclid, because, as he says, there would be three cases to consider in following Euclid's method. Simson however uses the third Axiom in a peculiar manner, when he first takes a triangle from a trapezium, and then another triangle from the *same* trapezium, and infers that the remainders are equal. If the demonstration is to be conducted strictly after Euclid's manner, three cases must be made, by dividing the latter part of the demonstration into two. In the left-hand figure we may suppose the point of intersection of BE and DC to be denoted by G. Then, the triangle ABE is equal to the triangle DCF; take away the triangle DGE from each; then the figure $ABGD$ is equal to the figure $EGCF$; add the triangle GBC to each; then the parallelogram $ABCD$ is equal to the parallelogram $EBCF$. In the right-hand figure we have the triangle AEB equal to the triangle DFC; add the figure $BEDC$ to each; then the parallelogram $ABCD$ is equal to the parallelogram $EBCF$.

The equality of the parallelograms in I. 35 is an equality of area, and not an identity of figure. Legendre proposed to use the word *equivalent* to express the equality of area, and to restrict the word *equal* to the case in which magnitudes admit of superposition and coincidence. This distinction, however, has not been generally adopted, probably because there are few cases in which any ambiguity can arise; in such cases we may say especially, *equal in area,* to prevent misconception.

Cresswell, in his *Treatise of Geometry,* has given a demonstration of I. 35 which shews that the parallelograms may be divided into pairs of pieces admitting of superposition and coincidence; see also his Preface, page x.

I. 38. An important case of I. 38 is that in which the triangles are on equal bases and have a *common* vertex.

I. 40. We may demonstrate I. 40 without adopting the indirect method. Join *BD*, *CD*. The triangles *DBC* and *DEF* are equal, by I. 38; the triangles *ABC* and *DEF* are equal, by hypothesis; therefore the triangles *DBC* and *ABC* are equal, by the first Axiom. Therefore *AD* is parallel to *BC*, by I. 39. *Philosophical Magazine*, October 1850.

I. 44. In I. 44, Euclid does not shew that *AH* and *FG* will meet. "I cannot help being of opinion that the construction would have been more in Euclid's manner if he had made *GH* equal to *BA* and then joining *HA* had proved that *HA* was parallel to *GB* by the thirty-third proposition." *Williamson.*

I. 47. Tradition ascribed the discovery of I. 47 to Pythagoras. Many demonstrations have been given of this celebrated proposition; the following is one of the most interesting.

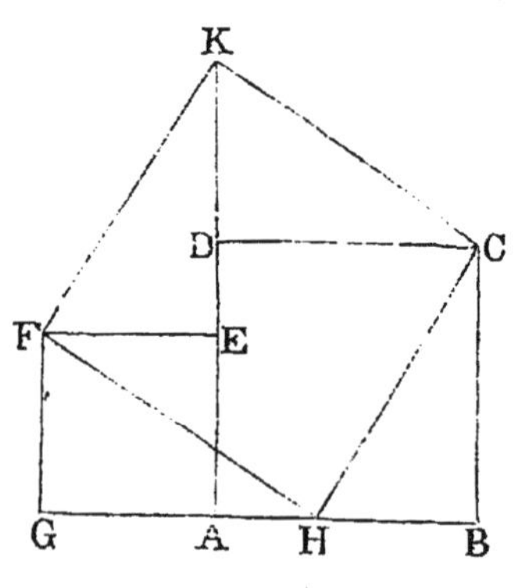

Let *ABCD*, *AEFG* be any two squares, placed so that their bases may join and form one straight line. Take *GH* and *EK* each equal to *AB*, and join *HC*, *CK*, *KF*, *FH*.

Then it may be shewn that the triangle *HBC* is equal in all respects to the triangle *FEK*, and the triangle *KDC* to the triangle *FGH*. Therefore the two squares are together equivalent to the figure *CKFH*. It may then be shewn, with the aid of I. 32, that the figure *CKFH* is a square. And the side *CH* is the hypotenuse of a right-angled triangle of which the sides *CB*, *BH* are equal to the sides of the two given squares. This demonstration requires no proposition of Euclid after I. 32, and it shews how two given squares may be cut into pieces which will fit together so as to form a third square. *Quarterly Journal of Mathematics*, Vol. I.

A large number of demonstrations of this proposition are collected in a dissertation by Joh. Jos. Ign. Hoffmann, entitled *Der Pythagorische Lehrsatz....Zweyte...Ausgabe. Mainz.* 1821.

THE SECOND BOOK.

The second book is devoted to the investigation of relations between the rectangles contained by straight lines divided into segments in various ways.

When a straight line is divided into two parts, each part is called a segment by Euclid. It is found convenient to extend the meaning of the word *segment*, and to lay down the following definition. When a point is taken in a straight line, or in the straight line produced, the distances of the point from the ends of the straight line are called segments of the straight line. When it is necessary to distinguish them, such segments are called *internal* or *external*, according as the point is in the straight line, or in the straight line produced.

The student cannot fail to notice that there is an analogy between the first ten propositions of this book and some elementary facts in Arithmetic and Algebra.

Let $ABCD$ represent a rectangle which is 4 inches long and 3 inches broad. Then, by drawing straight lines parallel to the sides, the figure may be divided into 12 squares, each square being described on a side which represents an inch in length. A square described on a side measuring an inch is called, for shortness, a *square inch.* Thus if a rectangle is 4 inches long and 3 inches broad it may be divided into 12 square inches; this is expressed by saying, that its area is equal to 12 square inches, or, more briefly, that it contains 12 square inches. And a similar result is easily seen to hold in all similar cases. Suppose, for example, that a rectangle is 12 feet long and 7 feet broad; then its area is equal to 12 times 7 square feet, that is to 84 square feet; this may be expressed briefly in common language thus; if a rectangle measures 12 feet by 7 it contains 84 square feet. It must be carefully observed that the sides of the rectangle are supposed to be measured by the same unit of length. Thus if a rectangle is a yard in length, and a foot and a half in breadth, we

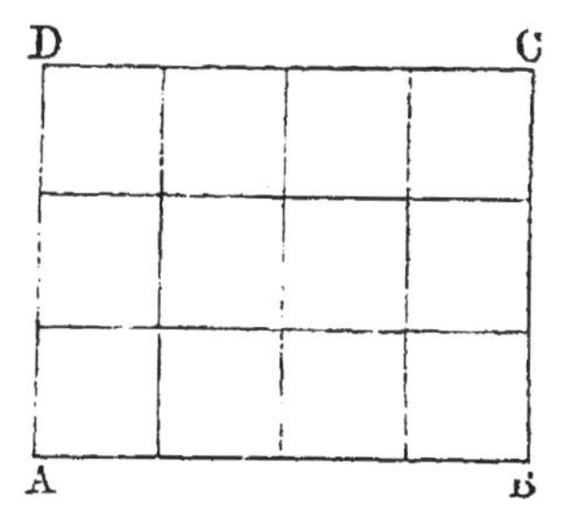

must express each of these dimensions in terms of the same unit; we may say that the rectangle measures 36 inches by 18 inches, and contains 36 times 18 square inches, that is, 648 square inches.

Thus universally, if one side of a rectangle contain a unit of length an exact number of times, and if an adjacent side of the rectangle also contain the same unit of length an exact number of times, the product of these numbers will be the number of square units contained in the rectangle.

Next suppose we have a *square*, and let its side be 5 inches in length. Then, by our rule, the area of the square is 5 times 5 square inches, that is 25 square inches. Now the number 25 is called in Arithmetic the square of the number 5. And universally, if a straight line contain a unit of length an exact number of times, the area of the square described *on* the straight line is denoted by the square *of* the number which denotes the length of the straight line.

Thus we see that there is in general a connexion between the product of two numbers and the rectangle contained by two straight lines, and in particular a connexion between the square *of* a number and the square *on* a straight line; and in consequence of this connexion the first ten propositions in Euclid's Second Book correspond to propositions in Arithmetic and Algebra.

The student will perceive that we speak of the square described *on* a straight line, when we refer to the geometrical figure, and of the square *of* a number when we refer to Arithmetic. The editors of Euclid generally use the words "square described *upon*" in I. 47 and I. 48, and afterwards speak of the square *of* a straight line. Euclid himself retains throughout the same form of expression, and we have imitated him.

Some editors of Euclid have added Arithmetical or Algebraical demonstrations of the propositions in the second book, founded on the connexion we have explained. We have thought it unnecessary to do this, because the student who is acquainted with the elements of Arithmetic and Algebra will find no difficulty in supplying such demonstrations himself, so far as they are usually given. We say *so far as they are usually given*, because these demonstrations usually imply that the sides of rectangles can always be expressed *exactly* in terms of some unit of length; whereas the student will find hereafter that this is not the case, owing to the existence of what are technically called *incommensurable* magnitudes. We do not enter on this subject, as it would lead us too far from Euclid's *Elements of Geometry*, with which we are here occupied.

The first ten propositions in the second book of Euclid may be arranged and enunciated in various ways; we will briefly indicate this, but we do not consider it of any importance to distract the attention of a beginner with these diversities.

II. 2 and II. 3 are particular cases of II. 1.

II. 4 is very important; the following particular case of it should be noticed; *the square described on a straight line made up of two equal straight lines is equal to four times the square described on one of the two equal straight lines.*

II. 5 and II. 6 may be included in one enunciation thus; *the rectangle under the sum and difference of two straight lines is equal to the difference of the squares described on those straight lines;* or thus, *the rectangle contained by two straight lines together with the square described on half their difference, is equal to the square described on half their sum.*

II. 7 may be enunciated thus; *the square described on a straight line which is the difference of two other straight lines is less than the sum of the squares described on those straight lines by twice the rectangle contained by those straight lines.* Then from this and II. 4, and the second Axiom, we infer that *the square described on the sum of two straight lines, and the square described on their difference, are together double of the sum of the squares described on the straight lines;* and this enunciation includes both II. 9 and II. 10, so that the demonstrations given of these propositions by Euclid might be superseded.

II. 8 coincides with the second form of enunciation which we have given to II. 5 and II. 6, bearing in mind the particular case of II. 4 which we have noticed.

II. 11. When the student is acquainted with the elements of Algebra he should notice that II. 11 gives a geometrical construction for the solution of a particular quadratic equation.

II. 12, II. 13. These are interesting in connexion with I. 47; and, as the student may see hereafter, they are of great importance in Trigonometry; they are however not required in any of the parts of Euclid's *Elements* which are usually read. The converse of I. 47 is proved in I. 48; and we can easily shew that converses of II. 12 and II. 13 are true.

Take the following, which is the converse of II. 12; *if the square described on one side of a triangle be greater than the sum of the squares described on the other two sides, the angle opposite to the first side is obtuse.*

For the angle cannot be a right angle, since the square described on the first side would then be equal to the sum of the squares described on the other two sides, by I. 47; and the angle cannot be acute, since the square described on the first side would then be less than the sum of the squares described on the other two sides, by II. 13; therefore the angle must be obtuse.

Similarly we may demonstrate the following, which is the converse of II. 13; *if the square described on one side of a triangle be less than the sum of the squares described on the other two sides, the angle opposite to the first side is acute.*

II. 13. Euclid enunciates II. 13 thus; *in acute-angled triangles*, &c.; and he gives only the first case in the demonstration. But, as Simson observes, the proposition holds for any triangle; and accordingly Simson supplies the second and third cases. It has, however, been often noticed that the same demonstration is applicable to the first and second cases; and it would be a great improvement as to brevity and clearness to take these two cases together. Then the whole demonstration will be as follows.

Let ABC be any triangle, and the angle at B one of its acute angles; and, if AC be not perpendicular to BC, let fall on BC, produced if necessary, the perpendicular AD from the opposite angle: the square on AC opposite to the angle B, shall be less than the squares on CB, BA, by twice the rectangle CB, BD.

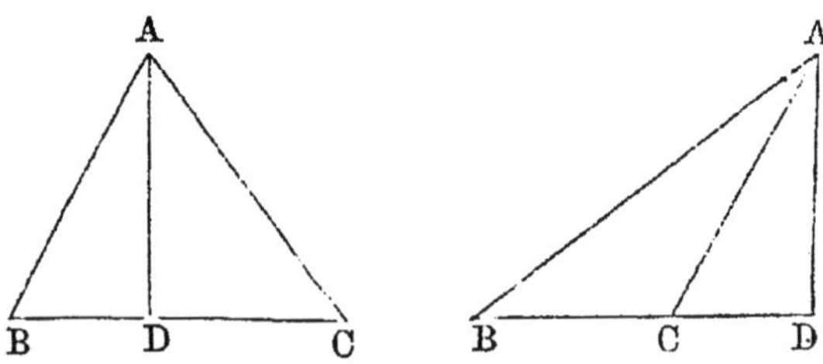

First, suppose AC not perpendicular to BC.

The squares on CB, BD are equal to twice the rectangle CB, BD, together with the square on CD. [II. 7.

To each of these equals add the square on DA.

Therefore the squares on CB, BD, DA are equal to twice the rectangle CB, BD, together with the squares on CD, DA.

But the square on AB is equal to the squares on BD, DA,

and the square on AC is equal to the squares on CD, DA, because the angle BDA is a right angle. [I. 47.
Therefore the squares on CB, BA are equal to the square on AC, together with twice the rectangle CB, BD;
that is, the square on AC alone is less than the squares on CB, BA, by twice the rectangle CB, BD.

Next, suppose AC perpendicular to BC.
Then BC is the straight line intercepted between the perpendicular and the acute angle at B. 
And the square on AB is equal to the squares on AC, CB. [I. 47.
Therefore the square on AC is less than the squares on AB, BC, by twice the square on BC.

A

B C

II. 14. This is not required in any of the parts of Euclid's *Elements* which are usually read; it is included in VI. 22.

THE THIRD BOOK.

THE third book of the Elements is devoted to properties of circles.

Different opinions have been held as to what is, or should be, included in the third definition of the third book. One opinion is that the definition only means that the circles do not cut in the *neighbourhood of the point of contact*, and that it must be shewn that they do not cut elsewhere. Another opinion is that the definition means that the circles do not cut at all; and this seems the correct opinion. The definition may therefore be presented more distinctly thus. Two circles are said to touch internally when their circumferences have one or more common points, and when every point in one circle is within the other circle, except the common point or points. Two circles are said to touch externally when their circumferences have one or more common points, and when every point in each circle is without the other circle, except the common point or points. It is then shewn in the third Book that the circumferences of two circles which touch can have only *one* common point.

A straight line which touches a circle is often called a *tangent* to the circle, or briefly, a *tangent*.

It is very convenient to have a word to denote a portion of

the boundary of a circle, and accordingly we use the word *arc*. Euclid himself uses *circumference* both for the whole boundary and for a portion of it.

III. 1. In the construction, DC is said to be *produced* to E; this assumes that D is within the circle, which Euclid demonstrates in III. 2.

III. 3. This consists of two parts, each of which is the converse of the other; and the whole proposition is the converse of the corollary in III. 1.

III. 5 and III. 6 should have been taken together. They amount to this, *if the circumferences of two circles meet at a point they cannot have the same centre*, so that circles which have the same centre and one point in their circumferences common, must coincide altogether. It would seem as if Euclid had made three cases, one in which the circles cut, one in which they touch internally, and one in which they touch externally, and had then omitted the last case as evident.

III. 7, III. 8. It is observed by Professor De Morgan that in III. 7 it is *assumed* that the angle FEB is greater than the angle FEC, the hypothesis being only that the angle DFB is greater than the angle DFC; and that in III. 8 it is *assumed* that K falls within the triangle DLM, and E without the triangle DMF. He intimates that these assumptions may be established by means of the following two propositions which may be given in order after I. 21.

The perpendicular is the shortest straight line which can be drawn from a given point to a given straight line; and of others that which is nearer to the perpendicular is less than the more remote, and the converse; and not more than two equal straight lines can be drawn from the given point to the given straight line, one on each side of the perpendicular.

Every straight line drawn from the vertex of a triangle to the base is less than the greater of the two sides, or than either of them if they be equal.

The following proposition is analogous to III. 7 and III. 8.

If any point be taken on the circumference of a circle, of all the straight lines which can be drawn from it to the circumference, the greatest is that in which the centre is; and of any others, that which is nearer to the straight line which passes through the centre is always greater than one more remote; and from the same point there can be drawn to the circumference two straight lines, and only two, which are equal to one another, one on each side of the greatest line.

The first two parts of this proposition are contained in III. 15; all three parts might be demonstrated in the manner of III. 7, and they should be demonstrated, for the third part is really required, as we shall see in the note on III. 10.

III. 9. The point E might be supposed to fall *within* the angle ADC. It cannot then be shewn that DC is greater than DB, and DB greater than DA, but only that either DC or DA is less than DB; this however is sufficient for establishing the proposition.

Euclid has given two demonstrations of III. 9, of which Simson has chosen the second. Euclid's other demonstration is as follows. Join D with the middle point of the straight line AB; then it may be shewn that this straight line is at right angles to AB; and therefore the centre of the circle must lie in this straight line, by III. 1, Corollary. In the same manner it may be shewn that the centre of the circle must lie in the straight line which joins D with the middle point of the straight line BC. The centre of the circle must therefore be at D, because two straight lines cannot have more than one common point.

III. 10. Euclid has given two demonstrations of III. 10, of which Simson has chosen the second. Euclid's first demonstration resembles his first demonstration of III. 9. He shews that the centre of each circle is on the straight line which joins K with the middle point of the straight line BG, and also on the straight line which joins K with the middle point of the straight line BH; therefore K must be the centre of each circle.

The demonstration which Simson has chosen requires some additions to make it complete. For the point K might be supposed to fall *without* the circle DEF, or *on* its circumference, or *within* it; and of these three suppositions Euclid only considers the last. If the point K be supposed to fall *without* the circle DEF we obtain a contradiction of III. 8; which is absurd. If the point K be supposed to fall *on* the circumference of the circle DEF we obtain a contradiction of the proposition which we have enunciated at the end of the note on III. 7 and III. 8; which is absurd.

What is demonstrated in III. 10 is that the circumferences of two circles cannot have more than *two* common points; there is nothing in the demonstration which assumes that the circles *cut* one another, but the enunciation refers to this case only because it is shewn in III. 13 that if two circles *touch* one another, their circumferences cannot have more than *one* common point.

III. 11, III. 12. The enunciations as given by Simson and others speak of *the* point of contact; it is however not shewn until III. 13 that there is only *one* point of contact. It should be observed that the demonstration in III. 11 will hold even if D and H be supposed to coincide, and that the demonstration in III. 12 will hold even if C and D be supposed to coincide. We may combine III. 11 and III. 12 in one enunciation thus.

If two circles touch one another their circumferences cannot have a common point out of the direction of the straight line which joins the centres.

III. 13 may be deduced from III. 7. For GH is the least line that can be drawn from G to the circumference of the circle whose centre is F, by III. 7. Therefore GH is less than GA, that is, less than GD; which is absurd. Similarly III. 12 may be deduced from III. 8.

III. 13. Simson observes, "As it is much easier to imagine that two circles may touch one another within in more points than one, upon the same side, than upon opposite sides, the figure of that case ought not to have been omitted; but the construction in the Greek text would not have suited with this figure so well, because the centres of the circles must have been placed near to the circumferences; on which account another construction and demonstration is given, which is the same with the second part of that which Campanus has translated from the Arabic, where, without any reason, the demonstration is divided into two parts."

It would not be obvious from this note which figure Simson himself supplied, because it is uncertain what he means by the "same side" and "opposite sides." It is the left-hand figure in the first part of the demonstration. Euclid, however, seems to be quite correct in omitting this figure, because he has shewn in III. 11 that if two circles touch internally there cannot be a point of contact out of the direction of the straight line which joins the centres. Thus, in order to shew that there is only *one* point of contact, it is sufficient to put the second supposed point of contact on the direction of the straight line which joins the centres. Accordingly in his own demonstration Euclid confines himself to the right-hand figure; and he shews that this case cannot exist, because the straight line BD would be a diameter of both circles, and would therefore be bisected at two different points; which is absurd.

Euclid might have used a similar method for the second part of the proposition; for as there cannot be a point of contact out of the straight line joining the centres, *it is obviously impossible* that there can be a second point of contact when the circles touch externally. It is easy to *see* this; but Euclid preferred a method in which there is more formal reasoning.

We may observe that Euclid's mode of dealing with the contact of circles has often been censured by commentators, but apparently not always with good reason. For example, Walker gives another demonstration of III. 13; and says that Euclid's is worth nothing, and that Simson fails; for it is not proved that two circles which touch cannot have any arc common to both circumferences. But it is shewn in III. 10 that this is impossible; Walker appears to have supposed that III. 10 is limited to the case of circles which *cut*. See the note on III. 10.

III. 17. It is obvious from the construction in III. 17 that *two* straight lines can be drawn from a given external point to touch a given circle; and these two straight lines are equal in length and equally inclined to the straight line which joins the given external point with the centre of the given circle.

After reading III. 31 the student will see that the problem in III. 17 may be solved in another way, as follows: describe a circle on AE as diameter; then the points of intersection of this circle with the given circle will be the points of contact of the two straight lines which can be drawn from A to touch the given circle.

III. 18. It does not appear that III. 18 adds anything to what we have already obtained in III. 16. For in III. 16 it is shewn, that there is only one straight line which touches a given circle at a given point, and that the angle between this straight line and the radius drawn to the point of contact is a right angle.

III. 20. There are two assumptions in the demonstration of III. 20. Suppose that A is double of B and C double of D; then in the first part it is assumed that the sum of A and C is double of the sum of B and D, and in the second part it is assumed that the difference of A and C is double of the difference of B and D. The former assumption is a particular case of V. 1, and the latter is a particular case of V. 5.

An important extension may be given to III. 20 by introducing angles greater than two right angles. For, in the first figure, suppose we draw the straight lines BF and CF. Then, the angle BEA is double of the angle BFA, and the angle CEA is double of the angle CFA; therefore the sum of the angles BEA and CEA is double of the angle BFC. The sum of the angles BEA and CEA is greater than two right angles; we will call the sum, the *re-entrant* angle BEC. Thus the re-entrant angle BEC is double of the angle BFC. (See note on I. 32). If this extension be used some of the demonstrations in the third book may be abbreviated. Thus III. 21 may be demonstrated without making two cases; III. 22 will follow immediately from the fact that the sum of the angles at the centre is equal to four right angles; and III. 31 will follow immediately from III. 20.

III. 21. In III. 21 Euclid himself has given only the first case; the second case has been added by Simson and others. In either of the figures of III. 21 if a point be taken on the same side of BD as A, the angle contained by the straight lines which join this point to the extremities of BD is *greater* or *less* than the angle BAD, according as the point is *within* or *without* the angle BAD; this follows from I. 21.

We shall have occasion to refer to IV. 5 in some of the remaining notes to the third Book; and the student is accordingly recommended to read that proposition at the present stage.

The following proposition is very important. *If any number of triangles be constructed on the same base and on the same side of it, with equal vertical angles, the vertices will all lie on the circumference of a segment of a circle.*

For take any one of these triangles, and describe a circle round it, by IV. 5; then the vertex of any other of the triangles must be on the circumference of the segment containing the assumed vertex, since, by the former part of this note, the vertex cannot be without the circle or within the circle.

III. 22. The converse of III. 22 is true and very important; namely, *if two opposite angles of a quadrilateral be together equal to two right angles, a circle may be circumscribed about the quadrilateral.* For, let $ABCD$ denote the quadrilateral. Describe a circle round the triangle ABC, by IV. 5. Take any point E, on the circumference of the segment cut off by AC, and on the same side of AC as D is. Then, the angles at B and E are together equal to two right angles, by III. 22; and the angles at B and D are together equal to two right angles, by hypothesis. Therefore the angle at E is equal to the angle at D. Therefore, by the preceding note D is on the circumference of the same segment as E.

III. 32. The converse of III. 32 is true and important; namely, *if a straight line meet a circle, and from the point of meeting a straight line be drawn cutting the circle, and the angle between the two straight lines be equal to the angle in the alternate segment of the circle, the straight line which meets the circle shall touch the circle.*

This may be demonstrated indirectly. For, if possible, suppose that the straight line which meets the circle does not touch it. Draw through the point of meeting a straight line to touch the circle. Then, by III. 32 and the hypothesis, it will follow that two different straight lines pass through the same point, and make the same angle, on the same side, with a third straight line which also passes through that point; but this is impossible.

III. 35, III. 36. The following proposition constitutes a large part of the demonstrations of III. 35 and III. 36. *If any point be taken in the base, or the base produced, of an isosceles triangle, the rectangle contained by the segments of the base is equal to the difference of the square on the straight line joining this point to the vertex and the square on the side of the triangle.*

This proposition is in fact demonstrated by Euclid, without using any property of the circle; if it were enunciated and demonstrated before III. 35 and III. 36 the demonstrations of these two propositions might be shortened and simplified.

The following converse of III. 35 and the Corollary of III. 36 may be noticed. *If two straight lines* AB, CD *intersect at* O, *and the rectangle* AO, OB *be equal to the rectangle* CO, OD, *the circumference of a circle will pass through the four points* A, B, C, D.

For a circle may be described round the triangle ABC, by IV. 5; and then it may be shewn indirectly, by the aid of III. 35 or the Corollary of III. 36 that the circumference of this circle will also pass through D.

THE FOURTH BOOK.

THE fourth Book of the Elements consists entirely of problems. The first five propositions relate to triangles of any kind; the remaining propositions relate to polygons which have all their sides equal and all their angles equal. A polygon which has all its sides equal and all its angles equal is called a *regular* polygon.

IV. 4. By a process similar to that in IV. 4 we can describe a circle which shall touch one side of a triangle and the other two sides produced. Suppose, for example, that we wish to describe a circle which shall touch the side BC, and the sides AB and AC produced: bisect the angle between AB produced and BC, and bisect the angle between AC produced and BC; then the point at which the bisecting straight lines meet will be the centre of the required circle. The demonstration will be similar to that in IV. 4.

A circle which touches one side of a triangle and the other two sides produced, is called an *escribed* circle of the triangle.

We can also describe a triangle equiangular to a given triangle, and such that one of its sides and the other two sides produced shall touch a given circle. For, in the figure of IV. 3 suppose AK produced to meet the circle again; and at the point of intersection draw a straight line touching the circle; this straight line with parts of NB and NC, will form a triangle, which will be equiangular to the triangle MLN, and therefore equiangular to the triangle EDF; and one of the sides of this triangle, and the other two sides produced, will touch the given circle.

IV. 5. Simson introduced into the demonstration of IV. 5 the part which shews that DF and EF will meet. It has also been proposed to shew this in the following way: join DE; then the angles EDF and DEF are together less than the angles ADF and AEF, that is, they are together less than two right angles; and therefore DF and EF will meet, by Axiom 12. This assumes that ADE and AED are *acute* angles; it may however be easily shewn that DE is parallel to BC, so that the triangle ADE is equiangular to the triangle ABC; and we must therefore select the two sides AB and AC such that ABC and ACB may be acute angles.

IV. 10. The vertical angle of the triangle in IV. 10 is easily seen to be the fifth part of two right angles; and as it may be bisected, we can thus divide a right angle geometrically into five equal parts.

It follows from what is given in the fourth Book of the Elements that the circumference of a circle can be divided into 3, 6, 12, 24, ... equal parts; and also into 4, 8, 16, 32, ... equal parts; and also into 5, 10, 20, 40, ... equal parts; and also into 15, 30, 60, 120, ... equal parts. Hence also regular polygons having as many sides as any of these numbers may be inscribed in a circle, or described about a circle. This however does not enable us to describe a regular polygon of any assigned number of sides; for example, we do not know how to describe geometrically a regular polygon of 7 sides.

It was first demonstrated by Gauss in 1801, in his *Disquisitiones Arithmeticæ*, that it is possible to describe geometrically a regular polygon of 2^n+1 sides, provided 2^n+1 be a prime number; the demonstration is not of an elementary character. As an example, it follows that a regular polygon of 17 sides can be described geometrically; this example is discussed in Catalan's *Théorèmes et Problèmes de Géométrie Elémentaire.*

For an approximate construction of a regular heptagon see the *Philosophical Magazine* for February and for April, 1864.

THE FIFTH BOOK.

THE fifth Book of the Elements is on *Proportion*. Much has been written respecting Euclid's treatment of this subject; besides the Commentaries on the Elements to which we have already referred, the student may consult the articles *Ratio* and *Proportion* in the *English Cyclopædia*, and the tract on the *Connexion of Number and Magnitude* by Professor De Morgan.

The fifth Book relates not merely to length and space, but to any kind of magnitude of which we can form multiples.

V. *Def.* 1. The word *part* is used in two senses in Geometry. Sometimes the word denotes any magnitude which is less than another of the same kind, as in the axiom, *the whole is greater than its part.* In this sense the word has been used up to the present point, but in the fifth Book Euclid confines the word to a more restricted sense. This restricted sense agrees with that which is given in Arithmetic and Algebra to the term *aliquot part*, or to the term *submultiple.*

V. *Def.* 3. Simson considers that the definitions 3 and 8 are "not Euclid's, but added by some unskilful editor." Other commentators also have rejected these definitions as useless. The last word of the third definition should be *quantuplicity,* not *quantity;* so that the definition indicates that ratio refers to the *number of times* which one magnitude contains another. See De Morgan's *Differential and Integral Calculus,* page 18.

V. *Def.* 4. This definition amounts to saying that the quantities must be of the *same kind.*

V. *Def.* 5. The fifth definition is the foundation of Euclid's doctrine of proportion. The student will find in works on Algebra a comparison of Euclid's definition of proportion with the simpler definitions which are employed in Arithmetic and Algebra. Euclid's definition is applicable to *incommensurable* quantities, as well as to *commensurable* quantities.

We should recommend the student to read the first proposition of the sixth Book immediately after the fifth definition of the fifth Book; he will there see how Euclid applies his definition, and will thus obtain a better notion of its meaning and importance.

Compound Ratio. The definition of compound ratio was supplied by Simson. The Greek text does not give any definition of compound ratio here, but gives one as the fifth definition of the sixth Book, which Simson rejects as absurd and useless.

V. *Defs.* 18, 19, 20. The definitions 18, 19, 20 are not presented by Simson precisely as they stand in the original. The last sentence in definition 18 was supplied by Simson. Euclid does not connect definitions 19 and 20 with definition 18. In 19 he defines *ordinate proportion,* and in 20 he defines *perturbate proportion.* Nothing would be lost if Euclid's definition 18 were entirely omitted, and the term *ex æquali* never employed. Euclid employs such a term in the enunciations of V. 20, 21, 22, 23; but it seems quite useless, and is accordingly neglected by Simson and others in their translations.

The axioms given after the definitions of the fifth Book are not in Euclid; they were supplied by Simson.

The propositions of the fifth Book might be divided into four sections. Propositions 1 to 6 relate to the properties of equimultiples. Propositions 7 to 10 and 13 and 14 connect the notion of the *ratio* of magnitudes with the ordinary notions of *greater*, *equal*, and *less*. Propositions 11, 12, 15 and 16 may be considered as introduced to shew that, *if four quantities of the same kind be proportionals they will also be proportionals when taken alternately*. The remaining propositions shew that magnitudes are proportional by *composition*, by *division*, and *ex æquo*.

In this division of the fifth Book propositions 13 and 14 are supposed to be placed immediately after proposition 10; and they might be taken in this order without any change in Euclid's demonstrations.

The propositions headed *A*, *B*, *C*, *D*, *E* were supplied by Simson.

V. 1, 2, 3, 5, 6. These are simple propositions of Arithmetic, though they are here expressed in terms which make them appear less familiar than they really are. For example, V. 1 "states no more than that *ten* acres and *ten* roods make *ten* times as much as one acre and one rood." *De Morgan.*

In V. 5 Simson has substituted another construction for that given by Euclid, because Euclid's construction assumes that we can divide a given straight line into any assigned number of equal parts, and this problem is not solved until VI. 9.

V. 18. This demonstration is Simson's. We will give here Euclid's demonstration.

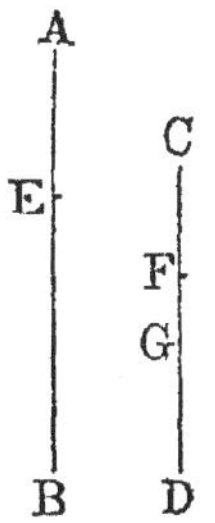

Let *AE* be to *EB* as *CF* is to *FD*: *AB* shall be to *BE* as *CD* is to *DF*.

A E B C F G D

For, if not, *AB* will be to *BE* as *CD* is to some magnitude less than *DF*, or greater than *DF*.

First, suppose that *AB* is to *BE* as *CD* is to *DG*, which is less than *DF*.

Then, because *AB* is to *BE* as *CD* is to *DG*,
therefore *AE* is to *EB* as *CG* is to *GD*. [V. 17.
But *AE* is to *EB* as *CF* is to *FD*, [*Hypothesis.*
therefore *CG* is to *GD* as *CF* is to *FD*. [V. 11.
But *CG* is greater than *CF*; [*Hypothesis.*
therefore *GD* is greater than *FD*. [V. 14.
But *GD* is less than *FD*; which is impossible.

In the same manner it may be shewn that *AB* is not to *BE* as *CD* is to a magnitude greater than *DF*.
Therefore *AB* is to *BE* as *CD* is to *DF*.

The objection urged by Simson against Euclid's demonstration is that "it depends upon this hypothesis, that to any three magnitudes, two of which, at least, are of the same kind, there may be a fourth proportional: . . . . . . Euclid does not demonstrate it, nor does he shew how to find the fourth proportional, before the 12th Proposition of the 6th Book. . . . ."

The following demonstration is given by Austin in his *Examination of the first six books of Euclid's Elements.*

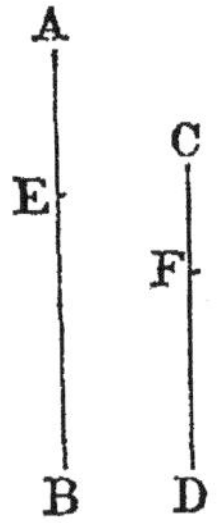

Let AE be to EB as CF is to FD: AB shall be to BE as CD is to DF.

For, because AE is to EB as CF is to FD, therefore, alternately, AE is to CF as EB is to FD. [V. 16.
And as one of the antecedents is to its consequent so is the sum of the antecedents to the sum of the consequents; [V. 12.
therefore as EB is to FD so are AE and EB together to CF and FD together,
that is, AB is to CD as EB is to FD.
Therefore, alternately, AB is to EB as CD is to FD. [V. 16.

A E B C F D

V. 25. The first step in the demonstration of this proposition is "take AG equal to E and CH equal to F"; and here a reference is sometimes given to I. 3. But the magnitudes in the proposition are not necessarily *straight lines*, so that this reference to I. 3 should not be given; it must however be assumed that we can perform on the magnitudes considered, an operation similar to that which is performed on straight lines in I. 3. Since the fifth Book of the Elements treats of magnitudes generally, and not merely of lengths, areas, and angles, there is no reference made in it to any proposition of the first four Books.

Simson adds four propositions relating to compound ratio, which he distinguishes by the letters F, G, H, K; it seems however unnecessary to reproduce them as they are now rarely read and never required.

THE SIXTH BOOK.

THE sixth Book of the Elements consists of the application of the theory of proportion to establish properties of geometrical figures.

VI. *Def.* 1. For an important remark bearing on the first definition, see the note on VI. 5.

VI. *Def.* 2. The second definition is useless, for Euclid makes no mention of reciprocal figures.

VI. *Def.* 4. The fourth definition is strictly only applicable to a triangle, because no other figure has a point which can be exclusively called its *vertex.* The altitude of a parallelogram is the perpendicular drawn to the base from any point in the opposite side.

VI. 2. The enunciation of this important proposition is open to objection, for the manner in which the sides may be cut is not sufficiently limited. Suppose, for example, that AD is double of DB, and CE double of EA; the sides are then cut proportionally, for each side is divided into two parts, one of which is double of the other; but DE is not parallel to BC. It should therefore be stated in the enunciation that *the segments terminated at the vertex of the triangle are to be homologous terms in the ratios, that is, are to be the antecedents or the consequents of the ratios.*

It will be observed that there are three figures corresponding to three cases which may exist; for the straight line drawn parallel to one side may cut the other sides, or may cut the other sides when they are produced through the extremities of the base, or may cut the other sides when they are produced through the vertex. In all these cases the triangles which are shewn to be equal have their vertices at the extremities of the base of the given triangle, and have for their common base the straight line which is, either by hypothesis or by demonstration, parallel to the base of the triangle. The triangle with which these two triangles are compared has the same base as they have, and has its vertex coinciding with the vertex of the given triangle.

VI. *A.* This proposition was supplied by Simson.

VI. 4. We have preferred to adopt the term "triangles which are equiangular to one another," instead of "equiangular triangles," when the words are used in the sense they bear in this proposition. Euclid himself does not use the term *equiangular triangle* in the sense in which the modern editors use it in the Corollary to I. 5, so that he is not prevented from using the term in the sense it bears in the enunciation of VI. 4 and elsewhere; but modern editors, having already employed the term in one sense ought to keep to that sense. In the demonstrations, where Euclid uses such language as "the triangle ABC is equiangular to the triangle DEF," the modern editors sometimes adopt it, and sometimes change it to "the triangles ABC and DEF are equiangular."

In VI. 4 the manner in which the two triangles are to be

placed is very imperfectly described; their bases are to be in the same straight line and contiguous, their vertices are to be on the same side of the base, and each of the two angles which have a common vertex is to be equal to the remote angle of the other triangle.

By superposition we might deduce VI. 4 immediately from VI. 2.

VI. 5. The hypothesis in VI. 5 involves more than is directly asserted; the enunciation should be, "if the sides of two triangles, *taken in order*, about each of their angles . . . . . ;" that is, some restriction equivalent to the words *taken in order* should be introduced. It is quite possible that there should be two triangles ABC, DEF, such that AB is to BC as DE is to EF, and BC to CA as DF is to ED, and therefore, by V. 23, AB to AC as DF is to EF; in this case the sides of the triangles about each of their angles are proportionals, but not in the same order, and the triangles are not necessarily equiangular to one another. For a numerical illustration we may suppose the sides of one triangle to be 3, 4 and 5 feet respectively, and those of another to be 12, 15 and 20 feet respectively. *Walker.*

Each of the two propositions VI. 4 and VI. 5 is the converse of the other. They shew that if two triangles have either of the two properties involved in the definition of similar figures they will have the other also. This is a special property of triangles. In other figures either of the properties may exist alone. For example, any rectangle and a square have their angles equal, but not their sides proportional; while a square and any rhombus have their sides proportional, but not their angles equal.

VI. 7. In VI. 7 the enunciation is imperfect; it should be, "if two triangles have one angle of the one equal to one angle of the other, and the sides about two other angles proportionals, *so that the sides subtending the equal angles are homologous;* then if each . . . . ." The imperfection is of the same nature as that which is pointed out in the note on VI. 5. *Walker.*

The proposition might be conveniently broken up and the essential part of it presented thus: *if two triangles have two sides of the one proportional to two sides of the other, and the angles opposite to one pair of homologous sides equal, the angles which are opposite to the other pair of homologous sides shall either be equal, or be together equal to two right angles.*

For, the angles included by the proportional sides must be

either equal or unequal. If they are equal, then since the triangles have two angles of the one equal to two angles of the other, each to each, they are equiangular to one another. We have therefore only to consider the case in which the angles included by the proportional sides are unequal.

Let the triangles ABC, DEF have the angle at A equal to the angle at D, and AB to BC as DE is to EF, but the angle ABC not equal to the angle DEF: the angles ACB and DFE shall be together equal to two right angles.

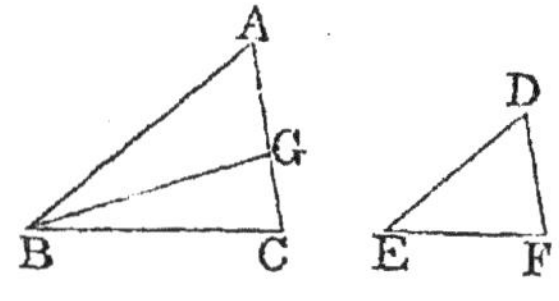

For, one of the angles ABC, DEF must be greater than the other; suppose ABC the greater; and make the angle ABG equal to the angle DEF. Then it may be shewn, as in VI. 7, that BG is equal to BC, and the angle BGA equal to the angle EFD. Therefore the angles ACB and DFE are together equal to the angles BGC and AGB, that is, to two right angles.

Then the results enunciated in VI. 7 will readily follow. For if the angles ACB and DFE are both greater than a right angle, or both less than a right angle, or if one of them be a right angle, they must be equal.

VI. 8. In the demonstration of VI. 8, as given by Simson, it is inferred that two triangles which are similar to a third triangle are similar to each other; this is a particular case of VI. 21, which the student should consult, in order to see the validity of the inference.

VI. 9. The word *part* is here used in the restricted sense of the first definition of the fifth Book. VI. 9 is a particular case of VI. 10.

VI. 10. The most important case of this proposition is that in which a straight line is to be divided either *internally or externally* into two parts which shall be in a given ratio.

The case in which the straight line is to be divided *internally* is given in the text; suppose, for example, that the given ratio is that of AE to EC; then AB is divided at G in the given ratio.

Suppose, however, that AB is to be divided *externally* in a given ratio; that is, suppose that AB is to be produced so that the whole straight line made up of AB and the part produced may be to the part produced in a given ratio. Let the given ratio be that of AC to CE. Join EB; through C draw a straight line parallel to EB; then this straight line will meet AB, produced through B, at the required point.

VI. 11. This is a particular case of VI. 12.

VI. 14. The following is a full exhibition of the steps which lead to the result that FB and BG are in one straight line.

The angle DBF is equal to the angle GBE; [*Hypothesis.*
add to each the angle FBE;
therefore the angles DBF, FBE are together equal to the angles GBE, FBE. [*Axiom* 2.
But the angles DBF, FBE are together equal to two right angles; [I. 13.
therefore the angles GBE, FBE are together equal to two right angles; [*Axiom* 1.
therefore FB and BG are in one straight line. [I. 14.

VI. 15. This may be inferred from VI. 14, since a triangle is half of a parallelogram with the same base and altitude.

It is not difficult to establish a third proposition conversely connected with the two involved in VI. 14, and a third proposition similarly conversely connected with the two involved in VI. 15. These propositions are the following.

Equal parallelograms which have their sides reciprocally proportional, have their angles equal, each to each.

Equal triangles which have the sides about a pair of angles reciprocally proportional, have those angles equal or together equal to two right angles.

We will take the latter proposition.

Let ABC, ADE be equal triangles; and let CA be to AD as AE is to AB: either the angle BAC shall be equal to the angle DAE, or the angles BAC and DAE shall be together equal to two right angles.

[The student can construct the figure for himself.]

Place the triangles so that CA and AD may be in one straight line; then if EA and AB are in one straight line the angle BAC is equal to the angle DAE. [I. 15.
If EA and AB are not in one straight line, produce BA through A to F, so that AF may be equal to AE; join DF and EF.

Then because CA is to AD as AE is to AB, [*Hypothesis.*
and AF is equal to AE, [*Construction.*
therefore CA is to AD as AF is to AB. [V. 9, V. 11.
Therefore the triangle DAF is equal to the triangle BAC. [VI. 15.

But the triangle DAE is equal to the triangle BAC. [*Hypothesis.*
Therefore the triangle DAE is equal to the triangle DAF. [*Ax.* 1.
Therefore EF is parallel to AD. [I. 39.

Suppose now that the angle DAE is greater than the angle DAF.
Then the angle CAE is equal to the angle AEF, [I. 29.
and therefore the angle CAE is equal to the angle AFE, [I. 5.
and therefore the angle CAE is equal to the angle BAC. [I. 29.
Therefore the angles BAC and DAE are together equal to two right angles.

Similarly the proposition may be demonstrated if the angle DAE is less than the angle DAF.

VI. 16. This is a particular case of VI. 14.

VI. 17. This is a particular case of VI. 16.

VI. 22. There is a step in the second part of VI. 22 which requires examination. After it has been shewn that the figure SR is equal to the similar and similarly situated figure NH, it is added "therefore PR is equal to GH." In the Greek text reference is here made to a *lemma* which follows the proposition. The word *lemma* is occasionally used in mathematics to denote an auxiliary proposition. From the unusual circumstance of a reference to something following, Simson probably concluded that the lemma could not be Euclid's, and accordingly he takes no notice of it.

The following is the substance of the lemma.

If PR be not equal to GH, one of them must be greater than the other; suppose PR greater than GH.

Then, because SR and NH are similar figures, PR is to PS as GH is to GN. [VI. *Definition* 1.
But PR is greater than GH, [*Hypothesis.*
therefore PS is greater than GN. [V. 14.
Therefore the triangle RPS is greater than the triangle HGN. [I. 4, *Axiom* 9.
But, because SR and NH are similar figures, the triangle RPS is equal to the triangle HGN; [VI. 20.
which is impossible.
Therefore PR is equal to GH.

VI. 23. In the figure of VI. 23 suppose BD and GE drawn. Then the triangle BCD is to the triangle GCE as the parallelogram AC is to the parallelogram CF. Hence the result may be extended to triangles, and we have the following theorem.

triangles which have one angle of the one equal to one angle of the other, have to one another the ratio which is compounded of the ratios of their sides.

Then VI. 19 is an immediate consequence of this theorem. For let ABC and DEF be similar triangles, so that AB is to BC as DE is to EF; and therefore, alternately, AB is to DE as BC is to EF. Then, by the theorem, the triangle ABC has to the triangle DEF the ratio which is compounded of the ratios of AB to DE and of BC to EF, that is, the ratio which is compounded of the ratios of BC to EF and of BC to EF. And, from the definitions of duplicate ratio and of compound ratio, it follows that the ratio compounded of the ratios of BC to EF and of BC to EF is the duplicate ratio of BC to EF.

VI. 25. It will be easy for the student to exhibit in detail the process of shewing that BC and CF are in one straight line, and also LE and EM; the process is exactly the same as that in I. 45, by which it is shewn that KH and HM are in one straight line, and also FG and GL.

It seems that VI. 25 is out of place, since it separates propositions so closely connected as VI. 24 and VI. 26. We may enunciate VI. 25 in familiar language thus: *to make a figure which shall have the form of one figure and the size of another.*

VI. 26. This proposition is the converse of VI. 24; it might be extended to the case of two similar and similarly situated parallelograms which have a pair of angles *vertically opposite.*

We have omitted in the sixth Book Propositions 27, 28, 29, and the first solution which Euclid gives of Proposition 30, as they appear now to be never required, and have been condemned as useless by various modern commentators; see Austin, Walker, and Lardner. Some idea of the nature of these propositions may be obtained from the following statement of the problem proposed by Euclid in VI. 29. AB is a given straight line; it has to be produced through B to a point O, and a parallelogram described on AO subject to the following conditions; the parallelogram is to be equal to a given rectilineal figure, and the parallelogram on the base BO which can be cut off by a straight line through B is to be similar to a given parallelogram.

VI. 32. This proposition seems of no use. Moreover the enunciation is imperfect. For suppose ED to be produced through D to a point F, such that DF is equal to DE; and join CF. Then the triangle CDF will satisfy all the conditions in Euclid's enunciation, as well as the triangle CDE; but CF and CB are not in one straight line. It should be stated that the bases must lie on corresponding sides of both the parallels; the bases CF and BC do not lie on corresponding sides of the parallels AB and DC, and so the triangle CDF would not fulfil all the conditions, and would therefore be excluded.

VI. 33. In VI. 33 Euclid implicitly gives up the restriction, which he seems to have adopted hitherto, that no angle is to be considered greater than two right angles. For in the demonstration the angle BGL may be any multiple whatever of the angle BGC, and so may be greater than any number of right angles.

VI. B, C, D. These propositions were introduced by Simson. The important proposition VI. D occurs in the Μεγάλη Σύνταξις of Ptolemy.

THE ELEVENTH BOOK.

IN addition to the first six Books of the Elements it is usual to read part of the eleventh Book. For an account of the contents of the other Books of the Elements the student is referred to the article *Eucleides* in Dr Smith's *Dictionary of Greek and Roman Biography*, and to the article *Irrational Quantities* in the *English Cyclopædia.* We may state briefly that Books VII, VIII, IX treat on Arithmetic, Book X on Irrational Quantities, and Books XI, XII on Solid Geometry.

XI. *Def.* 10. This definition is omitted by Simson, and justly, because, as he shews, it is not true that solid figures contained by the same number of similar and equal plane figures are equal to one another. For, conceive two pyramids, which have their bases similar and equal, but have different altitudes. Suppose one of these bases applied exactly on the other; then if the vertices be put on opposite sides of the base a certain solid is formed, and if the vertices be put on the same side of the base another solid is formed. The two solids thus formed are contained by the same number of similar and equal plane figures, but they are not equal.

It will be observed that in this example one of the solids has a *re-entrant* solid angle; see page 264. It is however true that two *convex* solid figures are equal if they are contained by equal plane figures similarly arranged; see Catalan's *Théorèmes et Problèmes de Géométrie Elémentaire.* This result was first demonstrated by Cauchy, who turned his attention to the point at the request of Legendre and Malus; see the *Journal de l'École Polytechnique*, Cahier 16.

XI. *Def.* 26. The word tetrahedron is now often used to denote a solid bounded by any four triangular faces, that is, a pyramid on a triangular base; and when the tetrahedron is to be such as Euclid defines, it is called a regular tetrahedron.

Two other definitions may conveniently be added.

A straight line is said to be parallel to a plane when they do not meet if produced.

The angle made by two straight lines which do not meet is the angle contained by two straight lines parallel to them, drawn through any point.

XI. 21. In XI. 21 the first case only is given in the original. In the second case a certain condition must be introduced, or the proposition will not be true; the polygon *BCDEF* must have no *re-entrant* angle. See note on I. 32.

The propositions in Euclid on Solid Geometry which are now not read, contain some very important results respecting the volumes of solids. We will state these results, as they are often of use; the demonstrations of them are now usually given as examples of the *Integral Calculus.*

We have already explained in the notes to the second Book how the area of a figure is measured by the number of square inches or square feet which it contains. In a similar manner the volume of a solid is measured by the number of *cubic inches* or *cubic feet* which it contains; a *cubic inch* is a cube in which each of the faces is a square inch, and a *cubic foot* is similarly defined.

The volume of a prism is found by multiplying the number of square inches in its base by the number of inches in its altitude; the volume is thus expressed in cubic inches. Or we may multiply the number of square feet in the base by the number of feet in the altitude; the volume is thus expressed in cubic feet. By the *base* of a prism is meant either of the *two equal, similar, and parallel figures* of XI. *Definition* 13; and the *altitude* of the prism is the perpendicular distance between these two planes.

The rule for the volume of a prism involves the fact that *prisms on equal bases and between the same parallels are equal in volume.*

A parallelepiped is a particular case of a prism. The volume of a pyramid is one third of the volume of a prism on the same base and having the same altitude.

For an account of what are called the *five regular solids* the student is referred to the chapter on *Polyhedrons* in the Treatise on *Spherical Trigonometry.*

THE TWELFTH BOOK.

Two propositions are given from the twelfth Book, as they are very important, and are required in the University Examinations. The Lemma is the first proposition of the tenth Book, and is required in the demonstration of the second proposition of the twelfth Book.

APPENDIX.

THIS Appendix consists of a collection of important propositions which will be found useful, both as affording geometrical exercises, and as exhibiting results which are often required in mathematical investigations. The student will have no difficulty in drawing for himself the requisite figures in the cases where they are not given.

1. *The sum of the squares on the sides of a triangle is equal to twice the square on half the base, together with twice the square on the straight line which joins the vertex to the middle point of the base.*

Let ABC be a triangle; and let D be the middle point of the base AB. Draw CE perpendicular to the base

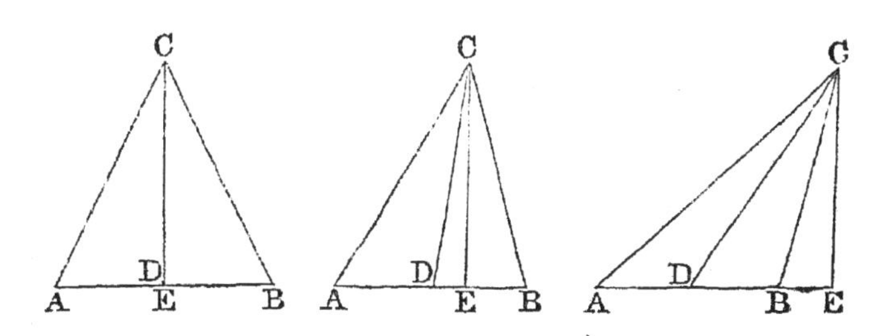

meeting it at E; then E may be either in AB or in AB produced.

First, let E coincide with D; then the proposition follows immediately from I. 47.

Next, let E not coincide with D; then of the two angles ADC and BDC, one must be obtuse and one acute. Suppose the angle ADC obtuse. Then, by II. 12, the square on AC is equal to the squares on AD, DC, together with twice the rectangle AD, DE; and, by II. 13, the square on BC together with twice the rectangle BD, DE is equal to the squares on BD, DC. Therefore, by Axiom 2, the squares on AC, BC, together with twice the rectangle BD, DE are equal to the squares on AD, DB, and twice the square on DC, together with twice the rectangle AD, DE. But AD is equal to DB. Therefore the squares on AC, BC are equal to twice the squares on AD, DC.

2. *If two chords intersect within a circle, the angle which they include is measured by half the sum of the intercepted arcs.*

Let the chords AB and CD of a circle intersect at E; join AD.

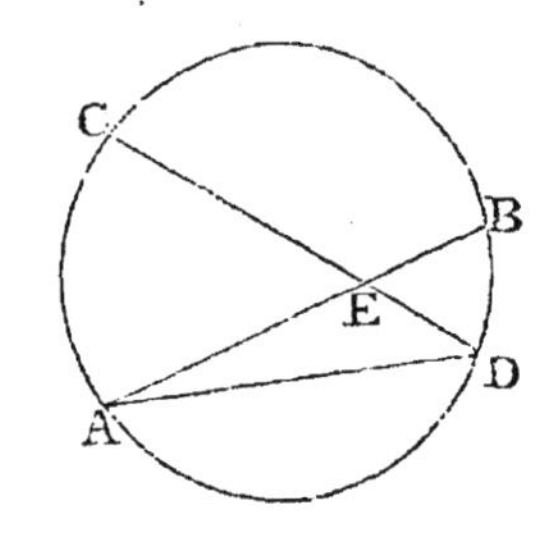

The angle AEC is equal to the angles ADE, and DAE, by I. 32; that is, to the angles standing on the arcs AC and BD. Thus the angle AEC is equal to an angle at the *circumference* of the circle standing on the sum of the arcs AC and BD; and is therefore equal to an angle at the *centre* of the circle standing on half the sum of these arcs.

Similarly the angle CEB is measured by half the sum of the arcs CB and AD.

3. *If two chords produced intersect without a circle, the angle which they include is measured by half the difference of the intercepted arcs.*

Let the chords AB and CD of a circle, produced, intersect at E; join AD.

The angle ADC is equal to the angles EAD and AED, by I. 32. Thus the angle AEC is equal to the difference of the angles ADC and BAD; that is, to an angle at the *circumference* of the circle standing on an arc which is the

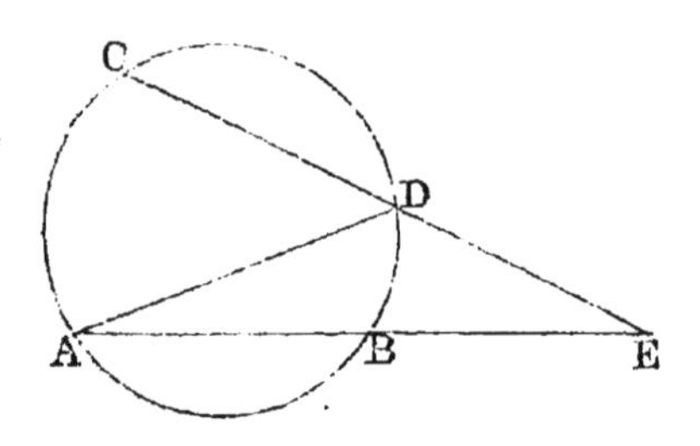

difference of AC and BD; and is therefore equal to an angle at the *centre* of the circle standing on half the difference of these arcs.

4. *To draw a straight line which shall touch two given circles.*

Let A be the centre of the greater circle, and B the centre of the less circle. With centre A, and radius equal to the difference of the radii of the given circles, describe a circle; from B draw a straight line touching the circle

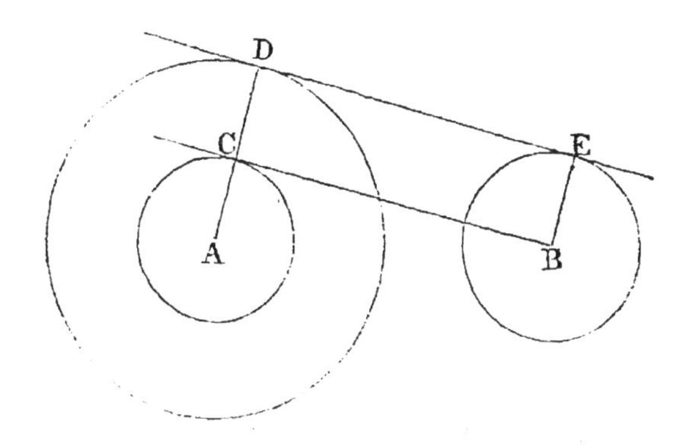

so described at C. Join AC and produce it to meet the circumference at D. Draw the radius BE parallel to AD, and on the same side of AB; and join DE. Then DE shall touch both circles.

See I. 33, I. 29, and III. 16 Corollary.

Since two straight lines can be drawn from B to touch the described circle, two solutions can be obtained; and the two straight lines which are thus drawn to touch the two given circles can be shewn to meet AB, produced through B, at the same point. The construction is applicable when each of the given circles is without the other, and also when they intersect.

When each of the given circles is without the other we can obtain two other solutions. For, describe a circle with A as a centre and radius equal to the sum of the radii of the given circles; and continue as before, except that BE and AD will now be on *opposite* sides of AB. The two straight lines which are thus drawn to touch the two given circles can be shewn to intersect AB at the same point.

5. *To describe a circle which shall pass through three given points not in the same straight line.*

This is solved in Euclid IV. 5.

6. *To describe a circle which shall pass through two given points on the same side of a given straight line, and touch that straight line.*

Let A and B be the given points; join AB and produce it to meet the given straight line at C. Make a square equal to the rectangle CA, CB (II. 14), and on the

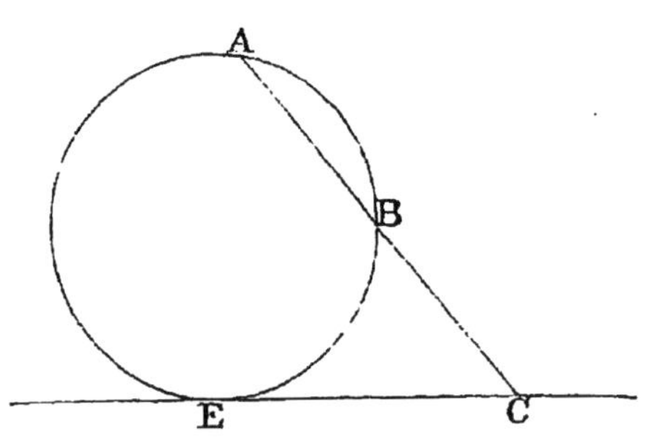

given straight line take CE equal to a side of this square. Describe a circle through A, B, E (5); this will be the circle required (III. 37).

Since E can be taken on either side of C, there are two solutions.

The construction fails if AB is parallel to the given straight line. In this case bisect AB at D, and draw DC at right angles to AB, meeting the given straight line at C. Then describe a circle through A, B, C.

7. *To describe a circle which shall pass through a given point and touch two given straight lines.*

Let A be the given point; produce the given straight lines to meet at B, and join AB. Through B draw a straight line, bisecting that angle included by the given straight lines within which A lies; and in this bisecting straight line take any point C. From C draw a perpendicular on one of the given straight lines, meeting it at D; with centre C, and radius CD, describe a circle, meeting AB, produced if necessary, at E. Join CE; and through A draw a straight line parallel to CE, meeting BC, produced if

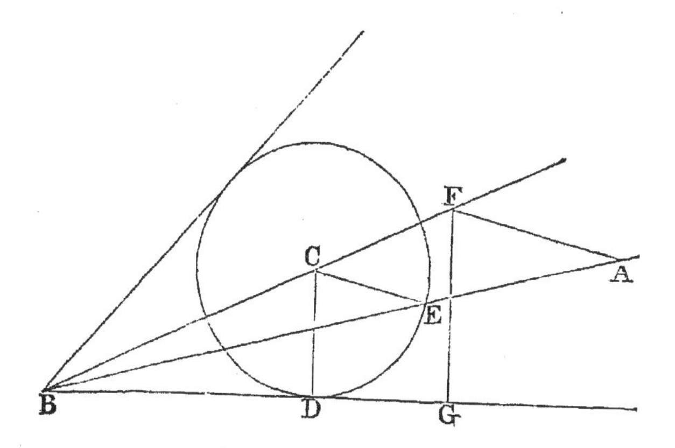

necessary, at F. The circle described from the centre F, with radius FA, will touch the given straight lines.

For, draw a perpendicular from F on the straight line BD, meeting it at G. Then CE is to FA as BC is to BF, and CD is to FG as BC is to BF (VI. 4, V. 16). Therefore CE is to FA as CD is to FG (V. 11). Therefore CE is to CD as FA is to FG (V. 16). But CE is equal to CD; therefore FA is equal to FG (V. *A*).

If A is on the straight line BC we determine E as before; then join ED, and draw a straight line through A parallel to ED meeting BD produced if necessary at G; from G draw a straight line at right angles to BG, and the point of intersection of this straight line with BC, produced if necessary, is the required centre.

As the circle described from the centre C, with the radius CD, will meet AB at two points, there are two solutions.

If A is on one of the given straight lines, draw from A a straight line at right angles to this given straight line; the point of intersection of this straight line with either of the two straight lines which bisect the angles made by the given straight lines may be taken for the centre of the required circle.

If the two given straight lines are parallel, instead of drawing a straight line BC to bisect the angle between them, we must draw it parallel to them, and equidistant from them.

8. *To describe a circle which shall touch three given straight lines, not more than two of which are parallel.*

Proceed as in Euclid IV. 4. If the given straight lines form a triangle, four circles can be described, namely, one as in Euclid, and three others each touching one side of the triangle and the other two sides produced. If two of the given straight lines are parallel, two circles can be described, namely, one on each side of the third given straight line.

9. *To describe a circle which shall touch a given circle, and touch a given straight line at a given point.*

Let A be the given point in the given straight line, and C be the centre of the given circle. Through C draw a straight line perpendicular to the given straight line,

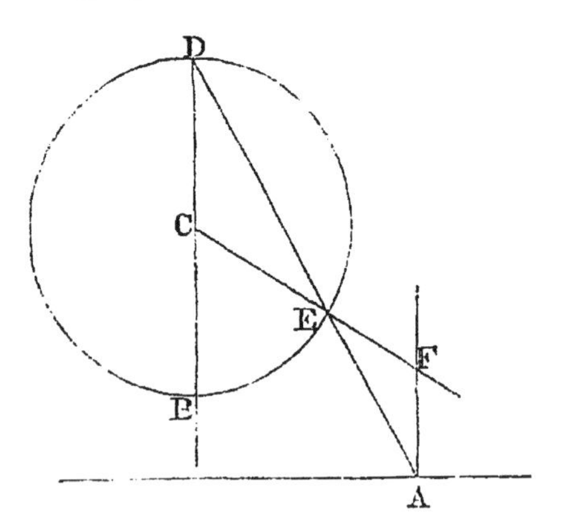

and meeting the circumference of the circle at B and D, of which D is the more remote from the given straight line. Join AD, meeting the circumference of the circle at E. From A draw a straight line at right angles to the given straight line, meeting CE produced at F. Then F shall be the centre of the required circle, and FA its radius.

For the angle AEF is equal to the angle CED (I. 15); and the angle EAF is equal to the angle CDE (I. 29); therefore the angle AEF is equal to the angle EAF; therefore AF is equal to EF (I. 6).

In a similar manner another solution may be obtained by joining AB. If the given straight line falls without the given circle, the circle obtained by the first solution touches the given circle externally, and the circle obtained by the second solution touches the given circle internally. If the given straight line cuts the given circle, both the circles obtained touch the given circle externally.

10. *To describe a circle which shall pass through two given points and touch a given circle.*

Let A and B be the given points. Take any point C on the circumference of the given circle, and describe a circle through A, B, C. If this described circle touches the given circle, it is the required circle. But if not, let D

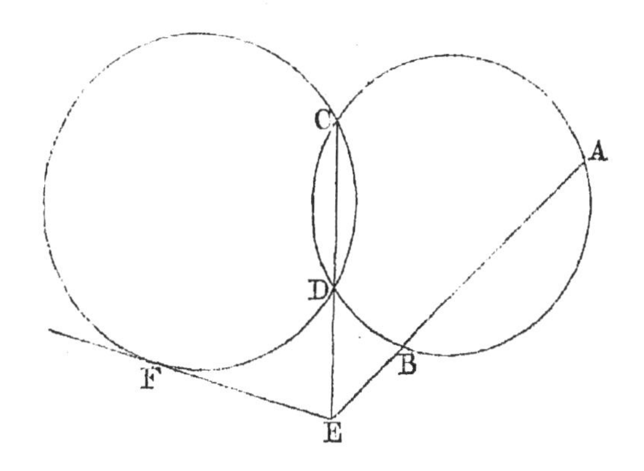

be the other point of intersection of the two circles. Let AB and CD be produced to meet at E; from E draw a straight line touching the given circle at F. Then a circle described through A, B, F shall be the required circle. See III. 35 and III. 37.

There are two solutions, because two straight lines can be drawn from E to touch the given circle.

If the straight line which bisects AB at right angles passes through the centre of the given circle, the construction fails, for AB and CD are parallel. In this case F must be determined by drawing a straight line parallel to AB so as to touch the given circle.

11. *To describe a circle which shall touch two given straight lines and a given circle.*

Draw two straight lines parallel to the given straight lines, at a distance from them equal to the radius of the given circle, and on the sides of them remote from the centre of the given circle. Describe a circle touching the straight lines thus drawn, and passing through the centre of the given circle (7). A circle having the same centre as the circle thus described, and a radius equal to the excess of its radius over that of the given circle, will be the required circle.

Two solutions will be obtained, because there are two solutions of the problem in 7; the circles thus obtained touch the given circle externally.

We may obtain two circles which touch the given circle internally, by drawing the straight lines parallel to the given straight lines on the sides of them adjacent to the centre of the given circle.

12. *To describe a circle which shall pass through a given point and touch a given straight line and a given circle.*

We will suppose the given point and the given straight line without the circle; other cases of the problem may be treated in a similar manner.

Let A be the given point, and B the centre of the given circle. From B draw a perpendicular to the given straight line, meeting it at C, and meeting the circumference of the given circle at D and E, so that D is between B and C. Join EA and determine a point F in EA, produced if necessary, such that the rectangle EA, EF may be equal to the rectangle EC, ED; this can be done by describing a circle through A, C, D, which will meet EA at the required point (III. 36, *Corollary*). Describe a circle to pass through A and F and touch the given straight line (6); this shall be the required circle.

For, let the circle thus described touch the given straight line at G; join EG meeting the given circle at H,

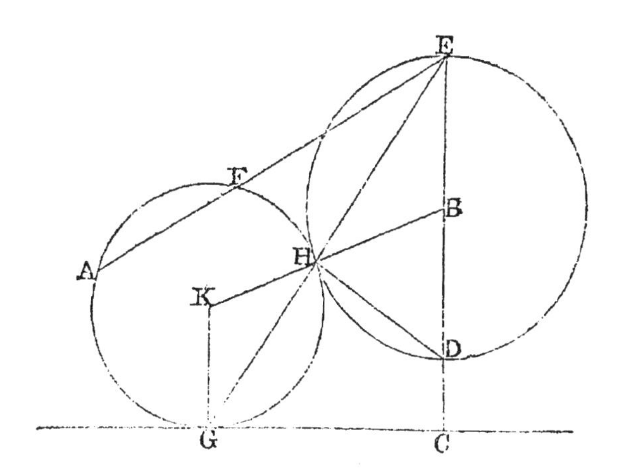

and join DH. Then the triangles EHD and ECG are similar; and therefore the rectangle EC, ED is equal to the rectangle EG, EH (III. 31, VI. 4, VI. 16). Thus the rectangle EA, EF is equal to the rectangle EH, EG; and therefore H is on the circumference of the described circle (III. 36, *Corollary*). Take K the centre of the described circle; join KG, KH, and BH. Then it may be shewn that the angles KHG and EHB are equal (I. 29, I. 5). Therefore KHB is a straight line; and therefore the described circle touches the given circle.

Two solutions will be obtained, because there are two solutions of the problem in 6; the circles thus described touch the given circle externally.

By joining DA instead of EA we can obtain two solutions in which the circles described touch the given circle internally.

13. *To describe a circle which shall touch a given straight line and two given circles.*

Let A be the centre of the larger circle and B the centre of the smaller circle. Draw a straight line parallel to the given straight line, at a distance from it equal to the radius of the smaller circle, and on the side of it remote from A. Describe a circle with A as centre, and radius equal to the difference of the radii of the given circles. Describe a circle which shall pass through B, touch externally the circle just described, and also touch the straight line which has been drawn parallel to the given straight line (12). Then a circle having the same centre as the second described circle, and a radius equal to the excess of its radius over the radius of the smaller given circle, will be the required circle.

Two solutions will be obtained, because there are two solutions of the problem in 12; the circles thus described touch the given circles externally.

We may obtain in a similar manner circles which touch the given circles internally, and also circles which touch one of the given circles internally and the other externally.

14. *Let* A *be the centre of a circle, and* B *the centre of a larger circle; let a straight line be drawn touching the former circle at* C *and the latter circle at* D, *and meeting* AB *produced through* A *at* T. *From* T *draw any straight line meeting the smaller circle at* K *and* L, *and the larger circle at* M *and* N; *so that the five letters* T, K, L, M, N *are in this order. Then the straight lines* AK, KC, CL, LA *shall be respectively parallel to the straight lines* BM, MD, DN, NB; *and the rectangle* TK, TN *shall be equal to the rectangle* TL, TM, *and equal to the rectangle* TC, TD.

Join AC, BD. Then the triangles TAC and TBD are

equiangular; and therefore TA is to TB as AC is to BD (VI. 4, V. 16), that is, as AK is to BM.

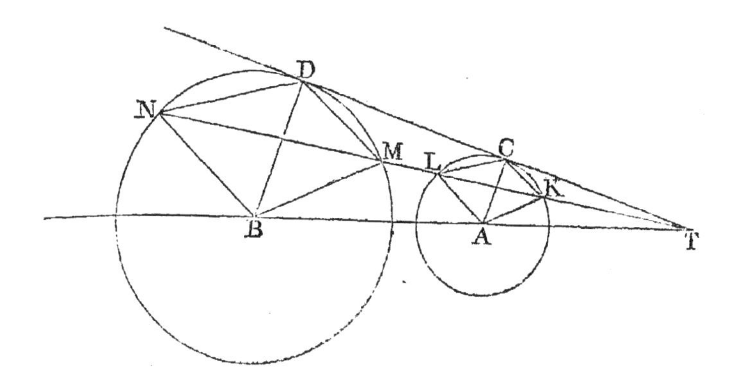

Therefore the triangles TAK and TBM are similar (VI. 7); therefore the angle TAK is equal to the angle TBM; and therefore AK is parallel to BM. Similarly AL is parallel to BN. And because AK is parallel to BM and AC parallel to BD, the angle CAK is equal to the angle DBM; and therefore the angle CLK is equal to the angle DNM (III. 20); and therefore CL is parallel to DN. Similarly CK is parallel to DM.

Now TM is to TD as TD is to TN (III. 37, VI. 16); and TM is to TD as TK is to TC (VI. 4); therefore TK is to TC as TD is to TN; and therefore the rectangle TK, TN is equal to the rectangle TC, TD. Similarly the rectangle TL, TM is equal to the rectangle TC, TD.

If each of the given circles is without the other we may suppose the straight line which touches both circles to meet AB at T *between* A and B, and the above results will all hold, provided we interchange the letters K and L; so that the five letters are now to be in the following order, L, K, T, M, N.

The point T is called a *centre of similitude* of the two circles.

15. *To describe a circle which shall pass through a given point and touch two given circles.*

Let A be the centre of the smaller circle and B the centre of the larger circle; and let E be the given point.

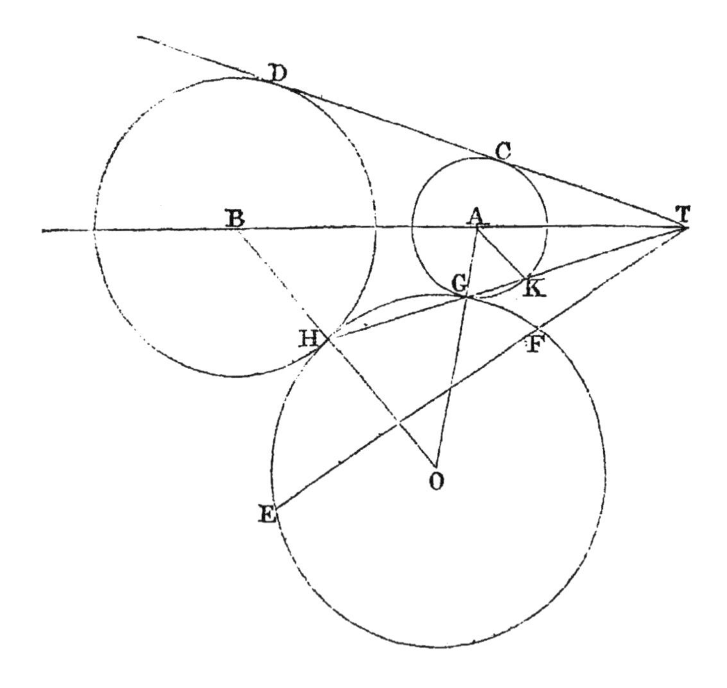

Draw a straight line touching the former circle at C and the latter at D, and meeting the straight line AB, produced through A, at T. Join TE and divide it at F so that the rectangle TE, TF may be equal to the rectangle TC, TD. Then describe a circle to pass through E and F and touch either of the given circles (10); this shall be the required circle.

For suppose that the circle is described so as to touch the smaller given circle; let G be the point of contact; we have then to shew that the described circle will also touch the larger given circle. Join TG, and produce it to meet the larger given circle at H. Then the rectangle TG, TH is equal to the rectangle TC, TD (14); therefore the rectangle TG, TH is equal to the rectangle TE, TF; and therefore the described circle passes through H.

Let O be the centre of this circle, so that OGA is a straight line; we have to shew that OHB is a straight line.

Let TG intersect the smaller circle again at K; then AK is parallel to BH (14); therefore the angle AKT is equal to the angle BHG; and the angle AKG is equal to the angle AGK, which is equal to the angle OGH, which is equal to the angle OHG. Therefore the angles BHG and OHG together are equal to AKT and AKG together; that is, to two right angles. Therefore OHB is a straight line.

Two solutions will be obtained, because there are two solutions of the problem in 10. Also, if each of the given circles is without the other, two other solutions can be obtained by taking for T the point between A and B where a straight line touching the two given circles meets AB. The various solutions correspond to the circumstance that the contact of circles may be external or internal.

16. *To describe a circle which shall touch three given circles.*

Let A be the centre of that circle which is not greater than either of the other circles; let B and C be the centres of the other circles. With centre B, and radius equal to the excess of the radius of the circle with centre B over the radius of the circle with centre A, describe a circle. Also with centre C, and radius equal to the excess of the radius of the circle with centre C over the radius of the circle with centre A, describe a circle. Describe a circle to touch externally these two described circles and to pass through A (15). Then a circle having the same centre as the last described circle, and having a radius equal to the excess of its radius over the radius of the circle with centre A, will touch externally the three given circles.

In a similar way we may describe a circle touching internally the three given circles, or touching one of them externally and the two others internally, or touching one of them internally and the two others externally.

17. *In a given indefinite straight line it is required to find a point such that the sum of its distances from two given points on the same side of the straight line shall be the least possible.*

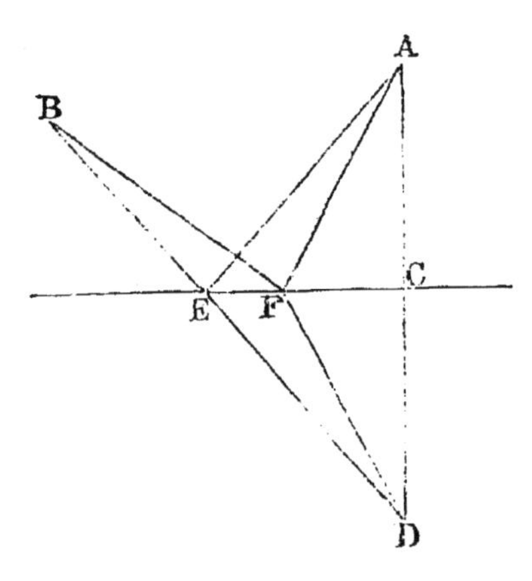

Let A and B be the two given points. From A draw a perpendicular to the given straight line meeting it at C; and produce AC to D so that CD may be equal to AC. Join DB meeting the given straight line at E. Then E shall be the required point.

For, let F be any other point in the given straight line. Then, because AC is equal to DC, and EC is common to the two triangles ACE, DCE; and that the right angle ACE is equal to the right angle DCE; therefore AE is equal to DE. Similarly, AF is equal to DF. And the sum of DF and FB is greater than BD (I. 20): therefore the sum of AF and FB is greater than BD; that is, the sum of AF and FB is greater than the sum of DE and EB; therefore the sum of AF and FB is greater than the sum of AE and EB.

18. *The perimeter of an isosceles triangle is less than that of any other triangle of equal area standing on the same base.*

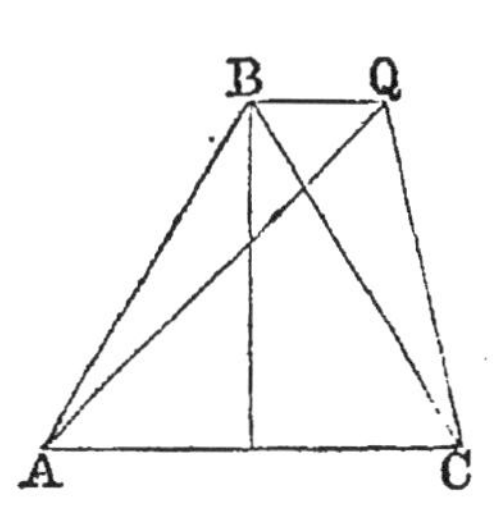

Let ABC be an isosceles triangle; AQC any other triangle equal in area and standing on the same base AC.

Join BQ; then BQ is parallel to AC (I. 39).

And it will follow from 17 that the sum of AQ and QC is greater than the sum of AB and BC.

19. *If a polygon be not equilateral a polygon may be found of the same number of sides, and equal in area, but having a less perimeter.*

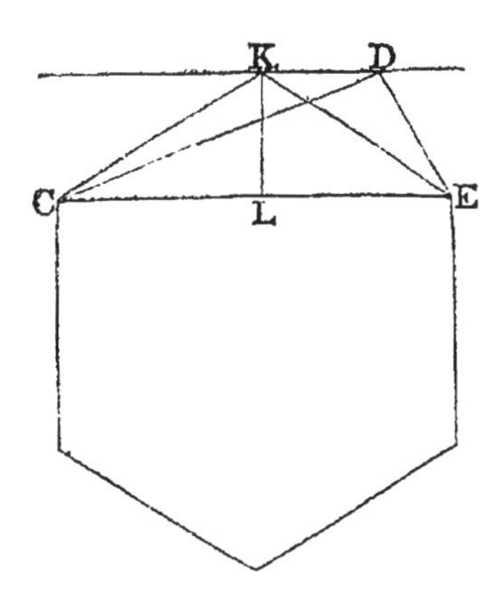

For, let CD, DE be two adjacent unequal sides of the polygon. Join CE. Through D draw a straight line parallel to CE. Bisect CE at L; from L draw a straight line at right angles to CE meeting the straight line drawn through D at K. Then by removing from the given polygon the triangle CDE and applying the triangle CKE, we obtain a polygon having the same number of sides as the given polygon, and equal to it in area, but having a less perimeter (18).

20. A *and* B *are two given points on the same side of a given straight line, and* AB *produced meets the given straight line at* C; *of all points in the given straight line on each side of* C, *it is required to determine that at which* AB *subtends the greatest angle.*

Describe a circle to pass through A and B, and to touch the given straight line on that side of C which is to be considered (6). Let D be the point of contact: D shall be the required point.

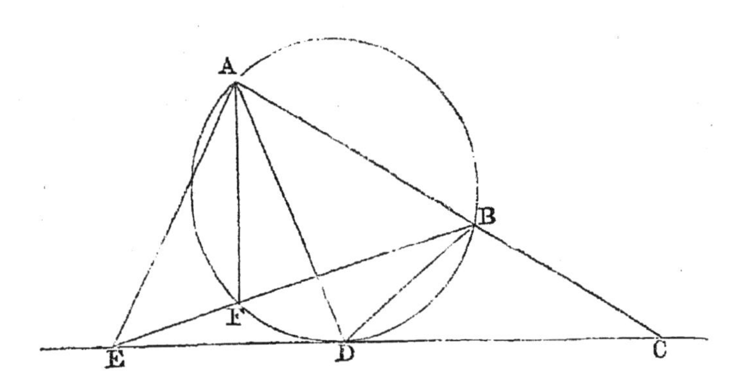

For, take any other point E in the given straight line, on the same side of C as D is; draw EA, EB; then one at least of these straight lines will cut the circumference ADB.

Suppose that BE cuts the circumference at F; join AF. Then the angle AFB is equal to the angle ADB (III. 21); and the angle AFB is greater than the angle AEB (I. 16); therefore the angle ADB is greater than the angle AEB.

21. A *and* B *are two given points within a circle; and* AB *is drawn and produced both ways so as to divide the whole circumference into two arcs; it is required to determine the point in each of these arcs at which* AB *subtends the greatest angle.*

Describe a circle to pass through A and B and to touch the circumference considered (10): the point of contact will be the required point. The demonstration is similar to that in the preceding proposition.

22. A *and* B *are two given points without a given circle; it is required to determine the points on the circumference of the given circle at which* AB *subtends the greatest and least angles.*

Suppose that neither AB nor AB produced cuts the given circle.

Describe two circles to pass through A and B, and to touch the given circle (10): the point of contact of the circle which touches the given circle externally will be the point where the angle is greatest, and the point of contact of the circle which touches the given circle internally will be the point where the angle is least. The demonstration is similar to that in 20.

If AB cuts the given circle, both the circles obtained by 10 touch the given circle internally; in this case the angle subtended by AB at a point of contact is less than the angle subtended at any other point of the circumference of the given circle which is on the same side of AB. Here the angle is greatest at the points where AB cuts the circle, and is there equal to two right angles.

If AB *produced* cuts the given circle, both the circles obtained by 10 touch the given circle externally; in this case the angle subtended by AB at a point of contact is greater than the angle subtended at any other point of the circumference of the given circle which is on the same side of AB. Here the angle is least at the points where AB produced cuts the circle, and is there zero.

23. *If there be four magnitudes such that the first is to the second as the third is to the fourth; then shall the first together with the second be to the excess of the first above the second as the third together with the fourth is to the excess of the third above the fourth.*

For, the first together with the second is to the second as the third together with the fourth is to the fourth (V. 18). Therefore, alternately, the first together with the second is to the third together with the fourth as the second is to the fourth (V. 16).

Similarly, by V. 17 and V. 16, the excess of the first above the second is to the excess of the third above the fourth as the second is to the fourth.
Therefore, by V. 11, the first together with the second is to the excess of the first above the second as the third together with the fourth is to the excess of the third above the fourth.

24. *The straight lines drawn at right angles to the sides of a triangle from the points of bisection of the sides meet at the same point.*

Let ABC be a triangle; bisect BC at D, and bisect CA at E; from D draw a straight line at right angles to BC, and from E draw a straight line at right angles to CA;

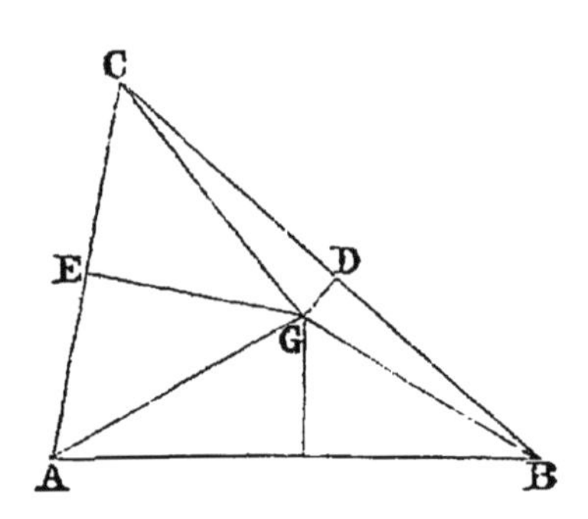

let these straight lines meet at G: we have then to shew that the straight line which bisects AB at right angles also passes through G. From the triangles BDG and

CDG we can shew that BG is equal to CG; and from the triangles CEG and AEG we can shew that CG is equal to AG; therefore BG is equal to AG. Then if we draw a straight line from G to the middle point of AB we can shew that this straight line is at right angles to AB: that is, the line which bisects AB at right angles passes through G.

25. *The straight lines drawn from the angles of a triangle to the points of bisection of the opposite sides meet at the same point.*

Let ABC be a triangle; bisect BC at D, bisect CA at E, and bisect AB at F; join BE and CF meeting at G;

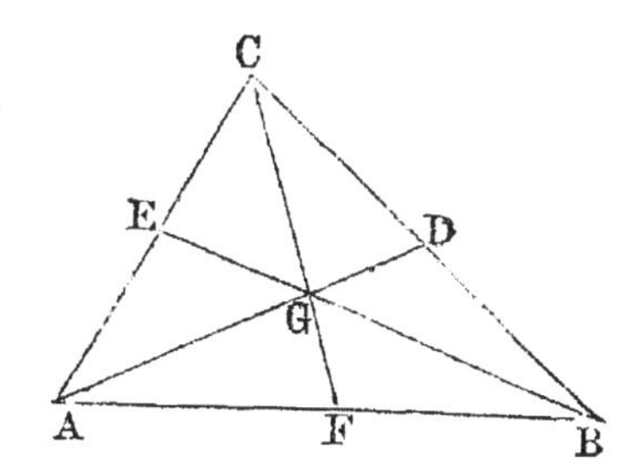

join AG and GD: then AG and GD shall lie in a straight line.

The triangle BEA is equal to the triangle BEC, and the triangle GEA is equal to the triangle GEC (I. 38); therefore, by the third Axiom, the triangle BGA is equal to the triangle BGC.

Similarly, the triangle CGA is equal to the triangle CGB.

Therefore the triangle BGA is equal to the triangle CGA. And the triangle BGD is equal to the triangle CGD (I. 38); therefore the triangles BGA and BGD together are equal to the triangles CGA and CGD together. Therefore the triangles BGA and BGD together are equal to half the triangle ABC. Therefore G must fall on the straight line AD; that is, AG and GD lie in a straight line.

26. *The straight lines which bisect the angles of a triangle meet at the same point.*

Let ABC be a triangle; bisect the angles at B and C

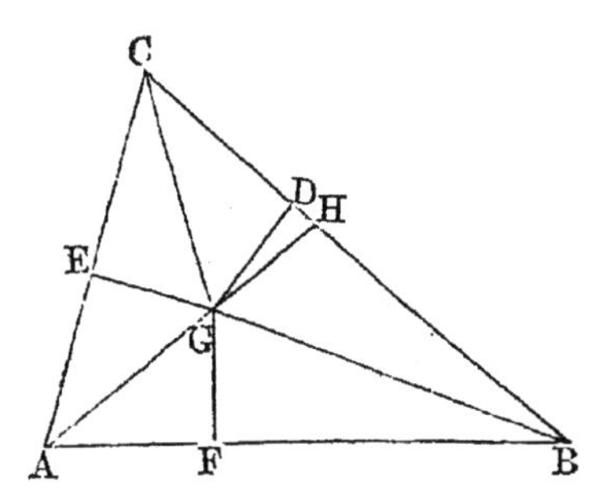

by straight lines meeting at G; join AG: then AG shall bisect the angle at A.

From G draw GD perpendicular to BC, GE perpendicular to CA, and GF perpendicular to AB.

From the triangles BGF and BGD we can shew that GF is equal to GD; and from the triangles CGE and CGD we can shew that GE is equal to GD; therefore GF is equal to GE. Then from the triangles AFG and AEG we can shew that the angle FAG is equal to the angle EAG.

The theorem may also be demonstrated thus. Produce AG to meet BC at H. Then AB is to BH as AG is to GH, and AC is to CH as AG is to GH (VI. 3); therefore AB is to BH as AC is to CH (V. 11); therefore AB is to AC as BH is to CH (V. 16); therefore the straight line AH bisects the angle at A (VI. 3).

27. *Let two sides of a triangle be produced through the base; then the straight lines which bisect the two exterior angles thus formed, and the straight line which bisects the vertical angle of the triangle, meet at the same point.*

This may be shewn like 26: if we adopt the second method we shall have to use VI. A.

23. *The perpendiculars drawn from the angles of a triangle on the opposite sides meet at the same point.*

Let ABC be a triangle; and first suppose that it is not obtuse angled. From B draw BE perpendicular to CA;

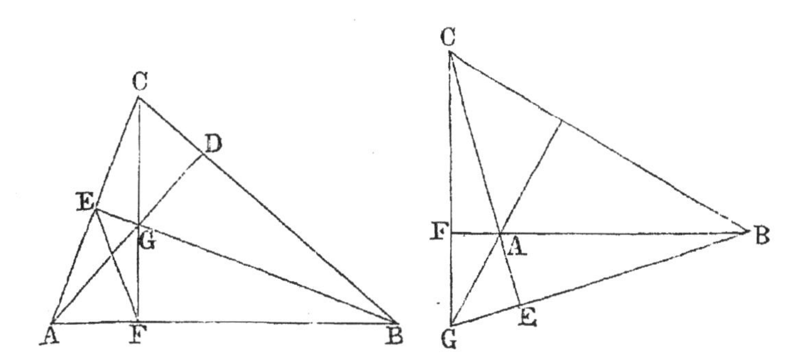

from C draw CF perpendicular to AB; let these perpendiculars meet at G; join AG, and produce it to meet BC at D: then AD shall be perpendicular to BC.

For a circle will go round $AEGF$ (*Note on* III. 22); therefore the angle FAG is equal to the angle FEG (III. 21). And a circle will go round $BCEF$ (III. 31, *Note on* III. 21); therefore the angle FEB is equal to the angle FCB. Therefore the angle BAD is equal to the angle BCF. And the angle at B is common to the two triangles BAD and BCF. Therefore the third angle BDA is equal to the third angle BFC (*Note on* I. 32). But the angle BFC is a right angle, by construction; therefore the angle BDA is a right angle.

In the same way the theorem may be demonstrated when the triangle is obtuse angled. Or this case may be deduced from what has been already shewn. For suppose the angle at A obtuse, and let the perpendicular from B on the opposite side meet that side *produced* at E, and let the perpendicular from C on the opposite side meet that side *produced* at F; and let BE and CF be produced to meet at G. Then in the triangle BCG the perpendiculars BF and CE meet at A; therefore by the former case the straight line GA produced will be perpendicular to BC.

29. *If from any point in the circumference of the circle described round a triangle perpendiculars be drawn to the sides of the triangle, the three points of intersection are in the same straight line.*

Let ABC be a triangle, P any point on the circumference of the circumscribing circle; from P draw PD,

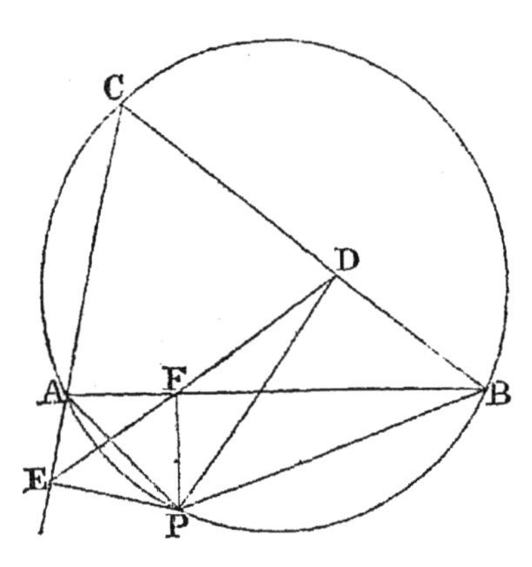

PE, PF perpendiculars to the sides BC, CA, AB respectively: D, E, F shall be in the same straight line.

[We will suppose that P is on the arc cut off by AB, on the opposite side from C, and that E is on CA produced through A; the demonstration will only have to be slightly modified for any other figure.]

A circle will go round $PEAF$ (*Note on* III. 22); therefore the angle PFE is equal to the angle PAE (III. 21). But the angles PAE and PAC are together equal to two right angles (I. 13); and the angles PAC and PBC are together equal to two right angles (III. 22). Therefore the angle PAE is equal to the angle PBC; therefore the angle PFE is equal to the angle PBC.

Again, a circle will go round $PFDB$ (*Note on* III. 21); therefore the angles PFD and PBD are together equal to two right angles (III. 22). But the angle PBD has been shewn equal to the angle PFE. Therefore the angles PFD and PFE are together equal to two right angles. Therefore EF and FD are in the same straight line.

30. ABC *is a triangle, and* O *is the point of intersection of the perpendiculars from* A, B, C *on the opposite sides of the triangle: the circle which passes through the middle points of* OA, OB, OC *will pass through the feet of the perpendiculars and through the middle points of the sides of the triangle.*

Let D, E, F be the middle points of OA, OB, OC respectively; let G be the foot of the perpendicular from A on BC, and H the middle point of BC.

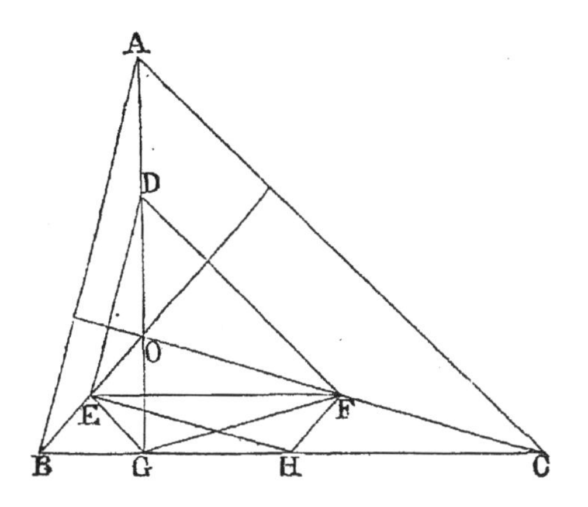

Then OBG is a right-angled triangle and E is the middle point of the hypotenuse OB; therefore EG is equal to EO; therefore the angle EGO is equal to the angle EOG. Similarly, the angle FGO is equal to the angle FOG. Therefore the angle FGE is equal to the angle FOE. But the angles FOE and BAC are together equal to two right angles; therefore the angles FGE and BAC are together equal to two right angles. And the angle BAC is equal to the angle EDF, because ED, DF are parallel to BA, AC (VI. 2). Therefore the angles FGE and EDF are together equal to two right angles. Hence G is on the circumference of the circle which passes through D, E, F (*Note on* III. 22).

Again, FH is parallel to OB, and EH parallel to OC; therefore the angle EHF is equal to the angle EGF. Therefore H is also on the circumference of the circle.

Similarly, the two points in each of the other sides of the triangle ABC may be shewn to be on the circumference of the circle.

The circle which is thus shewn to pass through these nine points may be called the *Nine points circle:* it has some curious properties, of which we will now give two.

The radius of the Nine points circle is half of the radius of the circle described round the original triangle.

For the triangle DEF has its sides respectively halves of the sides of the triangle ABC, so that the triangles are similar. Hence the radius of the circle described round DEF is half of the radius of the circle described round ABC.

If S *be the centre of the circle described round the triangle* ABC, *the centre of the Nine points circle is the middle point of* SO.

For HS is at right angles to BC, and therefore parallel to GO. Hence the straight line which bisects HG at right angles must bisect SO. And H and G are on the circumference of the Nine points circle, so that the straight line which bisects HG at right angles must pass through the centre of the Nine points circle. Similarly, from the other sides of the triangle ABC two other straight lines can be obtained, which pass through the centre of the Nine points circle and also bisect SO. Hence the centre of the Nine points circle must coincide with the middle point of SO.

We may state that the *Nine points circle* of any triangle touches the inscribed circle and the escribed circles of the triangle: a demonstration of this theorem will be found in the *Nouvelles Annales de Mathématiques* for 1842, page 196. For the history of this theorem see the volume of the same Journal for 1863, page 562.

31. *If two straight lines bisecting two angles of a triangle and terminated at the opposite sides be equal, the bisected angles shall be equal.*

Let ABC be a triangle; let the straight line BD bisect the angle at B, and be terminated at the side AC; and let the straight line CE bisect the angle at C, and be terminated at the side AB; and let the straight line BD be equal to the straight line CE: then the angle at B shall be equal to the angle at C.

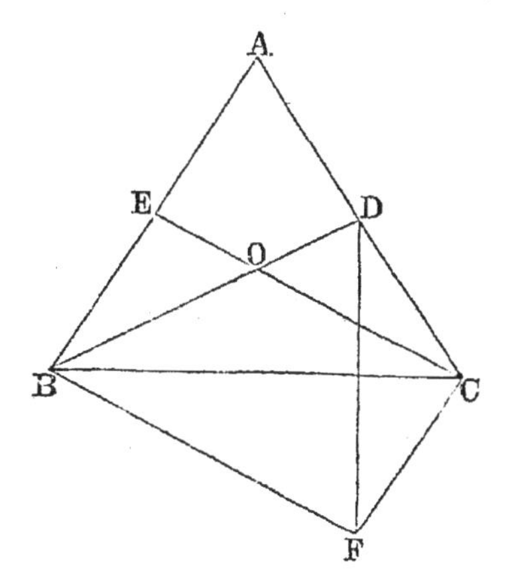

For, let BD and CE meet at O; then if the angle OBC be not equal to the angle OCB, one of them must be greater than the other; let the angle OBC be the greater. Then, because CB and BD are equal to BC and CE, each to each; but the angle CBD is greater than the angle BCE; therefore CD is greater than BE (I. 24).

On the other side of the base BC make the triangle BCF equal to the triangle CBE, so that BF may be equal to CE, and CF equal to BE (I. 22); and join DF.

Then because BF is equal to BD, the angle BFD is equal to the angle BDF. And the angle OCD is, by hypothesis, less than the angle OBE; and the angle COD is equal to the angle BOE; therefore the angle ODC is greater than the angle OEB (I. 32), and therefore the angle ODC is greater than the angle BFC.

Hence, by taking away the equal angles BDF and BFD, the angle FDC is greater than the angle DFC; and therefore CF is greater than CD (I. 19); therefore BE is greater than CD.

But it was shewn that CD is greater than BE; which is absurd.

Therefore the angles OBC and OCB are not unequal, that is, they are equal; and therefore the angle at B is equal to the angle at C.

[For the history of this theorem see *Lady's and Gentleman's Diary* for 1859, page 88.]

32. *If a quadrilateral figure does not admit of having a circle described round it, the sum of the rectangles contained by the opposite sides is greater than the rectangle contained by the diagonals.*

Let $ABCD$ be a quadrilateral figure which does not admit of having a circle described round it; then the rectangle AB, DC, together with the rectangle BC, AD, shall be greater than the rectangle AC, BD.

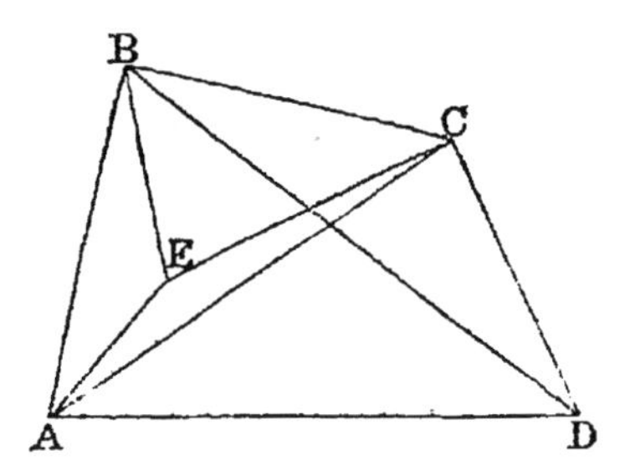

For, make the angle ABE equal to the angle DBC, and the angle BAE equal to the angle BDC; then the triangle ABE is similar to the triangle BDC (VI. 4); therefore AB is to AE as DB is to DC; and therefore the rectangle AB, DC is equal to the rectangle AE, DB.

Join EC. Then, since the angle ABE is equal to the angle DBC, the angle CBE is equal to the angle DBA. And because the triangles ABE and DBC are similar, AB is to DB as BE is to BC; therefore the triangles ABD and EBC are similar (VI. 6); therefore CB is to CE as DB is to DA; and therefore the rectangle CB, DA is equal to the rectangle CE, DB.

Therefore the rectangle AB, DC, together with the rectangle BC, AD is equal to the rectangle AE, BD together with the rectangle CE, BD; that is, equal to the rectangle contained by BD and the sum of AE and EC. But the sum of AE and EC is greater than AC (I. 20); therefore the rectangle AB, DC, together with the rectangle BC, AD is greater than the rectangle AC, BD.

33. *If the rectangle contained by the diagonals of a quadrilateral be equal to the sum of the rectangles contained by the opposite sides, a circle can be described round the quadrilateral.*

This is the converse of VI. D; it can be demonstrated indirectly with the aid of 32.

34. *It is required to find a point in a given straight line, such that the rectangle contained by its distances from two given points in the straight line may be equal to the rectangle contained by its distances from two other given points in the straight line.*

Let A, B, C, D be four given points in the same straight line: it is required to find a point in the straight

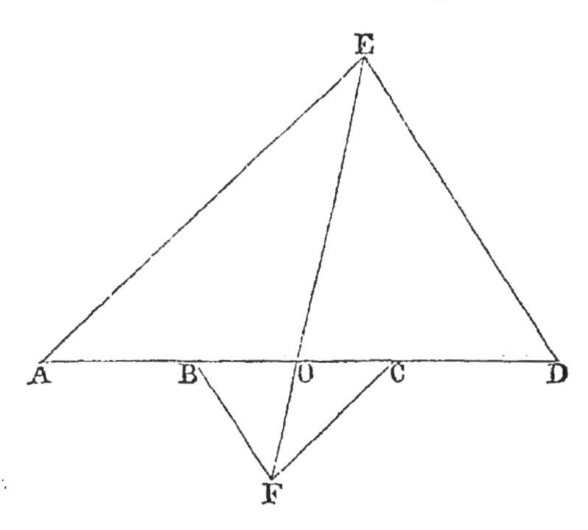

line, such that the rectangle contained by its distances from A and B may be equal to the rectangle contained by its distances from C and D.

On AD describe any triangle AED; and on CB describe a similar triangle CFB, so that CF is parallel to AE, and BF to DE; join EF, and let it meet the given straight line at O. Then O shall be the required point.

For, OE is to OA as OF is to OC (VI. 4); therefore OE is to OF as OA is to OC (V. 16). Similarly OE is to OF as OD is to OB. Therefore OA is to OC as OD is to OB (V. 11). Therefore the rectangle OA, OB is equal to the rectangle OC, OD.

The figure will vary slightly according to the situation of the four given points, but corresponding to an assigned situation there will be only *one* point such as is required. For suppose there could be such a point P, besides the point O which is determined by the construction given above; and that the points are in the order A, C, D, B, O, P. Join PE, and let it meet CF, produced at G; join BG. Then the rectangle PA, PB is, by hypothesis, equal to the rectangle PC, PD; and therefore PA is to PC as PD is to PB. But PA is to PC as PE is to PG (VI. 2); therefore PD is to PB as PE is to PG (V. 11); therefore BG is parallel to DE.

But, by the construction, BF is parallel to ED; therefore BG and BF are themselves parallel (I. 30); which is absurd. Therefore P is not such a point as is required.

ON GEOMETRICAL ANALYSIS.

35. The substantives *analysis* and *synthesis*, and the corresponding adjectives *analytical* and *synthetical*, are of frequent occurrence in mathematics. In general *analysis* means decomposition, or the separating a whole into its parts, and *synthesis* means composition, or making a whole out of its parts. In Geometry however these words are used in a more special sense. In synthesis we begin with results already established, and end with some new result; thus, by the aid of theorems already demonstrated, and problems already solved, we demonstrate some new theorem, or solve some new problem. In analysis we begin with assuming the truth of some theorem or the solution of some problem, and we deduce from the assumption consequences which we can compare with results already established, and thus test the validity of our assumption.

36. The propositions in Euclid's Elements are all exhibited synthetically; the student is only employed in examining the soundness of the reasoning by which each successive addition is made to the collection of geometrical truths already obtained; and there is no hint given as to the manner in which the propositions were originally discovered. Some of the constructions and demonstrations appear rather artificial, and we are thus naturally induced to enquire whether any rules can be discovered by which we may be guided easily and naturally to the investigation of new propositions.

37. Geometrical analysis has sometimes been described in language which might lead to the expectation that directions could be given which would enable a student to proceed to the demonstration of any proposed theorem, or the solution of any proposed problem, with confidence of success; but no such directions can be given. We will state the exact extent of these directions. Suppose that a new theorem is proposed for investigation, or a new problem for trial. Assume the truth of the theorem or the solution of the problem, and deduce consequences from this assumption combined with results which have been already established. If a consequence can be deduced which contradicts some result already established, this amounts to a demonstration that our assumption is inadmissible; that is, the theorem is not true, or the problem cannot be solved. If a consequence can be deduced which coincides with some result already established, we cannot say that the assumption is inadmissible; and it *may happen* that by starting from the consequence which we deduced, and retracing our steps, we can succeed in giving a synthetical demonstration of the theorem, or solution of the problem. These directions however are very vague, because no certain rule can be prescribed by which we are to combine our assumption with results already established; and moreover no test exists by which we can ascertain whether a valid consequence which we have drawn from an assumption will enable us to establish the assumption itself. That a proposition may be false and yet furnish consequences which are true, can be seen from a simple example. Suppose a theorem were proposed for investigation in the following words; *one angle of a triangle is to another as the side opposite to the first angle is to the side opposite to the other.* If this be assumed to be true we can immediately deduce Euclid's result in I. 19; but from Euclid's result in I. 19 we cannot retrace our steps and establish the proposed theorem, and in fact the proposed theorem is false.

Thus the only definite statement in the directions respecting Geometrical analysis is, that if a consequence can be deduced from an assumed proposition which contradicts a result already established, that assumed proposition must be false.

38. We may mention, in particular, that a consequence would contradict results already established, if we could shew that it would lead to the solution of a problem already given up as impossible. There are three famous problems which are now admitted to be beyond the power of Geometry; namely, to find a straight line equal in length to the circumference of a given circle, to trisect any given angle, and to find two mean proportionals between two given straight lines. The grounds on which the geometrical solution of these problems is admitted to be impossible cannot be explained without a knowledge of the higher parts of mathematics; the student of the Elements may however be content with the fact that innumerable attempts have been made to obtain solutions, and that these attempts have been made in vain.

The first of these problems is usually referred to as the *Quadrature of the Circle*. For the history of it the student should consult the article in the *English Cyclopædia* under that head, and also a series of papers in the *Athenæum* for 1863 and 1864, entitled a *Budget of Paradoxes*, by Professor De Morgan.

For *approximate* solutions of the problem we may refer to Davies's edition of Hutton's *Course of Mathematics*, Vol. I. page 400, the *Lady's and Gentleman's Diary* for 1855, page 86, and the *Philosophical Magazine* for April, 1862.

The third of the three problems is often referred to as the *Duplication of the Cube*. See the note on VI. 13 in *Lardner's Euclid*, and a dissertation by C. H. Biering entitled *Historia Problematis Cubi Duplicandi*...Hauniæ, 1844.

We will now give some examples of Geometrical analysis.

39. *From two given points it is required to draw to the same point in a given straight line, two straight lines equally inclined to the given straight line.*

Let A and B be the given points, and CD the given straight line.

Suppose AE and EB to be the two straight lines equally inclined to CD. Draw BF perpendicular to CD, and produce AE and BF to meet at G. Then the angle

BED is equal to the angle AEC, by hypothesis; and the angle AEC is equal to the angle DEG (I. 15). Hence the

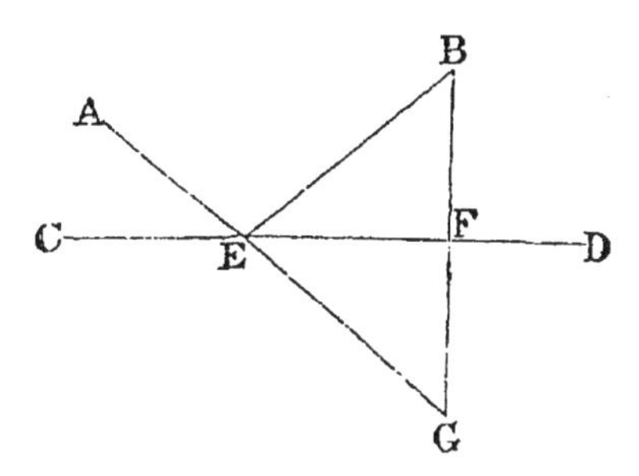

triangles BEF and GEF are equal in all respects (I. 26); therefore FG is equal to FB.

This result shews how we may synthetically solve the problem. Draw BF perpendicular to CD, and produce it to G, so that FG may be equal to FB; then join AG, and AG will intersect CD at the required point.

40. *To divide a given straight line into two parts such that the difference of the squares on the parts may be equal to a given square.*

Let AB be the given straight line, and suppose C the required point.

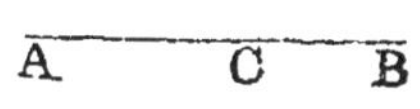

Then the difference of the squares on AC and BC is to be equal to a given square. But the difference of the squares on AC and BC is equal to the rectangle contained by their sum and difference; therefore this rectangle must be equal to the given square. Hence we have the following synthetical solution. On AB describe a rectangle equal to the given square (I. 45); then the difference of AC and CB will be equal to the side of the rectangle adjacent to AB, and is therefore known. And the sum of AC and CB is known. Thus AC and CB are known.

It is obvious that the given square must not exceed the square on AB, in order that the problem may be possible.

There are two positions of C, if it is not specified which of the two segments AC and CB is to be greater than the other; but only one position, if it is specified.

In like manner we may solve the problem, *to produce a given straight line so that the square on the whole straight line made up of the given straight line and the part produced, may exceed the square on the part produced by a given square, which is not less than the square on the given straight line.*

The two problems may be combined in one enunciation thus, *to divide a given straight line internally or externally so that the difference of the squares on the segments may be equal to a given square.*

41. *To find a point in the circumference of a given segment of a circle, so that the straight lines which join the point to the extremities of the straight line on which the segment stands may be together equal to a given straight line.*

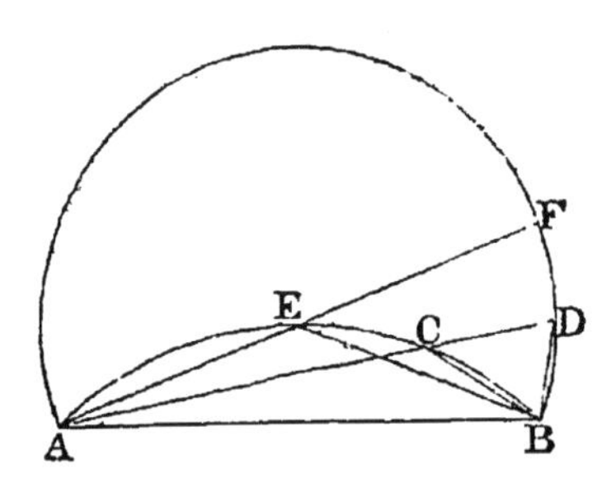

Let ACB be the circumference of the given segment, and suppose C the required point, so that the sum of AC and CB is equal to a given straight line.

Produce AC to D so that CD may be equal to CB; and join DB.

Then AD is equal to the given straight line. And the angle ACB is equal to the sum of the angles CDB and CBD (I. 32), that is, to twice the angle CDB (I. 5). Therefore the angle ADB is half of the angle in the given segment. Hence we have the following synthetical solution. Describe on AB a segment of a circle containing an angle equal to half the angle in the given segment. With A as centre, and a radius equal to the given straight line, describe a circle. Join A with a point of intersection of this circle and the segment which has been described; this joining straight line will cut the circumference of the given segment at a point which solves the problem.

The given straight line must exceed AB and it must not exceed a certain straight line which we will now determine. Suppose the circumference of the given segment bisected at E: join AE, and produce it to meet the circumference of the described segment at F. Then AE is equal to EB (III. 28), and EB is equal to EF for the same reason that CB is equal to CD. Thus EA, EB, EF are all equal; and therefore E is the centre of the circle of which ADB is a segment (III. 9). Hence AF is the longest straight line which can be drawn from A to the circumference of the described segment; so that the given straight line must not exceed twice AE.

42. *To describe an isosceles triangle having each of the angles at the base double of the third angle.*

This problem is solved in IV. 10; we may suppose the solution to have been discovered by such an analysis as the following.

Suppose the triangle ABD such a triangle as is required, so that each of the angles at B and D is double of the angle at A.

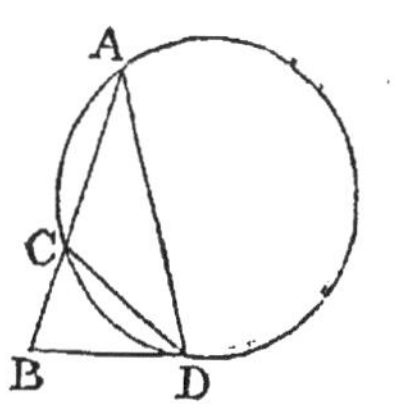

Bisect the angle at D by the straight line DC. Then the angle ADC is equal to the angle at A; therefore CA is equal to CD. The angle CBD is equal to the angle ADB, by hypothesis; the angle CDB is equal to the angle at A; therefore the third angle BCD is equal to the third angle ABD (I. 32). Therefore BD is equal to CD (I. 6); and therefore BD is equal to AC.

Since the angle BDC is equal to the angle at A, the straight line BD will touch at D the circle described round the triangle ACD (*Note on* III. 32). Therefore the rectangle AB, BC is equal to the square on BD (III. 36). Therefore the rectangle AB, BC is equal to the square on AC.

Therefore AB is divided at C in the manner required in II. 11.

Hence the synthetical solution of the problem is evident.

43. *To inscribe a square in a given triangle.*

Let ABC be the given triangle, and suppose $DEFG$ the required square.

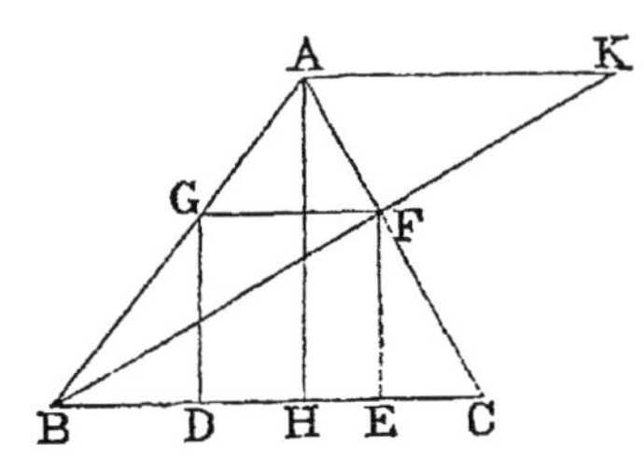

Draw AH perpendicular to BC, and AK parallel to BC; and let BF produced meet AK at K. Then BG is to GF as BA is to AK, and BG is to GD as BA is to AH (VI. 4). But GF is equal to GD, by hypothesis. Therefore BA is to AK as BA is to AH (V. 7, V. 11). Therefore AH is equal to AK (V. 7).

Hence we have the following synthetical solution. Draw AK parallel to BC, and equal to AH; and join BK. Then BK meets AC at one of the corners of the required square, and the solution can be completed.

44. *Through a given point between two given straight lines, it is required to draw a straight line, such that the rectangle contained by the parts between the given point and the given straight lines may be equal to a given rectangle.*

Let P be the given point, and AB and AC the given straight lines; suppose MPN the required straight line, so that the rectangle MP, PN is equal to a given rectangle.

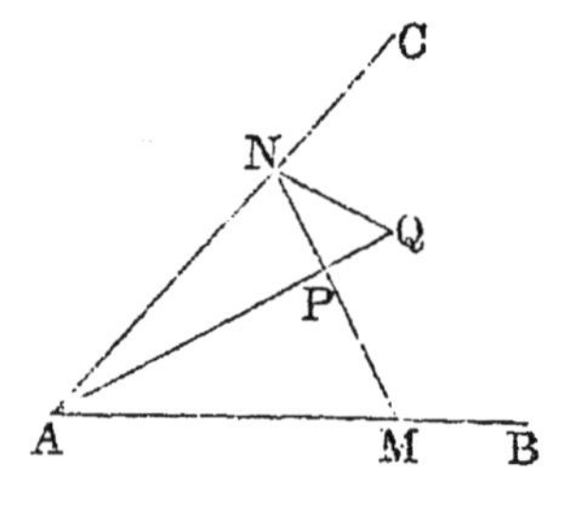

Produce AP to Q, so that the rectangle AP, PQ may be equal to the given rectangle. Then the rectangle MP, PN is equal to the rectangle AP, PQ. Therefore a circle will go round $AMQN$ (*Note on* III. 35). Therefore the angle PNQ is equal to the angle PAM (III. 21).

Hence we have the following synthetical solution. Produce AP to Q, so that the rectangle AP, PQ may be equal to the given rectangle; describe on PQ a segment of a circle containing an angle equal to the angle PAM; join P with a point of intersection of this circle and AC; the straight line thus drawn solves the problem.

45. *In a given circle it is required to inscribe a triangle so that two sides may pass through two given points, and the third side be parallel to a given straight line.*

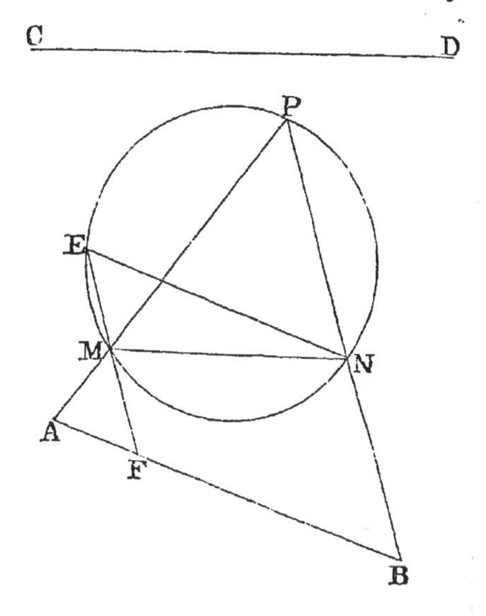

Let A and B be the given points, and CD the given straight line. Suppose PMN to be the required triangle inscribed in the given circle.

Draw NE parallel to AB; join EM, and produce it if necessary to meet AB at F.

If the point F were known the problem might be considered solved. For ENM is a known angle, and therefore the chord EM is known in magnitude. And then, since F is a known point, and EM is a known magnitude, the position of M becomes known.

We have then only to shew how F is to be determined. The angle MEN is equal to the angle MFA (I. 29). The angle MEN is equal to the angle MPN (III. 21). Hence MAF and BAP are similar triangles (VI. 4). Therefore MA is to AF as BA is to AP. Therefore the rectangle MA, AP is equal to the rectangle BA, AF (VI. 16). But since A is a *given* point the rectangle MA, AP is known; and AB is known; thus AF is determined.

46. *In a given circle it is required to inscribe a triangle so that the sides may pass through three given points.*

Let A, B, C be the three given points. Suppose PMN to be the required triangle inscribed in the given circle.

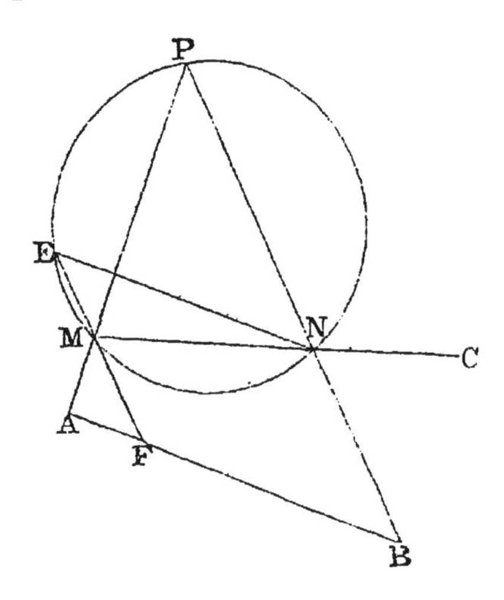

Draw NE parallel to AB, and determine the point F as in the preceding problem. We shall then have to describe in the given circle a triangle EMN so that two of its sides may pass through given points, F and C, and the third side be parallel to a given straight line AB. This can be done by the preceding problem.

This example and the preceding are taken from the work of Catalan already cited. The present problem is sometimes called *Castillon's* and sometimes *Cramer's;* the history of the general researches to which it has given rise will be found in a series of papers in the *Mathematician*, Vol. III. by the late T. S. Davies.

ON LOCI.

47. A *locus* consists of all the points which satisfy certain conditions and of those points alone. Thus, for example, the locus of the points which are at a given distance

from a given point is the surface of the sphere described from the given point as centre, with the given distance as radius; for all the points on this surface, and no other points, are at the given distance from the given point. If we restrict ourselves to all the points in a fixed plane which are at a given distance from a given point, the locus is the circumference of the circle described from the given point as centre, with the given distance as radius. In future we shall restrict ourselves to loci which are situated in a fixed plane, and which are properly called *plane loci*.

Several of the propositions in Euclid furnish good examples of loci. Thus the locus of the vertices of all triangles which are on the same base and on the same side of it, and which have the same area, is a straight line parallel to the base; this is shewn in I. 37 and I. 39.

Again, the locus of the vertices of all triangles which are on the same base and on the same side of it, and which have the same vertical angle, is a segment of a circle described on the base; for it is shewn in III. 21, that all the points thus determined satisfy the assigned conditions, and it is easily shewn that no other points do.

We will now give some examples. In each example we ought to shew not only that all the points which we indicate as the locus do fulfil the assigned conditions, but that no other points do. This second part however we leave to the student in all the examples except the last two; in these, which are more difficult, we have given the complete investigation.

48. *Required the locus of points which are equidistant from two given points.*

Let A and B be the two given points; join AB; and draw a straight line through the middle point of AB at right angles to AB; then it may be easily shewn that this straight line is the required locus.

49. *Required the locus of the vertices of all triangles on a given base* AB, *such that the square on the side terminated at* A *may exceed the square on the side terminated at* B, *by a given square.*

Suppose C to denote a point on the required locus; from C draw a perpendicular on the given base, meeting it, produced if necessary, at D. Then the square on AC is equal to the squares on AD and CD, and the square on BC is equal to the squares on BD and CD (I. 47); therefore the square on AC exceeds the square on BC by as much as the square on AD exceeds the square on BD. Hence D is a fixed point either in AB or in AB produced through B (40). And the required locus is the straight line drawn through D, at right angles to AB.

50. *Required the locus of a point such that the straight lines drawn from it to touch two given circles may be equal.*

Let A be the centre of the greater circle, B the centre of a smaller circle; and let P denote any point on the required locus. Since the straight lines drawn from P to touch the given circles are equal, the squares on these straight lines are equal. But the squares on PA and PB exceed these equal squares by the squares on the radii of the respective circles. Hence the square on PA exceeds the square on PB, by a known square, namely a square equal to the excess of the square on the radius of the circle of which A is the centre over the square on the radius of the circle of which B is the centre. Hence, the required locus is a certain straight line which is at right angles to AB (49).

This straight line is called the *radical axis* of the two circles.

If the given circles intersect, it follows from III. 36, that the straight line which is the locus coincides with the produced parts of the common chord of the two circles.

51. *Required the locus of the middle points of all the chords of a circle which pass through a fixed point.*

Let A be the centre of the given circle; B the fixed

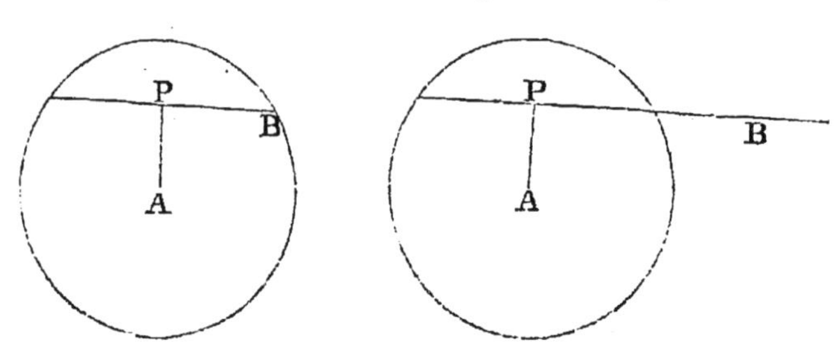

point; let any chord of the circle be drawn so that, produced if necessary, it may pass through B. Let P be the middle point of this chord, so that P is a point on the required locus.

The straight line AP is at right angles to the chord of which P is the middle point (III. 3); therefore P is on the circumference of a circle of which AB is a diameter. Hence if B be within the given circle the locus is the circumference of the circle described on AB as diameter; if B be without the given circle the locus is that part of the circumference of the circle described on AB as diameter, which is within the given circle.

52. *O is a fixed point from which any straight line is drawn meeting a fixed straight line at* P; *in* OP *a point* Q *is taken such that* OQ *is to* OP *in a fixed ratio: determine the locus of* Q.

We shall shew that the locus of Q is a straight line.

For draw a perpendicular from O on the fixed straight line, meeting it at C; in OC take a point D such that OD is to OC in the fixed ratio; draw from O any straight line OP meeting the fixed straight line at P, and in OP take a point Q such that OQ is to OP in the fixed ratio; join

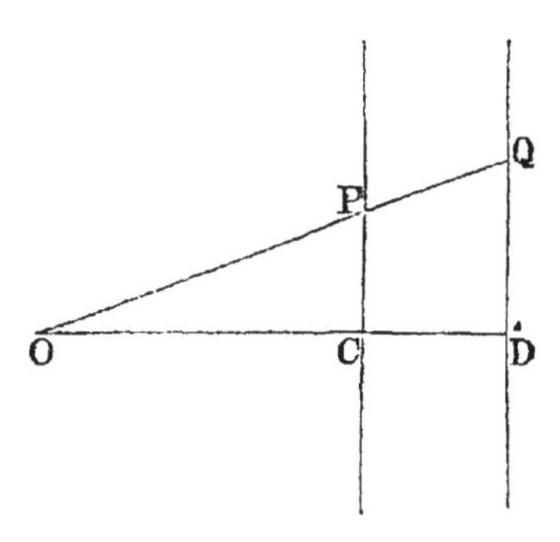

QD. The triangles ODQ and OCP are similar (VI. 6); therefore the angle ODQ is equal to the angle OCP, and is therefore a right angle. Hence Q lies in the straight line drawn through D at right angles to OD.

53. *O is a fixed point from which any straight line is drawn meeting the circumference of a fixed circle at* P; *in* OP *a point* Q *is taken such that* OQ *is to* OP *in a fixed ratio: determine the locus of* Q.

We shall shew that the locus is the circumference of a circle.

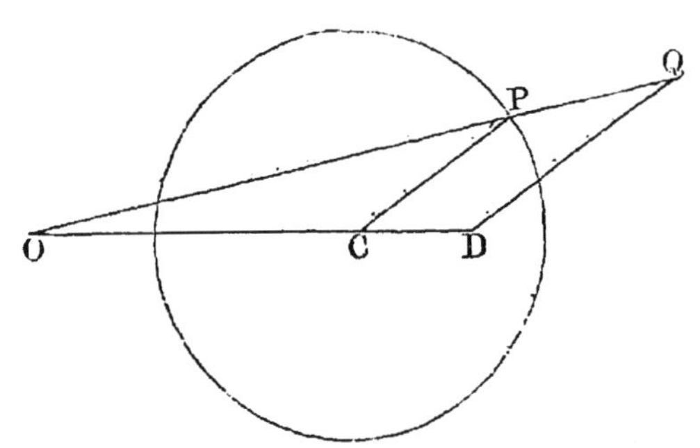

For let C be the centre of the fixed circle; in OC take a point D such that OD is to OC in the fixed ratio, and draw any radius CP of the fixed circle; draw DQ parallel to CP meeting OP, produced if necessary, at Q. Then the triangles OCP and ODQ are similar (VI. 4), and therefore OQ is to OP as OD is to OC, that is, in the fixed ratio. Therefore Q is a point on the locus. And DQ is to CP in the fixed ratio, so that DQ is of constant length. Hence the locus is the circumference of a circle of which D is the centre.

54. *There are four given points* A, B, C, D *in a straight line; required the locus of a point at which* AB *and* CD *subtend equal angles.*

Find a point O in the straight line, such that the rectangle OA, OD may be equal to the rectangle OB, OC (34), and take OK such that the square on OK may be equal to either of these rectangles (II. 14): the circumference of the circle described from O as centre, with radius OK, shall be the required locus.

[We will take the case in which the points are in the following order, O, A, B, C, D.]

For let P be any point on the circumference of this

circle. Describe a circle round PAD, and also a circle

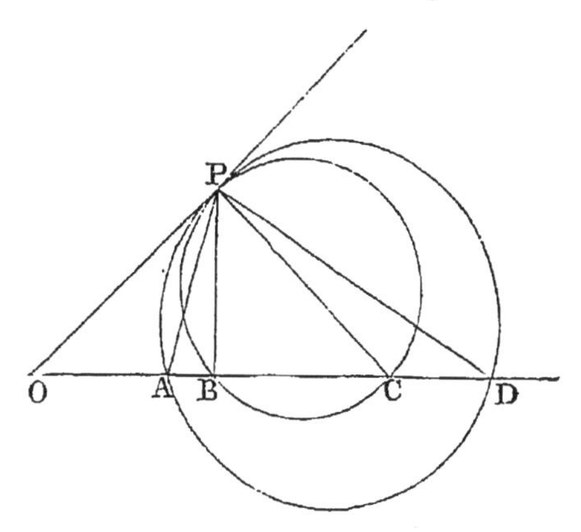

round PBC; then OP touches each of these circles (III. 37); therefore the angle OPA is equal to the angle PDA, and the angle OPB is equal to the angle PCB (III. 32). But the angle OPB is equal to the angles OPA and APB together, and the angle PCB is equal to the angles CPD and PDA together (I. 32). Therefore the angles OPA and APB together are equal to the angles CPD and PDA together; and the angle OPA has been shewn equal to the angle PDA; therefore the angle APB is equal to the angle CPD.

We have thus shewn that any point on the circumference of the circle satisfies the assigned conditions; we shall now shew that any point which satisfies the assigned conditions is on the circumference of the circle.

For take any point Q which satisfies the required conditions. Describe a circle round QAD, and also a circle round QBC. These circles will touch the same straight line at Q; for the angles AQB and CQD are equal, and the converse of III. 32 is true. Let this straight line which touches both circles at Q be drawn; and let it meet the straight line containing the four given points at R. Then the rectangle RA, RD is equal to the rectangle RB, RC; for each is equal to the square on RQ (III. 36). Therefore R must coincide with O (34); and therefore RQ must be equal to OK. Thus Q must be on the circumference of the circle of which O is the centre, and OK the radius.

55. *Required the locus of the vertices of all the triangles* ABC *which stand on a given base* AB, *and have the side* AC *to the side* BC *in a constant ratio.*

If the sides AC and BC are to be equal, the locus is

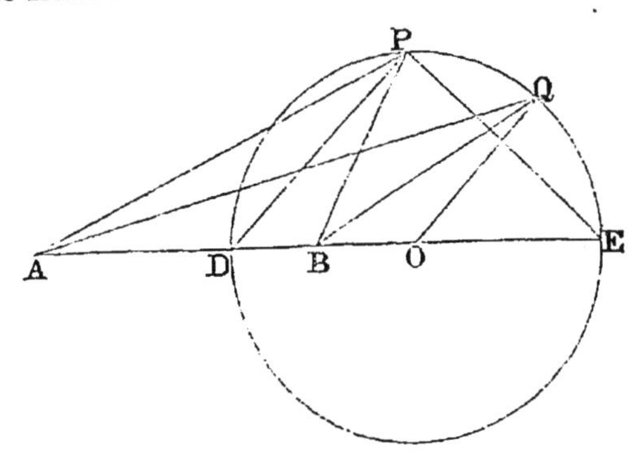

the straight line which bisects AB at right angles. We will suppose that the ratio is greater than a ratio of equality; so that AC is to be the greater side.

Divide AB at D so that AD is to DB in the given ratio (VI. 10); and produce AB to E, so that AE is to EB in the given ratio. Let P be any point in the required locus; join PD and PE. Then PD bisects the angle APB, and PE bisects the angle between BP and AP produced. Therefore the angle DPE is a right angle. Therefore P is on the circumference of a circle described on DE as diameter.

We have thus shewn that any point which satisfies the assigned conditions is on the circumference of the circle described on DE as diameter; we shall now shew that any point on the circumference of this circle satisfies the assigned conditions.

Let Q be any point on the circumference of this circle, QA shall be to QB in the assigned ratio. For, take O the centre of the circle; and join QO. Then, by construction, AE is to EB as AD is to DB, and therefore, alternately, AE is to AD as EB is to DB; therefore the sum of AE and AD is to their difference as the sum of EB and DB is to their difference (23); that is, twice AO is to twice DO as twice DO is to twice BO; therefore AO is to DO as DO is to BO; that is, AO is to OQ as QO is to OB. Therefore the triangles AOQ and QOB are similar triangles (VI. 6); and therefore AQ is to QB as QO is to BO. This shews that the ratio of AQ to BQ is *constant;* we have still to shew that this ratio is the same as the assigned ratio.

We have already shewn that AO is to DO as DO is to BO; therefore, the difference of AO and DO is to DO as the difference of DO and BO is to BO (V. 17); that is, AD is to DO as BD is to BO; therefore AD is to BD as DO is to BO; that is, AD is to DB as QO is to BO. This shews that the ratio of QO to BO is the same as the assigned ratio.

ON MODERN GEOMETRY.

56. We have hitherto restricted ourselves to Euclid's Elements, and propositions which can be demonstrated by strict adherence to Euclid's methods. In modern times various other methods have been introduced, and have led to numerous and important results. These methods may be called semi-geometrical, as they are not confined within the limits of the ancient pure geometry; in fact the power of the modern methods is obtained chiefly by combining arithmetic and algebra with geometry. The student who desires to cultivate this part of mathematics may consult Townsend's *Chapters on the Modern Geometry of the Point, Line, and Circle.*

We will give as specimens some important theorems, taken from what is called the theory of transversals.

Any line, straight or curved, which cuts a system of other lines is called a *transversal;* in the examples which we shall give, the lines will be straight lines, and the system will consist of three straight lines forming a triangle.

We will give a brief enunciation of the theorem which we are about to prove, for the sake of assisting the memory in retaining the result; but the enunciation will not be fully comprehended until the demonstration is completed.

57. *If a straight line cut the sides, or the sides produced, of a triangle, the product of three segments in order is equal to the product of the other three segments.*

Let ABC be a triangle, and let a straight line be drawn cutting the side BC at D, the side CA at E, and the side AB produced through B at F. Then BD and DC are

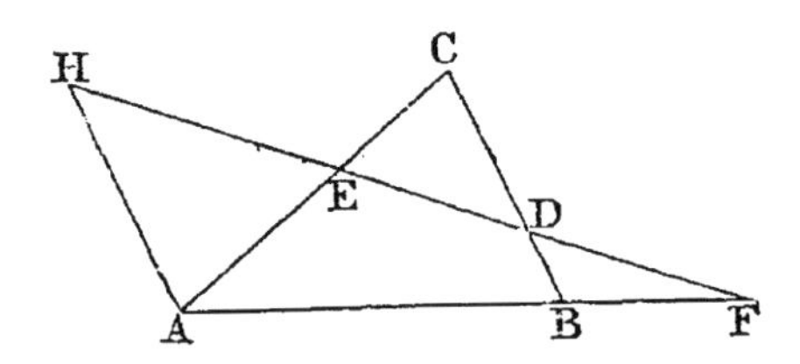

called *segments* of the side BC, and CE and EA are called *segments* of the side CA, and also AF and FB are called *segments* of the side AB.

Through A draw a straight line parallel to BC, meeting DF produced at H.

Then the triangles CED and EAH are equiangular to one another; therefore AH is to CD as AE is to EC (VI. 4).

Therefore the rectangle AH, EC is equal to the rectangle CD, AE (VI. 16).

Again, the triangles FAH and FBD are equiangular to one another; therefore AH is to BD as FA is to FB (VI. 4).

Therefore the rectangle AH, FB is equal to the rectangle BD, FA (VI. 16).

Now suppose the straight lines represented by numbers in the manner explained in the notes to the second Book of the Elements. We have then two results which we can express arithmetically: namely, the *product* $AH.EC$ is equal to the *product* $CD.AE$; and the *product* $AH.FB$ is equal to the *product* $BD.FA$.

Therefore, by the principles of arithmetic, the product $AH.EC.BD.FA$ is equal to the product $AH.FB.CD.AE$, and therefore, by the principles of arithmetic, the product $BD.CE.AF$ is equal to the product $DC.EA.FB$.

This is the result intended by the enunciation given above. Each product is made by three segments, one from every side of the triangle: and the two segments which terminated at any angular point of the triangle are never in the same product. Thus if we begin one product with the segment BD, the other segment of the side BC, namely DC, occurs in the other product; then the segment CE occurs in the first product, so that the two segments CD and CE, which terminate at C, do not occur in the same product; and so on.

The student should for exercise draw another figure for the case in which the transversal meets *all* the sides *produced*, and obtain the same result.

58. Conversely, it may be shewn by an indirect proof that if the product $BD.CE.AF$ be equal to the product $DC.EA.FB$, the three points D, E, F lie in the same straight line.

59. *If three straight lines be drawn through the angular points of a triangle to the opposite sides, and meet at the same point, the product of three segments in order is equal to the product of the other three segments.*

Let ABC be a triangle. From the angular points to the opposite sides let the straight lines AOD, BOE, COF be drawn, which meet at the point O: the product $AF.BD.CE$ shall be equal to the product $FB.DC.EA$.

For the triangle ABD is cut by the transversal FOC, and therefore by the theorem in 57 the following products are equal, $AF.BC.DO$, and $FB.CD.OA$.

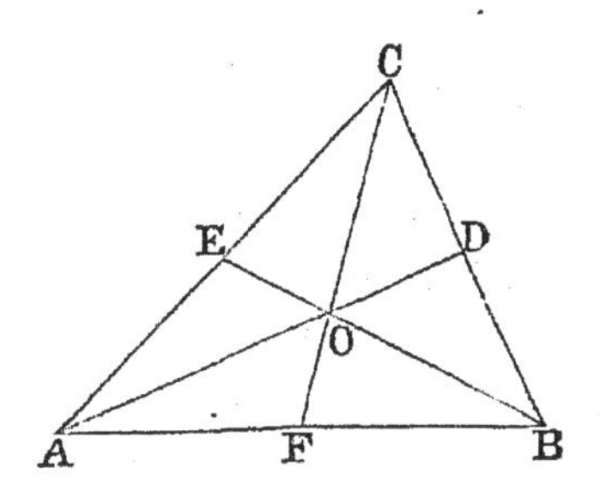

Again, the triangle ACD is cut by the transversal EOB, and therefore by the theorem in 57 the following products are equal, $AO.DB.CE$ and $OD.BC.EA$.

Therefore, by the principles of arithmetic, the following products are equal, $AF.BC.DO.AO.DB.CE$ and $FB.CD.OA.OD.BC.EA$. Therefore the following

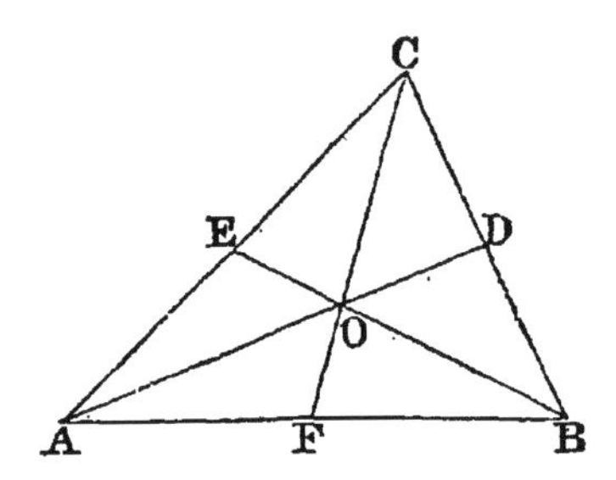

products are equal, $AF.BD.CE$ and $FB.DC.EA$.

We have supposed the point O to be within the triangle; if O be without the triangle two of the points D, E, F will fall on the sides produced.

60. Conversely, it may be shewn by an indirect proof that if the product $AF.BD.CE$ be equal to the product $FB.DC.EA$, the three straight lines AD, BE, CF meet at the same point.

61. We may remark that in geometrical problems the following terms sometimes occur, used in the same sense as in arithmetic; namely *arithmetical progression*, *geometrical progression*, and *harmonical progression*. A proposition respecting harmonical progression, which deserves notice, will now be given.

62. *Let ABC be a triangle; let the angle A be bisected by a straight line which meets BC at D, and let the exterior angle at A be bisected by a straight line which meets BC, produced through C, at E: then BD, BC, BE shall be in harmonical progression.*

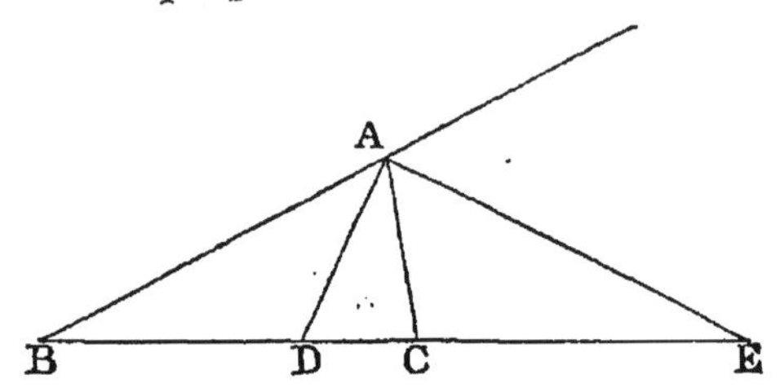

For BD is to DC as BA is to AC (VI. 3); and BE is to EC as BA is to AC (VI. A). Therefore BD is to DC as BE is to EC (V. 11). Therefore BD is to BE as DC is to EC (V. 16). Thus of the three straight lines BD, BC, BE, the first is to the third as the excess of the second over the first is to the excess of the third over the second. Therefore BD, BC, BE are in harmonical progression.

This result is sometimes expressed by saying that BE is divided harmonically at D and C.

EXERCISES IN EUCLID.

I. 1 to 15.

1. On a given straight line describe an isosceles triangle having each of the sides equal to a given straight line.

2. In the figure of I. 2 if the diameter of the smaller circle is the radius of the larger, shew where the given point and the vertex of the constructed triangle will be situated.

3. If two straight lines bisect each other at right angles, any point in either of them is equidistant from the extremities of the other.

4. If the angles ABC and ACB at the base of an isosceles triangle be bisected by the straight lines BD, CD, shew that DBC will be an isosceles triangle.

5. BAC is an isosceles triangle having each of the angles B and C double of the third angle A. If BD bisects the angle B and meets AC at D, shew that BD is equal to AD.

6. In the figure of I. 5 if FC and BG meet at H shew that FH and GH are equal.

7. In the figure of I. 5 if FC and BG meet at H, shew that AH bisects the angle BAC.

8. The sides AB, AD of a quadrilateral $ABCD$ are equal, and the diagonal AC bisects the angle BAD: shew that the sides CB and CD are equal, and that the diagonal AC bisects the angle BCD.

9. ACB, ADB are two triangles on the same side of AB, such that AC is equal to BD, and AD is equal to BC, and AD and BC intersect at O: shew that the triangle AOB is isosceles.

10. The opposite angles of a rhombus are equal.

11. A diagonal of a rhombus bisects each of the angles through which it passes.

12. If two isosceles triangles are on the same base the straight line joining their vertices, or that straight line produced, will bisect the base at right angles.

13. Find a point in a given straight line such that its distances from two given points may be equal.

14. Through two given points on opposite sides of a given straight line draw two straight lines which shall meet in that given straight line, and include an angle bisected by that given straight line.

15. A given angle BAC is bisected; if CA is produced to G and the angle BAG bisected, the two bisecting lines are at right angles.

16. If four straight lines meet at a point so that the opposite angles are equal, these straight lines are two and two in the same straight line.

I. 16 to 26.

17. ABC is a triangle and the angle A is bisected by a straight line which meets BC at D; shew that BA is greater than BD, and CA greater than CD.

18. In the figure of I. 17 shew that ABC and ACB are together less than two right angles, by joining A to any point in BC.

19. $ABCD$ is a quadrilateral of which AD is the longest side and BC the shortest; shew that the angle ABC is greater than the angle ADC, and the angle BCD greater than the angle BAD.

20. If a straight line be drawn through A one of the angular points of a square, cutting one of the opposite sides, and meeting the other produced at F, shew that AF is greater than the diagonal of the square.

21. The perpendicular is the shortest straight line that can be drawn from a given point to a given straight line; and of others, that which is nearer to the perpendicular is less than the more remote; and two, and only two, equal straight lines can be drawn from the given point to the given straight line, one on each side of the perpendicular.

22. The sum of the distances of any point from the three angles of a triangle is greater than half the sum of the sides of the triangle.

23. The four sides of any quadrilateral are together greater than the two diagonals together.

24. The two sides of a triangle are together greater than twice the straight line drawn from the vertex to the middle point of the base.

25. If one angle of a triangle is equal to the sum of the other two, the triangle can be divided into two isosceles triangles.

26. If the angle C of a triangle is equal to the sum of the angles A and B, the side AB is equal to twice the straight line joining C to the middle point of AB.

27. Construct a triangle, having given the base, one of the angles at the base, and the sum of the sides.

28. The perpendiculars let fall on two sides of a triangle from any point in the straight line bisecting the angle between them are equal to each other.

29. In a given straight line find a point such that the perpendiculars drawn from it to two given straight lines shall be equal.

30. Through a given point draw a straight line such that the perpendiculars on it from two given points may be on opposite sides of it and equal to each other.

31. A straight line bisects the angle A of a triangle ABC; from B a perpendicular is drawn to this bisecting line, meeting it at D, and BD is produced to meet AC or AC produced at E: shew that BD is equal to DE.

32. AB, AC are any two straight lines meeting at A: through any point P draw a straight line meeting them at E and F, such that AE may be equal to AF.

33. Two right-angled triangles have their hypotenuses equal, and a side of one equal to a side of the other: shew that they are equal in all respects.

I. 27 to 31.

34. Any straight line parallel to the base of an isosceles triangle makes equal angles with the sides.

35. If two straight lines A and B are respectively parallel to two others C and D, shew that the inclination of A to B is equal to that of C to D.

36. A straight line is drawn terminated by two parallel straight lines; through its middle point any straight line is

drawn and terminated by the parallel straight lines. Shew that the second straight line is bisected at the middle point of the first.

37. If through any point equidistant from two parallel straight lines, two straight lines be drawn cutting the parallel straight lines, they will intercept equal portions of these parallel straight lines.

38. If the straight line bisecting the exterior angle of a triangle be parallel to the base, shew that the triangle is isosceles.

39. Find a point B in a given straight line CD, such that if AB be drawn to B from a given point A, the angle ABC will be equal to a given angle.

40. If a straight line be drawn bisecting one of the angles of a triangle to meet the opposite side, the straight lines drawn from the point of section parallel to the other sides, and terminated by these sides, will be equal.

41. The side BC of a triangle ABC is produced to a point D; the angle ACB is bisected by the straight line CE which meets AB at E. A straight line is drawn through E parallel to BC, meeting AC at F, and the straight line bisecting the exterior angle ACD at G. Shew that EF is equal to FG.

42. AB is the hypotenuse of a right-angled triangle ABC: find a point D in AB such that DB may be equal to the perpendicular from D on AC.

43. ABC is an isosceles triangle: find points D, E in the equal sides AB, AC such that BD, DE, EC may all be equal.

44. A straight line drawn at right angles to BC the base of an isosceles triangle ABC cuts the side AB at D and CA produced at E: shew that AED is an isosceles triangle.

I. 32.

45. From the extremities of the base of an isosceles triangle straight lines are drawn perpendicular to the sides; shew that the angles made by them with the base are each equal to half the vertical angle.

46. On the sides of any triangle ABC equilateral triangles BCD, CAE, ABF are described, all external: shew that the straight lines AD, BE, CF are all equal.

47. What is the magnitude of the angles of a regular octagon?

48. Through two given points draw two straight lines forming with a straight line given in position an equilateral triangle.

49. If the straight lines bisecting the angles at the base of an isosceles triangle be produced to meet, they will contain an angle equal to an exterior angle of the triangle.

50. A is the vertex of an isosceles triangle ABC, and BA is produced to D, so that AD is equal to BA; and DC is drawn: shew that BCD is a right angle.

51. ABC is a triangle, and the exterior angles at B and C are bisected by the straight lines BD, CD respectively, meeting at D: shew that the angle BDC together with half the angle BAC make up a right angle.

52. Shew that any angle of a triangle is obtuse, right, or acute, according as it is greater than, equal to, or less than the other two angles of the triangle taken together.

53. Construct an isosceles triangle having the vertical angle four times each of the angles at the base.

54. In the triangle ABC the side BC is bisected at E and AB at G; AE is produced to F so that EF is equal to AE, and CG is produced to H so that GH is equal to CG: shew that FB and HB are in one straight line.

55. Construct an isosceles triangle which shall have one-third of each angle at the base equal to half the vertical angle.

56. AB, AC are two straight lines given in position: it is required to find in them two points P and Q, such that, PQ being joined, AP and PQ may together be equal to a given straight line, and may contain an angle equal to a given angle.

57. Straight lines are drawn through the extremities of the base of an isosceles triangle, making angles with it on the side remote from the vertex, each equal to one-third of one of the equal angles of the triangle and meeting the sides produced: shew that three of the triangles thus formed are isosceles.

58. AEB, CED are two straight lines intersecting at E; straight lines AC, DB are drawn forming two triangles ACE, BED; the angles ACE, DBE are bisected by the straight lines CF, BF, meeting at F. Shew that the angle CFB is equal to half the sum of the angles EAC, EDB.

59. The straight line joining the middle point of the hypotenuse of a right-angled triangle to the right angle is equal to half the hypotenuse.

60. From the angle A of a triangle ABC a perpendicular is drawn to the opposite side, meeting it, produced if necessary, at D; from the angle B a perpendicular is drawn to the opposite side, meeting it, produced if necessary, at E: shew that the straight lines which join D and E to the middle point of AB are equal.

61. From the angles at the base of a triangle perpendiculars are drawn to the opposite sides, produced if necessary: shew that the straight line joining the points of intersection will be bisected by a perpendicular drawn to it from the middle point of the base.

62. In the figure of I. 1, if C and H be the points of intersection of the circles, and AB be produced to meet one of the circles at K, shew that CHK is an equilateral triangle.

63. The straight lines bisecting the angles at the base of an isosceles triangle meet the sides at D and E: shew that DE is parallel to the base.

64. AB, AC are two given straight lines, and P is a given point in the former: it is required to draw through P a straight line to meet AC at Q, so that the angle APQ may be three times the angle AQP.

65. Construct a right-angled triangle, having given the hypotenuse and the sum of the sides.

66. Construct a right-angled triangle, having given the hypotenuse and the difference of the sides.

67. Construct a right-angled triangle, having given the hypotenuse and the perpendicular from the right angle on it.

68. Construct a right-angled triangle, having given the perimeter and an angle.

69. Trisect a right angle.

70. Trisect a given finite straight line.

71. From a given point it is required to draw to two parallel straight lines, two equal straight lines at right angles to each other.

72. Describe a triangle of given perimeter, having its angles equal to those of a given triangle.

I. 33, 34.

73. If a quadrilateral have two of its opposite sides parallel, and the two others equal but not parallel, any two of its opposite angles are together equal to two right angles.

74. If a straight line which joins the extremities of two equal straight lines, not parallel, make the angles on the same side of it equal to each other, the straight line which joins the other extremities will be parallel to the first.

75. No two straight lines drawn from the extremities of the base of a triangle to the opposite sides can possibly bisect each other.

76. If the opposite sides of a quadrilateral are equal it is a parallelogram.

77. If the opposite angles of a quadrilateral are equal it is a parallelogram.

78. The diagonals of a parallelogram bisect each other.

79. If the diagonals of a quadrilateral bisect each other it is a parallelogram.

80. If the straight line joining two opposite angles of a parallelogram bisect the angles the four sides of the parallelogram are equal.

81. Draw a straight line through a given point such that the part of it intercepted between two given parallel straight lines may be of given length.

82. Straight lines bisecting two adjacent angles of a parallelogram intersect at right angles.

83. Straight lines bisecting two opposite angles of a parallelogram are either parallel or coincident.

84. If the diagonals of a parallelogram are equal all its angles are equal.

85. Find a point such that the perpendiculars let fall from it on two given straight lines shall be respectively equal to two given straight lines. How many such points are there?

86. It is required to draw a straight line which shall be equal to one straight line and parallel to another, and be terminated by two given straight lines.

87. On the sides AB, BC, and CD of a parallelogram $ABCD$ three equilateral triangles are described, that on BC towards the same parts as the parallelogram, and those on AB, CD towards the opposite parts: shew that the

distances of the vertices of the triangles on AB, CD from that on BC are respectively equal to the two diagonals of the parallelograms.

88. If the angle between two adjacent sides of a parallelogram be increased, while their lengths do not alter, the diagonal through their point of intersection will diminish.

89. A, B, C are three points in a straight line, such that AB is equal to BC: shew that the sum of the perpendiculars from A and C on any straight line which does not pass between A and C is double the perpendicular from B on the same straight line.

90. If straight lines be drawn from the angles of any parallelogram perpendicular to any straight line which is outside the parallelogram, the sum of those from one pair of opposite angles is equal to the sum of those from the other pair of opposite angles.

91. If a six-sided plane rectilineal figure have its opposite sides equal and parallel, the three straight lines joining the opposite angles will meet at a point.

92. AB, AC are two given straight lines; through a given point E between them it is required to draw a straight line GEH such that the intercepted portion GH shall be bisected at the point E.

93. Inscribe a rhombus within a given rhombus, so that one of the angular points of the inscribed figure may bisect a side of the other.

94. $ABCD$ is a parallelogram, and E, F, the middle points of AB and BC respectively; shew that BE and DF will trisect the diagonal AC.

I. 35 to 45.

95. $ABCD$ is a quadrilateral having BC parallel to AD; shew that its area is the same as that of the parallelogram which can be formed by drawing through the middle point of DC a straight line parallel to AB.

96. $ABCD$ is a quadrilateral having BC parallel to AD, E is the middle point of DC; shew that the triangle AEB is half the quadrilateral.

97. Shew that any straight line passing through the middle point of the diameter of a parallelogram and terminated by two opposite sides, bisects the parallelogram.

98. Bisect a parallelogram by a straight line drawn through a given point within it.

99. Construct a rhombus equal to a given parallelogram.

100. If two triangles have two sides of the one equal to two sides of the other, each to each, and the sum of the two angles contained by these sides equal to two right angles, the triangles are equal in area.

101. A straight line is drawn bisecting a parallelogram $ABCD$ and meeting AD at E and BC at F: shew that the triangles EBF and CED are equal.

102. Shew that the four triangles into which a parallelogram is divided by its diagonals are equal in area.

103. Two straight lines AB and CD intersect at E, and the triangle AEC is equal to the triangle BED: shew that BC is parallel to AD.

104. $ABCD$ is a parallelogram; from any point P in the diagonal BD the straight lines PA, PC are drawn. Shew that the triangles PAB and PCB are equal.

105. If a triangle is described having two of its sides equal to the diagonals of any quadrilateral, and the included angle equal to either of the angles between these diagonals, then the area of the triangle is equal to the area of the quadrilateral.

106. The straight line which joins the middle points of two sides of any triangle is parallel to the base.

107. Straight lines joining the middle points of adjacent sides of a quadrilateral form a parallelogram.

108. D, E are the middle points of the sides AB, AC of a triangle, and CD, BE intersect at F: shew that the triangle BFC is equal to the quadrilateral $ADFE$.

109. The straight line which bisects two sides of any triangle is half the base.

110. In the base AC of a triangle take any point D; bisect AD, DC, AB, BC at the points E, F, G, H respectively: shew that EG is equal and parallel to FH.

111. Given the middle points of the sides of a triangle, construct the triangle.

112. If the middle points of any two sides of a triangle be joined, the triangle so cut off is one quarter of the whole.

113. The sides AB, AC of a given triangle ABC are bisected at the points E, F; a perpendicular is drawn from A to the opposite side, meeting it at D. Shew that the

angle FDE is equal to the angle BAC. Shew also that $AFDE$ is half the triangle ABC.

114. Two triangles of equal area stand on the same base and on opposite sides: shew that the straight line joining their vertices is bisected by the base or the base produced.

115. Three parallelograms which are equal in all respects are placed with their equal bases in the same straight line and contiguous; the extremities of the base of the first are joined with the extremities of the side opposite to the base of the third, towards the same parts: shew that the portion of the new parallelogram cut off by the second is one half the area of any one of them.

116. $ABCD$ is a parallelogram; from D draw any straight line DFG meeting BC at F and AB produced at G; draw AF and CG: shew that the triangles ABF, CFG are equal.

117. ABC is a given triangle: construct a triangle of equal area, having for its base a given straight line AD, coinciding in position with AB.

118. ABC is a given triangle: construct a triangle of equal area, having its vertex at a given point in BC and its base in the same straight line as AB.

119. $ABCD$ is a given quadrilateral: construct another quadrilateral of equal area having AB for one side, and for another a straight line drawn through a given point in CD parallel to AB.

120. $ABCD$ is a quadrilateral: construct a triangle whose base shall be in the same straight line as AB, vertex at a given point P in CD, and area equal to that of the given quadrilateral.

121. ABC is a given triangle: construct a triangle of equal area, having its base in the same straight line as AB, and its vertex in a given straight line parallel to AB.

122. Bisect a given triangle by a straight line drawn through a given point in a side.

123. Bisect a given quadrilateral by a straight line drawn through a given angular point.

124. If through the point O within a parallelogram $ABCD$ two straight lines are drawn parallel to the sides, and the parallelograms OB and OD are equal, the point O is in the diagonal AC.

I. 46 to 48.

125. On the sides AC, BC of a triangle ABC, squares $ACDE$, $BCFH$ are described: shew that the straight lines AF and BD are equal.

126. The square on the side subtending an acute angle of a triangle is less than the squares on the sides containing the acute angle.

127. The square on the side subtending an obtuse angle of a triangle is greater than the squares on the sides containing the obtuse angle.

128. If the square on one side of a triangle be less than the squares on the other two sides, the angle contained by these sides is an acute angle; if greater, an obtuse angle.

129. A straight line is drawn parallel to the hypotenuse of a right-angled triangle, and each of the acute angles is joined with the points where this straight line intersects the sides respectively opposite to them: shew that the squares on the joining straight lines are together equal to the square on the hypotenuse and the square on the straight line drawn parallel to it.

130. If any point P be joined to A, B, C, D, the angular points of a rectangle, the squares on PA and PC are together equal to the squares on PB and PD.

131. In a right-angled triangle if the square on one of the sides containing the right angle be three times the square on the other, and from the right angle two straight lines be drawn, one to bisect the opposite side, and the other perpendicular to that side, these straight lines divide the right angle into three equal parts.

132. If ABC be a triangle whose angle A is a right angle, and BE, CF be drawn bisecting the opposite sides respectively, shew that four times the sum of the squares on BE and CF is equal to five times the square on BC.

133. On the hypotenuse BC, and the sides CA, AB of a right-angled triangle ABC, squares $BDEC$, AF, and AG are described: shew that the squares on DG and EF are together equal to five times the square on BC.

II. 1 to 11.

134. A straight line is divided into two parts; shew that if twice the rectangle of the parts is equal to the sum of the squares described on the parts, the straight line is bisected.

135. Divide a given straight line into two parts such that the rectangle contained by them shall be the greatest possible.

136. Construct a rectangle equal to the difference of two given squares.

137. Divide a given straight line into two parts such that the sum of the squares on the two parts may be the least possible.

138. Shew that the square on the sum of two straight lines together with the square on their difference is double the squares on the two straight lines.

139. Divide a given straight line into two parts such that the sum of their squares shall be equal to a given square.

140. Divide a given straight line into two parts such that the square on one of them may be double the square on the other.

141. In the figure of II. 11 if CH be produced to meet BF at L, shew that CL is at right angles to BF.

142. In the figure of II. 11 if BE and CH meet at O, shew that AO is at right angles to CL.

143. Shew that in a straight line divided as in II. 11 the rectangle contained by the sum and difference of the parts is equal to the rectangle contained by the parts.

II. 12 to 14.

144. The square on the base of an isosceles triangle is equal to twice the rectangle contained by either side and by the straight line intercepted between the perpendicular let fall on it from the opposite angle and the extremity of the base.

145. In any triangle the sum of the squares on the sides is equal to twice the square on half the base together with twice the square on the straight line drawn from the vertex to the middle point of the base.

146. ABC is a triangle having the sides AB and AC equal; if AB is produced beyond the base to D so that BD is equal to AB, shew that the square on CD is equal to the square on AB, together with twice the square on BC.

147. The sum of the squares on the sides of a parallelogram is equal to the sum of the squares on the diagonals.

148. The base of a triangle is given and is bisected by the centre of a given circle: if the vertex be at any point of the circumference, shew that the sum of the squares on the two sides of the triangle is invariable.

149. In any quadrilateral the squares on the diagonals are together equal to twice the sum of the squares on the straight lines joining the middle points of opposite sides.

150. If a circle be described round the point of intersection of the diameters of a parallelogram as a centre, shew that the sum of the squares on the straight lines drawn from any point in its circumference to the four angular points of the parallelogram is constant.

151. The squares on the sides of a quadrilateral are together greater than the squares on its diagonals by four times the square on the straight line joining the middle points of its diagonals.

152. In AB the diameter of a circle take two points C and D equally distant from the centre, and from any point E in the circumference draw EC, ED: shew that the squares on EC and ED are together equal to the squares on AC and AD.

153. In BC the base of a triangle take D such that the squares on AB and BD are together equal to the squares on AC and CD, then the middle point of AD will be equally distant from B and C.

154. The square on any straight line drawn from the vertex of an isosceles triangle to the base is less than the square on a side of the triangle by the rectangle contained by the segments of the base.

155. A square $BDEC$ is described on the hypotenuse BC of a right-angled triangle ABC: shew that the squares on DA and AC are together equal to the squares on EA and AB.

156. ABC is a triangle in which C is a right angle, and DE is drawn from a point D in AC perpendicular to

AB: shew that the rectangle *AB*, *AE* is equal to the rectangle *AC*, *AD*.

157. If a straight line be drawn through one of the angles of an equilateral triangle to meet the opposite side produced, so that the rectangle contained by the whole straight line thus produced and the part of it produced is equal to the square on the side of the triangle, shew that the square on the straight line so drawn will be double the square on a side of the triangle.

158. In a triangle whose vertical angle is a right angle a straight line is drawn from the vertex perpendicular to the base: shew that the square on this perpendicular is equal to the rectangle contained by the segments of the base.

159. In a triangle whose vertical angle is a right angle a straight line is drawn from the vertex perpendicular to the base: shew that the square on either of the sides adjacent to the right angle is equal to the rectangle contained by the base and the segment of it adjacent to that side.

160. In a triangle *ABC* the angles *B* and *C* are acute: if *E* and *F* be the points where perpendiculars from the opposite angles meet the sides *AC*, *AB*, shew that the square on *BC* is equal to the rectangle *AB*, *BF*, together with the rectangle *AC*, *CE*.

161. Divide a given straight line into two parts so that the rectangle contained by them may be equal to the square described on a given straight line which is less than half the straight line to be divided.

III. 1 to 15.

162. Describe a circle with a given centre cutting a given circle at the extremities of a diameter.

163. Shew that the straight lines drawn at right angles to the sides of a quadrilateral inscribed in a circle from their middle points intersect at a fixed point.

164. If two circles cut each other, any two parallel straight lines drawn through the points of section to cut the circles are equal.

165. Two circles whose centres are *A* and *B* intersect at *C*; through *C* two chords *DCE* and *FCG* are drawn equally inclined to *AB* and terminated by the circles: shew that *DE* and *FG* are equal.

166. Through either of the points of intersection of two given circles draw the greatest possible straight line terminated both ways by the two circumferences.

167. If from any point in the diameter of a circle straight lines are drawn to the extremities of a parallel chord, the squares on these straight lines are together equal to the squares on the segments into which the diameter is divided.

168. A and B are two fixed points without a circle PQR; it is required to find a point P in the circumference, so that the sum of the squares described on AP and BP may be the least possible.

169. If in any two given circles which touch one another, there be drawn two parallel diameters, an extremity of each diameter, and the point of contact, shall lie in the same straight line.

170. A circle is described on the radius of another circle as diameter, and two chords of the larger circle are drawn, one through the centre of the less at right angles to the common diameter, and the other at right angles to the first through the point where it cuts the less circle. Shew that these two chords have the segments of the one equal to the segments of the other, each to each.

171. Through a given point within a circle draw the shortest chord.

172. O is the centre of a circle, P is any point in its circumference, PN a perpendicular on a fixed diameter: shew that the straight line which bisects the angle OPN always passes through one or the other of two fixed points.

173. Three circles touch one another externally at the points A, B, C; from A, the straight lines AB, AC are produced to cut the circle BC at D and E: shew that DE is a diameter of BC, and is parallel to the straight line joining the centres of the other circles.

174. Circles are described on the sides of a quadrilateral as diameters: shew that the common chord of any adjacent two is parallel to the common chord of the other two.

175. Describe a circle which shall touch a given circle, have its centre in a given straight line, and pass through a given point in the given straight line.

III. 16 to 19.

176. Shew that two tangents can be drawn to a circle from a given external point, and that they are of equal length.

177. Draw parallel to a given straight line a straight line to touch a given circle.

178. Draw perpendicular to a given straight line a straight line to touch a given circle.

179. In the diameter of a circle produced, determine a point so that the tangent drawn from it to the circumference shall be of given length.

180. Two circles have the same centre: shew that all chords of the outer circle which touch the inner circle are equal.

181. Through a given point draw a straight line so that the part intercepted by the circumference of a given circle shall be equal to a given straight line not greater than the diameter.

182. Two tangents are drawn to a circle at the opposite extremities of a diameter, and cut off from a third tangent a portion AB: if C be the centre of the circle shew that ACB is a right angle.

183. Describe a circle that shall have a given radius and touch a given circle and a given straight line.

184. A circle is drawn to touch a given circle and a given straight line. Shew that the points of contact are always in the same straight line with a fixed point in the circumference of the given circle.

185. Draw a straight line to touch each of two given circles.

186. Draw a straight line to touch one given circle so that the part of it contained by another given circle shall be equal to a given straight line not greater than the diameter of the latter circle.

187. Draw a straight line cutting two given circles so that the chords intercepted within the circles shall have given lengths.

188. A quadrilateral is described so that its sides touch a circle: shew that two of its sides are together equal to the other two sides.

189. Shew that no parallelogram can be described about a circle except a rhombus.

190. ABD, ACE are two straight lines touching a circle at B and C, and if DE be joined DE is equal to BD and CE together: shew that DE touches the circle.

191. If a quadrilateral be described about a circle the angles subtended at the centre of the circle by any two opposite sides of the figure are together equal to two right angles.

192. Two radii of a circle at right angles to each other when produced are cut by a straight line which touches the circle: shew that the tangents drawn from the points of section are parallel to each other.

193. A straight line is drawn touching two circles: shew that the chords are parallel which join the points of contact and the points where the straight line through the centres meets the circumferences.

194. If two circles can be described so that each touches the other and three of the sides of a quadrilateral figure, then the difference between the sums of the opposite sides is double the common tangent drawn across the quadrilateral.

195. AB is the diameter and C the centre of a semicircle: shew that O the centre of any circle inscribed in the semicircle is equidistant from C and from the tangent to the semicircle parallel to AB.

196. If from any point without a circle straight lines be drawn touching it, the angle contained by the tangents is double the angle contained by the straight line joining the points of contact and the diameter drawn through one of them.

197. A quadrilateral is bounded by the diameter of a circle, the tangents at its extremities, and a third tangent: shew that its area is equal to half that of the rectangle contained by the diameter and the side opposite to it.

198. If a quadrilateral, having two of its sides parallel, be described about a circle, a straight line drawn through the centre of the circle, parallel to either of the two parallel sides, and terminated by the other two sides, shall be equal to a fourth part of the perimeter of the figure.

199. A series of circles touch a fixed straight line at a fixed point: shew that the tangents at the points where they cut a parallel fixed straight line all touch a fixed circle.

200. Of all straight lines which can be drawn from two given points to meet in the convex circumference of a

given circle, the sum of the two is least which make equal angles with the tangent at the point of concourse.

201. C is the centre of a given circle, CA a radius, B a point on a radius at right angles to CA; join AB and produce it to meet the circle again at D, and let the tangent at D meet CB produced at E: shew that BDE is an isosceles triangle.

202. Let the diameter BA of a circle be produced to P, so that AP equals the radius; through A draw the tangent AED, and from P draw PEC touching the circle at C and meeting the former tangent at E; join BC and produce it to meet AED at D: then will the triangle DEC be equilateral.

III. 20 to 22.

203. Two tangents AB, AC are drawn to a circle; D is any point on the circumference outside of the triangle ABC: shew that the sum of the angles ABD and ACD is constant.

204. P, Q are any points in the circumferences of two segments described on the same straight line AB; the angles PAQ, PBQ are bisected by the straight lines AR, BR meeting at R: shew that the angle ARB is constant.

205. Two segments of a circle are on the same base AB, and P is any point in the circumference of one of the segments; the straight lines APD, BPC are drawn meeting the circumference of the other segment at D and C; AC and BD are drawn intersecting at Q. Shew that the angle AQB is constant.

206. APB is a fixed chord passing through P a point of intersection of two circles AQP, PBR; and QPR is any other chord of the circles passing through P: shew that AQ and RB when produced meet at a constant angle.

207. AOB is a triangle; C and D are points in BO and AO respectively, such that the angle ODC is equal to the angle OBA: shew that a circle may be described round the quadrilateral $ABCD$.

208. $ABCD$ is a quadrilateral inscribed in a circle, and the sides AB, CD when produced meet at O: shew that the triangles AOC, BOD are equiangular.

209. Shew that no parallelogram except a rectangle can be inscribed in a circle.

210. A triangle is inscribed in a circle: shew that the sum of the angles in the three segments exterior to the triangle is equal to four right angles.

211. A quadrilateral is inscribed in a circle: shew that the sum of the angles in the four segments of the circle exterior to the quadrilateral is equal to six right angles.

212. Divide a circle into two parts so that the angle contained in one segment shall be equal to twice the angle contained in the other.

213. Divide a circle into two parts so that the angle contained in one segment shall be equal to five times the angle contained in the other.

214. If the angle contained by any side of a quadrilateral and the adjacent side produced, be equal to the opposite angle of the quadrilateral, shew that any side of the quadrilateral will subtend equal angles at the opposite angles of the quadrilateral.

215. If any two consecutive sides of a hexagon inscribed in a circle be respectively parallel to their opposite sides, the remaining sides are parallel to each other.

216. A, B, C, D are four points taken in order on the circumference of a circle; the straight lines AB, CD produced intersect at P, and AD, BC at Q: shew that the straight lines which respectively bisect the angles APC, AQC are perpendicular to each other.

217. If a quadrilateral be inscribed in a circle, and a straight line be drawn making equal angles with one pair of opposite sides, it will make equal angles with the other pair.

218. A quadrilateral can have one circle inscribed in it and another circumscribed about it: shew that the straight lines joining the opposite points of contact of the inscribed circle are perpendicular to each other.

III. 23 to 30.

219. The straight lines joining the extremities of the chords of two equal arcs of a circle, towards the same parts are parallel to each other.

220. The straight lines in a circle which join the extremities of two parallel chords are equal to each other.

221. AB is a common chord of two circles; through C any point of one circumference straight lines CAD, CBE are drawn terminated by the other circumference: shew that the arc DE is invariable.

222. Through a point C in the circumference of a circle two straight lines ACB, DCE are drawn cutting the circle at B and E: shew that the straight line which bisects the angles ACE, DCB meets the circle at a point equidistant from B and E.

223. The straight lines bisecting any angle of a quadrilateral inscribed in a circle and the opposite exterior angle, meet in the circumference of the circle.

224. AB is a diameter of a circle, and D is a given point on the circumference: draw a chord DE so that one arc between the chord and diameter may be three times the other.

225. From A and B two of the angular points of a triangle ABC, straight lines are drawn so as to meet the opposite sides at P and Q in given equal angles: shew that the straight line joining P and Q will be of the same length in all triangles on the same base AB, and having vertical angles equal to C.

226. If two equal circles cut each other, and if through one of the points of intersection a straight line be drawn terminated by the circles, the straight lines joining its extremities with the other point of intersection are equal.

227. OA, OB, OC are three chords of a circle; the angle AOB is equal to the angle BOC, and OA is nearer to the centre than OB. From B a perpendicular is drawn on OA, meeting it at P, and a perpendicular on OC produced, meeting it at Q: shew that AP is equal to CQ.

228. AB is a given finite straight line; through A two indefinite straight lines are drawn equally inclined to AB; any circle passing through A and B meets these straight lines at L and M. Shew that if AB be between AL and AM the sum of AL and AM is constant; if AB be not between AL and AM the difference of AL and AM is constant.

229. AOB and COD are diameters of a circle at right angles to each other; E is a point in the arc AC, and EFG is a chord meeting COD at F, and drawn in such a

direction that EF is equal to the radius. Shew that the arc BG is equal to three times the arc AE.

230. The straight lines which bisect the vertical angles of all triangles on the same base and on the same side of it, and having equal vertical angles, all intersect at the same point.

231. If two circles touch each other internally, any chord of the greater circle which touches the less shall be divided at the point of its contact into segments which subtend equal angles at the point of contact of the two circles.

III. 31.

232. Right-angled triangles are described on the same hypotenuse: shew that the angular points opposite the hypotenuse all lie on a circle described on the hypotenuse as diameter.

233. The circles described on the equal sides of an isosceles triangle as diameters, will intersect at the middle point of the base.

234. The greatest rectangle which can be inscribed in a circle is a square.

235. The hypotenuse AB of a right-angled triangle ABC is bisected at D, and EDF is drawn at right angles to AB, and DE and DF are cut off each equal to DA; CE and CF are joined: shew that the last two straight lines will bisect the angle C and its supplement respectively.

236. On the side AB of any triangle ABC as diameter a circle is described; EF is a diameter parallel to BC: shew that the straight lines EB and FB bisect the interior and exterior angles at B.

237. If AD, CE be drawn perpendicular to the sides BC, AB of a triangle ABC, and DE be joined, shew that the angles ADE and ACE are equal to each other.

238. If two circles ABC, ABD intersect at A and B, and AC, AD be two diameters, shew that the straight line CD will pass through B.

239. If O be the centre of a circle and OA a radius and a circle be described on OA as diameter, the circum-

ference of this circle will bisect any chord drawn through it from A to meet the exterior circle.

240. Describe a circle touching a given straight line at a given point, such that the tangents drawn to it from two given points in the straight line may be parallel.

241. Describe a circle with a given radius touching a given straight line, such that the tangents drawn to it from two given points in the straight line may be parallel.

242. If from the angles at the base of any triangle perpendiculars are drawn to the opposite sides, produced if necessary, the straight line joining the points of intersection will be bisected by a perpendicular drawn to it from the centre of the base.

243. AD is a diameter of a circle; B and C are points on the circumference on the same side of AD; a perpendicular from D on BC produced through C, meets it at E: shew that the square on AD is greater than the sum of the squares on AB, BC, CD, by twice the rectangle BC, CE.

244. AB is the diameter of a semicircle, P is a point on the circumference, PM is perpendicular to AB; on AM, BM as diameters two semicircles are described, and AP, BP meet these latter circumferences at Q, R: shew that QR will be a common tangent to them.

245. AB, AC are two straight lines, B and C are given points in the same; BD is drawn perpendicular to AC, and DE perpendicular to AB; in like manner CF is drawn perpendicular to AB, and FG to AC. Shew that EG is parallel to BC.

246. Two circles intersect at the points A and B, from which are drawn chords to a point C in one of the circumferences, and these chords, produced if necessary, cut the other circumference at D and E: shew that the straight line DE cuts at right angles that diameter of the circle ABC which passes through C.

247. If squares be described on the sides and hypotenuse of a right-angled triangle, the straight line joining the intersection of the diagonals of the latter square with the right angle is perpendicular to the straight line joining the intersections of the diagonals of the two former.

248. C is the centre of a given circle, CA a straight line less than the radius; find the point of the circumference at which CA subtends the greatest angle.

249. AB is the diameter of a semicircle, D and E are any two points in its circumference. Shew that if the chords joining A and B with D and E each way intersect at F and G, then FG produced is at right angles to AB.

250. Two equal circles touch one another externally, and through the point of contact chords are drawn, one to each circle, at right angles to each other: shew that the straight line joining the other extremities of these chords is equal and parallel to the straight line joining the centres of the circles.

251. A circle is described on the shorter diagonal of a rhombus as a diameter, and cuts the sides; and the points of intersection are joined crosswise with the extremities of that diagonal: shew that the parallelogram thus formed is a rhombus with angles equal to those of the first.

252. If two chords of a circle meet at a right angle within or without a circle, the squares on their segments are together equal to the squares on the diameter.

III. 32 to 34.

253. B is a point in the circumference of a circle, whose centre is C; PA, a tangent at any point P, meets CB produced at A, and PD is drawn perpendicular to CB: shew that the straight line PB bisects the angle APD.

254. If two circles touch each other, any straight line drawn through the point of contact will cut off similar segments.

255. AB is any chord, and AD is a tangent to a circle at A. DPQ is any straight line parallel to AB, meeting the circumference at P and Q. Shew that the triangle PAD is equiangular to the triangle QAB.

256. Two circles $ABDH$, ABG, intersect each other at the points A, B; from B a straight line BD is drawn in the one to touch the other; and from A any chord whatever is drawn cutting the circles at G and H: shew that BG is parallel to DH.

257. Two circles intersect at A and B. At A the tangents AC, AD are drawn to each circle and terminated

by the circumference of the other. If CB, BD be joined, shew that AB or AB produced, if necessary, bisects the angle CBD.

258. Two circles intersect at A and B, and through P any point in the circumference of one of them the chords PA and PB are drawn to cut the other circle at C and D: shew that CD is parallel to the tangent at P.

259. If from any point in the circumference of a circle a chord and tangent be drawn, the perpendiculars dropped on them from the middle point of the subtended arc are equal to one another.

260. AB is any chord of a circle, P any point on the circumference of the circle; PM is a perpendicular on AB and is produced to meet the circle at Q; and AN is drawn perpendicular to the tangent at P: shew that the triangle NAM is equiangular to the triangle PAQ.

261. Two diameters AOB, COD of a circle are at right angles to each other; P is a point in the circumference; the tangent at P meets COD produced at Q, and AP, BP meet the same line at R, S respectively: shew that RQ is equal to SQ

262. Construct a triangle, having given the base, the vertical angle, and the point in the base on which the perpendicular falls.

263. Construct a triangle, having given the base, the vertical angle, and the altitude.

264. Construct a triangle, having given the base, the vertical angle, and the length of the straight line drawn from the vertex to the middle point of the base.

265. Having given the base and the vertical angle of a triangle, shew that the triangle will be greatest when it is isosceles.

266. From a given point A without a circle whose centre is O draw a straight line cutting the circle at the points B and C, so that the area BOC may be the greatest possible.

267. Two straight lines containing a constant angle always pass through two fixed points, their position being otherwise unrestricted: shew that the straight line bisecting the angle always passes through one or other of two fixed points.

268. Given one angle of a triangle, the side opposite

it, and the sum of the other two sides, construct the triangle.

III. 35 to 37.

269. If two circles cut one another, the tangents drawn to the two circles from any point in the common chord produced are equal.

270. Two circles intersect at A and B: shew that AB produced bisects their common tangent.

271. If AD, CE are drawn perpendicular to the sides BC, AB of a triangle ABC, shew that the rectangle contained by BC and BD is equal to the rectangle contained by BA and BE.

272. If through any point in the common chord of two circles which intersect one another, there be drawn any two other chords, one in each circle, their four extremities shall all lie in the circumference of a circle.

273. From a given point as centre describe a circle cutting a given straight line in two points, so that the rectangle contained by their distances from a fixed point in the straight line may be equal to a given square.

274. Two circles $ABCD$, $EBCF$, having the common tangents AE and DF, cut one another at B and C, and the chord BC is produced to cut the tangents at G and H: shew that the square on GH exceeds the square on AE or DF by the square on BC.

275. A series of circles intersect each other, and are such that the tangents to them from a fixed point are equal: shew that the straight lines joining the two points of intersection of each pair will pass through this point.

276. ABC is a right-angled triangle; from any point D in the hypotenuse BC a straight line is drawn at right angles to BC, meeting CA at E and BA produced at F: shew that the square on DE is equal to the difference of the rectangles BD, DC and AE, EC; and that the square on DF is equal to the sum of the rectangles BD, DC and AF, FB.

277. It is required to find a point in the straight line which touches a circle at the end of a given diameter, such that when a straight line is drawn from this point to the other extremity of the diameter, the rectangle contained

by the part of it without the circle and the part within the circle may be equal to a given square not greater than that on the diameter.

IV. 1 to 4.

278. In IV. 3 shew that the straight lines drawn through A and B to touch the circle will meet.

279. In IV. 4 shew that the straight lines which bisect the angles B and C will meet.

280. In IV. 4 shew that the straight line DA will bisect the angle at A.

281. If the circle inscribed in a triangle ABC touch the sides AB, AC at the points D, E, and a straight line be drawn from A to the centre of the circle meeting the circumference at G, shew that the point G is the centre of the circle inscribed in the triangle ADE.

282. Shew that the straight lines joining the centres of the circles touching one side of a triangle and the others produced, pass through the angular points of the triangle.

283. A circle touches the side BC of a triangle ABC and the other two sides produced: shew that the distance between the points of contact of the side BC with this circle and with the inscribed circle, is equal to the difference between the sides AB and AC.

284. A circle is inscribed in a triangle ABC, and a triangle is cut off at each angle by a tangent to the circle. Shew that the sides of the three triangles so cut off are together equal to the sides of ABC.

285. D is the centre of the circle inscribed in a triangle BAC, and AD is produced to meet the straight line drawn through B at right angles to BD at O: shew that O is the centre of the circle which touches the side BC and the sides AB, AC produced.

286. Three circles are described, each of which touches one side of a triangle ABC, and the other two sides produced. If D be the point of contact of the side BC, E that of AC, and F that of AB, shew that AE is equal to BD, BF to CE, and CD to AF.

287. Describe a circle which shall touch a given circle and two given straight lines which themselves touch the given circle.

288. If the three points be joined in which the circle inscribed in a triangle meets the sides, shew that the resulting triangle is acute angled.

289. Two opposite sides of a quadrilateral are together equal to the other two, and each of the angles is less than two right angles. Shew that a circle can be inscribed in the quadrilateral.

290. Two circles HPL, KPM, that touch each other externally, have the common tangents HK, LM; HL and KM being joined, shew that a circle may be inscribed in the quadrilateral $HKML$.

291. Straight lines are drawn from the angles of a triangle to the centres of the opposite escribed circles: shew that these straight lines intersect at the centre of the inscribed circle.

292. Two sides of a triangle whose perimeter is constant are given in position: shew that the third side always touches a certain circle.

293. Given the base, the vertical angle, and the radius of the inscribed circle of a triangle, construct it.

IV. 5 to 9.

294. In IV. 5 shew that the perpendicular from F on BC will bisect BC.

295. If DE be drawn parallel to the base BC of a triangle ABC, shew that the circles described about the triangles ABC and ADE have a common tangent.

296. If the inscribed and circumscribed circles of a triangle be concentric, shew that the triangle must be equilateral.

297. Shew that if the straight line joining the centres of the inscribed and circumscribed circles of a triangle passes through one of its angular points, the triangle is isosceles.

298. The common chord of two circles is produced to any point P; PA touches one of the circles at A, PBC is any chord of the other. Shew that the circle which passes through A, B, and C touches the circle to which PA is a tangent.

299. A quadrilateral $ABCD$ is inscribed in a circle, and AD, BC are produced to meet at E: shew that the circle described about the triangle ECD will have the tangent at E parallel to AB.

300. Describe a circle which shall touch a given straight line, and pass through two given points.

301. Describe a circle which shall pass through two given points and cut off from a given straight line a chord of given length.

302. Describe a circle which shall have its centre in a given straight line, and cut off from two given straight lines chords of equal given length.

303. Two triangles have equal bases and equal vertical angles: shew that the radius of the circumscribing circle of one triangle is equal to that of the other.

304. Describe a circle which shall pass through two given points, so that the tangent drawn to it from another given point may be of a given length.

305. C is the centre of a circle; CA, CB are two radii at right angles; from B any chord BP is drawn cutting CA at N: a circle being described round ANP, shew that it will be touched by BA.

306. AB and CD are parallel straight lines, and the straight lines which join their extremities intersect at E: shew that the circles described round the triangles ABE, CDE touch one another.

307. Find the centre of a circle cutting off three equal chords from the sides of a triangle.

308. If O be the centre of the circle inscribed in the triangle ABC, and AO be produced to meet the circumscribed circle at F, shew that FB, FO, and FC are all equal.

309. The opposite sides of a quadrilateral inscribed in a circle are produced to meet at P and Q, and about the triangles so formed without the quadrilateral, circles are described meeting again at R: shew that P, R, Q are in one straight line.

310. The angle ACB of any triangle is bisected, and the base AB is bisected at right angles, by straight lines which intersect at D: shew that the angles ACB, ADB are together equal to two right angles.

311. $ACDB$ is a semicircle, AB being the diameter, and the two chords AD, BC intersect at E: shew that if a circle be described round CDE it will cut the former at right angles.

312. The diagonals of a given quadrilateral $ABCD$ intersect at O: shew that the centres of the circles described about the triangles OAB, OBC, OCD, ODA, will lie in the angular points of a parallelogram.

313. A circle is described round the triangle ABC; the tangent at C meets AB produced at D; the circle whose centre is D and radius DC cuts AB at E: shew that EC bisects the angle ACB.

314. AB, AC are two straight lines given in position; BC is a straight line of given length; D, E are the middle points of AB, AC; DF, EF are drawn at right angles to AB, AC respectively. Shew that AF will be constant for all positions of BC.

315. A circle is described about an isosceles triangle ABC in which AB is equal to AC; from A a straight line is drawn meeting the base at D and the circle at E: shew that the circle which passes through B, D, and E, touches AB.

316. AC is a chord of a given circle; B and D are two given points in the chord, both within or both without the circle: if a circle be described to pass through B and D, and touch the given circle, shew that AB and CD subtend equal angles at the point of contact.

317. A and B are two points within a circle: find the point P in the circumference such that if PAH, PBK be drawn meeting the circle at H and K, the chord HK shall be the greatest possible.

318. The centre of a given circle is equidistant from two given straight lines: describe another circle which shall touch these two straight lines and shall cut off from the given circle a segment containing an angle equal to a given angle.

319. O is the centre of the circle circumscribing a triangle ABC; D, E, F the feet of the perpendiculars from A, B, C on the opposite sides: shew that OA, OB, OC are respectively perpendicular to EF, FD, DE.

320. If from any point in the circumference of a given circle straight lines be drawn to the four angular points of an inscribed square, the sum of the squares on the four straight lines is double the square on the diameter.

321. Shew that no rectangle except a square can be described about a circle.

322. Describe a circle about a given rectangle.

323. If tangents be drawn through the extremities of two diameters of a circle the parallelogram thus formed will be a rhombus.

IV. 10.

324. Shew that the angle ACD in the figure of IV. 10 is equal to three times the angle at the vertex of the triangle.

325. Shew that in the figure of IV. 10 there are two triangles which possess the required property: shew that there is also an isosceles triangle whose equal angles are each one third part of the third angle.

326. Shew that the base of the triangle in IV. 10 is equal to the side of a regular pentagon inscribed in the smaller circle of the figure.

327. On a given straight line as base describe an isosceles triangle having the third angle treble of each of the angles at the base.

328. In the figure of IV. 10 suppose the two circles to cut again at E: then DE is equal to DC.

329. If A be the vertex and BD the base of the constructed triangle in IV. 10, D being one of the two points of intersection of the two circles employed in the construction, and E the other, and AE be drawn meeting BD produced at G, shew that GAB is another isosceles triangle of the same kind.

330. In the figure of IV. 10 if the two equal chords of the smaller circle be produced to cut the larger, and these points of section be joined, another triangle will be formed having the property required by the proposition.

331. In the figure of IV. 10 suppose the two circles to cut again at E; join AE, CE, and produce AE, BD to meet at G: then $CDGE$ is a parallelogram.

332. Shew that the smaller of the two circles employed in the figure of IV. 10 is equal to the circle described round the required triangle.

333. In the figure of IV. 10 if AF be the diameter of the smaller circle, DF is equal to a radius of the circle which circumscribes the triangle BCD.

IV. 11 to 16.

334. The straight lines which connect the angular points of a regular pentagon which are not adjacent intersect at the angular points of another regular pentagon.

335. $ABCDE$ is a regular pentagon; join AC and BE, and let BE meet AC at F; shew that AC is equal to the sum of AB and BF.

336. Shew that each of the triangles made by joining the extremities of adjoining sides of a regular pentagon is less than a third and greater than a fourth of the whole area of the pentagon.

337. Shew how to derive a regular hexagon from an equilateral triangle inscribed in a circle, and from the construction shew that the side of the hexagon equals the radius of the circle, and that the hexagon is double of the triangle.

338. In a given circle inscribe a triangle whose angles are as the numbers 2, 5, 8.

339. If $ABCDEF$ is a regular hexagon, and AC, BD, CE, DF, EA, FB be joined, another hexagon is formed whose area is one third of that of the former.

340. Any equilateral figure which is inscribed in a circle is also equiangular.

VI. 1, 2.

341. Shew that one of the triangles in the figure of IV. 10 is a mean proportional between the other two.

342. Through D, any point in the base of a triangle ABC, straight lines DE, DF are drawn parallel to the sides AB, AC, and meeting the sides at E, F: shew that the triangle AEF is a mean proportional between the triangles FBD, EDC.

343. Perpendiculars are drawn from any point within an equilateral triangle on the three sides: shew that their sum is invariable.

344. Find a point within a triangle such that if straight lines be drawn from it to the three angular points the triangle will be divided into three equal triangles.

345. From a point E in the common base of two triangles ACB, ADB, straight lines are drawn parallel to AC, AD, meeting BC, BD at F, G: shew that FG is parallel to CD.

346. From any point in the base of a triangle straight lines are drawn parallel to the sides: shew that the intersection of the diagonals of every parallelogram so formed lies in a certain straight line.

347. In a triangle ABC a straight line AD is drawn perpendicular to the straight line BD which bisects the angle B: shew that a straight line drawn from D parallel to BC will bisect AC.

348. ABC is a triangle; any straight line parallel to BC meets AB at D and AC at E; join BE and CD meeting at F: shew that the triangle ADF is equal to the triangle AEF.

349. ABC is a triangle; any straight line parallel to BC meets AB at D and AC at E; join BE and CD meeting at F: shew that if AF be produced it will bisect BC.

350. If two sides of a quadrilateral figure be parallel to each other, any straight line drawn parallel to them will cut the other sides, or those sides produced, proportionally.

351. ABC is a triangle; it is required to draw from a given point P, in the side AB, or AB produced, a straight line to AC, or AC produced, so that it may be bisected by BC.

VI. 3, A.

352. The side BC of a triangle ABC is bisected at D, and the angles ADB, ADC are bisected by the straight lines DE, DF, meeting AB, AC at E, F respectively: shew that EF is parallel to BC.

353. AB is a diameter of a circle, CD is a chord at right angles to it, and E is any point in CD; AE and BE

are drawn and produced to cut the circle at F and G: shew that the quadrilateral $CFDG$ has any two of its adjacent sides in the same ratio as the remaining two.

354. Apply VI. 3 to solve the problem of the trisection of a finite straight line.

355. In the circumference of the circle of which AB is a diameter, take any point P; and draw PC, PD on opposite sides of AP, and equally inclined to it, meeting AB at C and D: shew that AC is to BC as AD is to BD.

356. AB is a straight line, and D is any point in it: determine a point P in AB produced such that PA is to PB as DA is to DB.

357. From the same point A straight lines are drawn making the angles BAC, CAD, DAE each equal to half a right angle, and they are cut by a straight line $BCDE$, which makes BAE an isosceles triangle: shew that BC or DE is a mean proportional between BE and CD.

358. The angle A of a triangle ABC is bisected by AD which cuts the base at D, and O is the middle point of BC: shew that OD bears the same ratio to OB that the difference of the sides bears to their sum.

359. AD and AE bisect the interior and exterior angles at A of a triangle ABC, and meet the base at D and E; and O is the middle point of BC: shew that OB is a mean proportional between OD and OE.

360. Three points D, E, F in the sides of a triangle ABC being joined form a second triangle, such that any two sides make equal angles with the side of the former at which they meet: shew that AD, BE, CF are at right angles to BC, CA, AB respectively.

VI. 4 to 6.

361. If two triangles be on equal bases and between the same parallels, any straight line parallel to their bases will cut off equal areas from the two triangles.

362. AB and CD are two parallel straight lines; E is the middle point of CD; AC and BE meet at F, and AE and BD meet at G: shew that FG is parallel to AB.

363. A, B, C are three fixed points in a straight line; any straight line is drawn through C; shew that the perpendiculars on it from A and B are in a constant ratio.

364. If the perpendiculars from two fixed points on a straight line passing between them be in a given ratio, the straight line must pass through a third fixed point.

365. Find a straight line such that the perpendiculars on it from three given points shall be in a given ratio to each other.

366. Through a given point draw a straight line, so that the parts of it intercepted between that point and perpendiculars drawn to the straight line from two other given points may have a given ratio.

367. A tangent to a circle at the point A intersects two parallel tangents at B, C, the points of contact of which with the circle are D, E respectively; and BE, CD intersect at F: shew that AF is parallel to the tangents BD, CE.

368. P and Q are fixed points; AB and CD are fixed parallel straight lines; any straight line is drawn from P to meet AB at M, and a straight line is drawn from Q parallel to PM meeting CD at N: shew that the ratio of PM to QN is constant, and thence shew that the straight line through M and N passes through a fixed point.

369. Shew that the diagonals of a quadrilateral, two of whose sides are parallel and one of them double of the other, cut one another at a point of trisection.

370. A and B are two points on the circumference of a circle of which C is the centre; draw tangents at A and B meeting at T; and from A draw AN perpendicular to CB: shew that BT is to BC as BN is to NA.

371. In the sides AB, AC of a triangle ABC are taken two points D, E, such that BD is equal to CE; DE, BC are produced to meet at F: shew that AB is to AC as EF is to DF.

372. If through the vertex and the extremities of the base of a triangle two circles be described intersecting each other in the base or base produced, their diameters are proportional to the sides of the triangle.

373. Find a point the perpendiculars from which on the sides of a given triangle shall be in a given ratio.

374. On AB, AC, two adjacent sides of a rectangle, two similar triangles are constructed, and perpendiculars are drawn to AB, AC from the angles which they subtend, intersecting at the point P. If AB, AC be homologous

sides, shew that P is in all cases in one of the diagonals of the rectangle.

375. In the figure of I. 43 shew that if EG and FH be produced they will meet on AC produced.

376. APB and CQD are parallel straight lines, and AP is to PB as DQ is to QC: shew that the straight lines PQ, AC, BD, produced if necessary, will meet at a point: shew also that the straight lines PQ, AD, BC, produced if necessary, will meet at a point.

377. ACB is a triangle, and the side AC is produced to D so that CD is equal to AC, and BD is joined: if any straight line drawn parallel to AB cuts the sides AC, CB, and from the points of section straight lines be drawn parallel to DB, shew that these straight lines will meet AB at points equidistant from its extremities.

378. If a circle be described touching externally two given circles, the straight line passing through the points of contact will intersect the straight line passing through the centres of the given circles at a fixed point.

379. D is the middle point of the side BC of a triangle ABC, and P is any point in AD; through P the straight lines BPE, CPF are drawn meeting the other sides at E, F: shew that EF is parallel to BC.

380. AB is the diameter of a circle, E the middle point of the radius OB; on AE, EB as diameters circles are described; PQL is a common tangent meeting the circles at P and Q, and AB produced at L: shew that BL is equal to the radius of the smaller circle.

381. $ABCDE$ is a regular pentagon, and AD, BE intersect at O: shew that a side of the pentagon is a mean proportional between AO and AD.

382. $ABCD$ is a parallelogram; P and Q are points in a straight line parallel to AB; PA and QB meet at R, and PD and QC meet at S; shew that RS is parallel to AD.

383. A and B are two given points; AC and BD are perpendicular to a given straight line CD; AD and BC intersect at E, and EF is perpendicular to CD: shew that AF and BF make equal angles with CD.

384. From the angular points of a parallelogram $ABCD$ perpendiculars are drawn on the diagonals meeting them at E, F, G, H respectively: shew that $EFGH$ is a parallelogram similar to $ABCD$.

385. If at a given point two circles intersect, and their centres lie on two fixed straight lines which pass through that point, shew that whatever be the magnitude of the circles their common tangents will always meet in one of two fixed straight lines which pass through the given point.

VI. 7 to 18.

386. If two circles touch each other, and also touch a given straight line, the part of the straight line between the points of contact is a mean proportional between the diameters of the circles.

387. Divide a given arc of a circle into two parts, so that the chords of these parts shall be to each other in a given ratio.

388. In a given triangle draw a straight line parallel to one of the sides, so that it may be a mean proportional between the segments of the base.

389. ABC is a triangle, and a perpendicular is drawn from A to the opposite side, meeting it at D between B and C: shew that if AD is a mean proportional between BD and CD the angle BAC is a right angle.

390. ABC is a triangle, and a perpendicular is drawn from A on the opposite side, meeting it at D between B and C: shew that if BA is a mean proportional between BD and BC, the angle BAC is a right angle.

391. C is the centre of a circle, and A any point within it; CA is produced through A to a point B such that the radius is a mean proportional between CA and CB: shew that if P be any point on the circumference, the angles CPA and CBP are equal.

392. O is a fixed point in a given straight line OA, and a circle of given radius moves so as always to be touched by OA; a tangent OP is drawn from O to the circle, and in OP produced PQ is taken a third proportional to OP and the radius: shew that as the circle moves along OA, the point Q will move in a straight line.

393. Two given parallel straight lines touch a circle, and SPT is another tangent cutting the two former tangents at S and T, and meeting the circle at P: shew

that the rectangle SP, PT is constant for all positions of P.

394. Find a point in a side of a triangle, from which two straight lines drawn, one to the opposite angle, and the other parallel to the base, shall cut off towards the vertex and towards the base, equal triangles.

395. ACB is a triangle having a right angle at C; from A a straight line is drawn at right angles to AB, cutting BC produced at E; from B a straight line is drawn at right angles to AB, cutting AC produced at D: shew that the triangle ECD is equal to the triangle ACB.

396. The straight line bisecting the angle ABC of the triangle ABC meets the straight lines drawn through A and C, parallel to BC and AB respectively, at E and F: shew that the triangles CBE, ABF are equal.

397. Shew that the diagonals of any quadrilateral figure inscribed in a circle divide the quadrilateral into four triangles which are similar two and two; and deduce the theorem of III. 35.

398. AB, CD are any two chords of a circle passing through a point O; EF is any chord parallel to OB; join CE, DF meeting AB at the points G and H, and DE, CF meeting AB at the points K and L: shew that the rectangle OG, OH is equal to the rectangle OK, OL.

399. $ABCD$ is a quadrilateral in a circle; the straight lines CE, DE which bisect the angles ACB, ADB cut BD and AC at F and G respectively: shew that EF is to EG as ED is to EC.

400. From an angle of a triangle two straight lines are drawn, one to any point in the side opposite to the angle, and the other to the circumference of the circumscribing circle, so as to cut from it a segment containing an angle equal to the angle contained by the first drawn line and the side which it meets: shew that the rectangle contained by the sides of the triangle is equal to the rectangle contained by the straight lines thus drawn.

401. The vertical angle C of a triangle is bisected by a straight line which meets the base at D, and is produced to a point E, such that the rectangle contained by CD and CE is equal to the rectangle contained by AC and CB: shew that if the base and vertical angle be given, the position of E is invariable.

402. A square is inscribed in a right-angled triangle ABC, one side DE of the square coinciding with the hypotenuse AB of the triangle: shew that the area of the square is equal to the rectangle AD, BE.

403. $ABCD$ is a parallelogram; from B a straight line is drawn cutting the diagonal AC at F, the side DC at G, and the side AD produced at E: shew that the rectangle EF, FG is equal to the square on BF.

404. If a straight line drawn from the vertex of an isosceles triangle to the base, be produced to meet the circumference of a circle described about the triangle, the rectangle contained by the whole line so produced, and the part of it between the vertex and the base shall be equal to the square on either of the equal sides of the triangle.

405. Two straight lines are drawn from a point A to touch a circle of which the centre is E; the points of contact are joined by a straight line which cuts EA at H; and on HA as diameter a circle is described: shew that the straight lines drawn through E to touch this circle will meet it on the circumference of the given circle.

VI. 19 to *D*.

406. An isosceles triangle is described having each of the angles at the base double of the third angle: if the angles at the base be bisected, and the points where the lines bisecting them meet the opposite sides be joined, the triangle will be divided into two parts in the proportion of the base to the side of the triangle.

407. Any regular polygon inscribed in a circle is a mean proportional between the inscribed and circumscribed regular polygons of half the number of sides.

408. In the figure of VI. 24 shew that EG and KH are parallel.

409. Divide a triangle into two equal parts by a straight line at right angles to one of the sides.

410. Through any point P in the diagonal AC of a parallelogram $ABCD$ a straight line is drawn meeting BC at E, and AD at F; and through P another straight line is drawn meeting AB at G, and CD at H: shew that GF is parallel to EH.

411. Through a given point draw a chord in a given circle so that it shall be divided at the point in a given ratio.

412. From a point without a circle draw a straight line cutting the circle, so that the two segments shall be equal to each other.

413. In the figure of II. 11 shew that four other straight lines, besides the given straight line are divided in the required manner.

414. Construct a triangle, having given the base, the vertical angle, and the rectangle contained by the sides.

415. A circle is described round an equilateral triangle, and from any point in the circumference straight lines are drawn to the angular points of the triangle: shew that one of these straight lines is equal to the other two together.

416. From the extremities B, C of the base of an isosceles triangle ABC, straight lines are drawn at right angles to AB, AC respectively, and intersecting at D: shew that the rectangle BC, AD is double of the rectangle AB, DB.

417. ABC is an isosceles triangle, the side AB being equal to AC; F is the middle point of BC; on any straight line through A perpendiculars FG and CE are drawn: shew that the rectangle AC, EF is equal to the sum of the rectangles FC, EG and FA, FG.

XI. 1 to 12.

418. Shew that equal straight lines drawn from a given point to a given plane are equally inclined to the plane.

419. If two straight lines in one plane be equally inclined to another plane, they will be equally inclined to the common section of these planes.

420. From a point A a perpendicular is drawn to a plane meeting it at B; from B a perpendicular is drawn on a straight line in the plane meeting it at C: shew that AC is perpendicular to the straight line in the plane.

421. ABC is a triangle; the perpendiculars from A and B on the opposite sides meet at D; through D a straight line is drawn perpendicular to the plane of the triangle, and E is any point in this straight line: shew that

the straight line joining E to any angular point of the triangle is at right angles to the straight line drawn through that angular point parallel to the opposite side of the triangle.

422. Straight lines are drawn from two given points without a given plane meeting each other in that plane: find when their sum is the least possible.

423. Three straight lines not in the same plane meet at a point, and a plane cuts these straight lines at equal distances from the point of intersection: shew that the perpendicular from that point on the plane will meet it at the centre of the circle inscribed in the triangle formed by the portion of the plane intercepted by the planes passing through the straight lines.

424. Give a geometrical construction for drawing a straight line which shall be equally inclined to three straight lines meeting at a point.

425. From a point E draw EC, ED perpendicular to two planes CAB, DAB which intersect in AB, and from D draw DF perpendicular to the plane CAB meeting it at F: shew that the straight line CF, produced if necessary, is perpendicular to AB.

426. Perpendiculars are drawn from a point to a plane, and to a straight line in that plane: shew that the straight line joining the feet of the perpendiculars is perpendicular to the former straight line.

XI. 13 to 21.

427. BCD is the common base of two pyramids, whose vertices A and E lie in a plane passing through BC; and AB, AC are respectively perpendicular to the faces BED, CED: shew that one of the angles at A together with the angles at E make up four right angles.

428. Within the area of a given triangle is inscribed another triangle: shew that the sum of the angles subtended by the sides of the interior triangle at any point not in the plane of the triangles is less than the sum of the angles subtended at the same point by the sides of the exterior angle.

429. From the extremities of the two parallel straight

lines AB, CD parallel straight lines Aa, Bb, Cc, Dd are drawn meeting a plane at a, b, c, d: shew that AB is to CD as ab to cd.

430. Shew that the perpendicular drawn from the vertex of a regular tetrahedron on the opposite face is three times that drawn from its own foot on any of the other faces.

431. A triangular pyramid stands on an equilateral base and the angles at the vertex are right angles: shew that the sum of the perpendiculars on the faces from any point of the base is constant.

432. Three straight lines not in the same plane intersect at a point, and through their point of intersection another straight line is drawn within the solid angle formed by them: shew that the angles which this straight line makes with the first three are together less than the sum, but greater than half the sum, of the angles which the first three make with each other.

433. Three straight lines which do not all lie in one plane, are cut in the same ratio by three planes, two of which are parallel: shew that the third will be parallel to the other two, if its intersections with the three straight lines are not all in the same straight line.

434. Draw two parallel planes, one through one straight line, and the other through another straight line which does not meet the former.

435. If two planes which are not parallel be cut by two parallel planes, the lines of section of the first two by the last two will contain equal angles.

436. From a point A in one of two planes are drawn AB at right angles to the first plane, and AC perpendicular to the second plane, and meeting the second plane at B, C: shew that BC is perpendicular to the line of intersection of the two planes.

437. Polygons formed by cutting a prism by parallel planes are equal.

438. Polygons formed by cutting a pyramid by parallel planes are similar.

439. The straight line $PBbp$ cuts two parallel planes at B, b, and the points P, p are equidistant from the planes; PAa, pcC are other straight lines drawn from P, p to cut the planes: shew that the triangles ABC, abc are equal.

440. Perpendiculars AE, BF are drawn to a plane

from two points A, B above it; a plane is drawn through A perpendicular to AB: shew that its line of intersection with the given plane is perpendicular to EF.

I. 1 to 48.

441. ABC is a triangle, and P is any point within it: shew that the sum of PA, PB, PC is less than the sum of the sides of the triangle.

442. From the centres A and B of two circles parallel radii AP, BQ are drawn; the straight line PQ meets the circumferences again at R and S: shew that AR is parallel to BS.

443. If any point be taken within a parallelogram the sum of the triangles formed by joining the point with the extremities of a pair of opposite sides is half the parallelogram.

444. If a quadrilateral figure be bisected by one diagonal the second diagonal is bisected by the first.

445. Any quadrilateral figure which is bisected by both of its diagonals is a parallelogram.

446. In the figure of I. 5 if the equal sides of the triangle be produced upwards through the vertex, instead of downwards through the base, a demonstration of I. 15 may be obtained without assuming any proposition beyond I. 5.

447. A is a given point, and B is a given point in a given straight line: it is required to draw from A to the given straight line, a straight line AP, such that the sum of AP and PB may be equal to a given length.

448. Shew that by superposition the first case of I. 26 may be immediately demonstrated, and also the second case with the aid of I. 16.

449. A straight line is drawn terminated by one of the sides of an isosceles triangle, and by the other side produced, and bisected by the base: shew that the straight lines thus intercepted between the vertex of the isosceles triangle and this straight line, are together equal to the two equal sides of the triangle.

450. Through the middle point M of the base BC of a triangle a straight line DME is drawn, so as to cut off equal parts from the sides AB, AC, produced if necessary: shew that BD is equal to CE.

451. Of all parallelograms which can be formed with diameters of given lengths the rhombus is the greatest.

452. Shew from I. 18 and I. 32 that if the hypotenuse BC of a right-angled triangle ABC be bisected at D, then AD, BD, CD are all equal.

453. If two equal straight lines intersect each other any where at right angles, the quadrilateral formed by joining their extremities is equal to half the square on either straight line.

454. Inscribe a parallelogram in a given triangle, in such a manner that its diagonals shall intersect at a given point within the triangle.

455. Construct a triangle of given area, and having two of its sides of given lengths.

456. Construct a triangle, having given the base, the difference of the sides, and the difference of the angles at the base.

457. AB, AC are two given straight lines: it is required to find in AB a point P, such that if PQ be drawn perpendicular to AC, the sum of AP and AQ may be equal to a given straight line.

458. The distance of the vertex of a triangle from the bisection of its base, is equal to, greater than, or less than half of the base, according as the vertical angle is a right, an acute, or an obtuse angle.

459. If in the sides of a given square, at equal distances from the four angular points, four other points be taken, one on each side, the figure contained by the straight lines which join them, shall also be a square.

460. On a given straight line as base, construct a triangle, having given the difference of the sides and a point through which one of the sides is to pass.

461. ABC is a triangle in which BA is greater than CA; the angle A is bisected by a straight line which meets BC at D; shew that BD is greater than CD.

462. If one angle of a triangle be triple another the triangle may be divided into two isosceles triangles.

463. If one angle of a triangle be double another, an isosceles triangle may be added to it so as to form together with it a single isosceles triangle.

464. Let one of the equal sides of an isosceles triangle be bisected at D, and let it also be doubled by being produced through the extremity of the base to E, then the distance of the other extremity of the base from E is double its distance from D.

465. Determine the locus of a point whose distance from one given point is double its distance from another given point.

466. A straight line AB is bisected at C, and on AC and CB as diagonals any two parallelograms $ADCE$ and $CFBG$ are described; let the parallelogram whose adjacent sides are CD and CF be completed, and also that whose adjacent sides are CE and CG: shew that the diagonals of these latter parallelograms are in the same straight line.

467. $ABCD$ is a rectangle of which A, C are opposite angles; E is any point in BC and F is any point in CD: shew that twice the area of the triangle AEF, together with the rectangle BE, DF is equal to the rectangle $ABCD$.

468. ABC, DBC are two triangles on the same base, and ABC has the side AB equal to the side AC; a circle passing through C and D has its centre E on CA, produced if necessary; a circle passing through B and D has its centre F on BA, produced if necessary: shew that the quadrilateral $AEDF$ has the sum of two of its sides equal to the sum of the other two.

469. Two straight lines AB, AC are given in position:

it is required to find in AB a point P, such that a perpendicular being drawn from it to AC, the straight line AP may exceed this perpendicular by a proposed length.

470. Shew that the opposite sides of any equiangular hexagon are parallel, and that any two sides which are adjacent are together equal to the two to which they are parallel.

471. From D and E, the corners of the square $BDEC$ described on the hypotenuse BC of a right-angled triangle ABC, perpendiculars DM, EN are let fall on AC, AB respectively: shew that AM is equal to AB, and AN equal to AC.

472. AB and AC are two given straight lines, and P is a given point: it is required to draw through P a straight line which shall form with AB and AC the least possible triangle.

473. ABC is a triangle in which C is a right angle: shew how to draw a straight line parallel to a given straight line, so as to be terminated by CA and CB, and bisected by AB.

474. ABC is an isosceles triangle having the angle at B four times either of the other angles; AB is produced to D so that BD is equal to twice AB, and CD is joined: shew that the triangles ACD and ABC are equiangular to one another.

475. Through a point K within a parallelogram $ABCD$ straight lines are drawn parallel to the sides: shew that the difference of the parallelograms of which KA and KC are diagonals is equal to twice the triangle BKD.

476. Construct a right-angled triangle, having given one side and the difference between the other side and the hypotenuse.

477. The straight lines AD, BE bisecting the sides BC, AC of a triangle intersect at G: shew that AG is double of GD.

478. BAC is a right-angled triangle; one straight line is drawn bisecting the right angle A, and another bisecting the base BC at right angles; these straight lines intersect at E: if D be the middle point of BC, shew that DE is equal to DA.

479. On AC the diagonal of a square $ABCD$, a rhombus $AEFC$ is described of the same area as the square,

and having its acute angle at A: if AF be joined, shew that the angle BAC is divided into three equal angles.

480. AB, AC are two fixed straight lines at right angles; D is any point in AB, and E is any point in AC; on DE as diagonal a half square is described with its vertex at G: shew that the locus of G is the straight line which bisects the angle BAC.

481. Shew that a square is greater than any other parallelogram of the same perimeter.

482. Inscribe a square of given magnitude in a given square.

483. ABC is a triangle; AD is a third of AB, and AE is a third of AC; CD and BE intersect at F: shew that the triangle BFC is half the triangle BAC, and that the quadrilateral $ADFE$ is equal to either of the triangles CFE or BDF.

484. ABC is a triangle, having the angle C a right angle; the angle A is bisected by a straight line which meets BC at D, and the angle B is bisected by a straight line which meets AC at E; AD and BE intersect at O: shew that the triangle AOB is half the quadrilateral $ABDE$.

485. Shew that a scalene triangle cannot be divided by a straight line into two parts which will coincide.

486. $ABCD$, $ACED$ are parallelograms on equal bases BC, CE, and between the same parallels AD, BE; the straight lines BD and AE intersect at F: shew that BF is equal to twice DF.

487. Parallelograms $AFGC$, $CBKH$ are described on AC, BC outside the triangle ABC; FG and KH meet at Z; ZC is joined, and through A and B straight lines AD and BE are drawn, both parallel to ZC, and meeting FG and KH at D and E respectively: shew that the figure $ADEB$ is a parallelogram, and that it is equal to the sum of the parallelograms FC, CK.

488. If a quadrilateral have two of its sides parallel shew that the straight line drawn parallel to these sides through the intersection of the diagonals is bisected at that point.

489. Two triangles are on equal bases and between the same parallels: shew that the sides of the triangles intercept equal lengths of any straight line which is parallel to their bases.

490. In a right-angled triangle, right-angled at A, if the side AC be double of the side AB, the angle B is more than double of the angle C.

491. Trisect a parallelogram by straight lines drawn through one of its angular points.

492. AHK is an equilateral triangle; $ABCD$ is a rhombus, a side of which is equal to a side of the triangle, and the sides BC and CD of which pass through H and K respectively: shew that the angle A of the rhombus is ten-ninths of a right angle.

493. Trisect a given triangle by straight lines drawn from a given point in one of its sides.

494. In the figure of I. 35 if two diagonals be drawn to the two parallelograms respectively, one from each extremity of the base, and the intersection of the diagonals be joined with the intersection of the sides (or sides produced) in the figure, shew that the joining straight line will bisect the base.

II. 1 to 14.

495. Produce one side of a given triangle so that the rectangle contained by this side and the produced part may be equal to the difference of the squares on the other two sides.

496. Produce a given straight line so that the sum of the squares on the given straight line and on the part produced may be equal to twice the rectangle contained by the whole straight line thus produced and the part produced.

497. Produce a given straight line so that the sum of the squares on the given straight line and on the whole straight line thus produced may be equal to twice the rectangle contained by the whole straight line thus produced and the part produced.

498. Produce a given straight line so that the rectangle contained by the whole straight line thus produced and the part produced may be equal to the square on the given straight line.

499. Describe an isosceles obtuse-angled triangle such that the square on the largest side may be equal to three times the square on either of the equal sides.

500. Find the obtuse angle of a triangle when the

square on the side opposite to the obtuse angle is greater than the sum of the squares on the sides containing it, by the rectangle of the sides.

501. Construct a rectangle equal to a given square when the sum of two adjacent sides of the rectangle is equal to a given quantity.

502. Construct a rectangle equal to a given square when the difference of two adjacent sides of the rectangle is equal to a given quantity.

503. The least square which can be inscribed in a given square is that which is half of the given square.

504. Divide a given straight line into two parts so that the squares on the whole line and on one of the parts may be together double of the square on the other part.

505. Two rectangles have equal areas and equal perimeters: shew that they are equal in all respects.

506. $ABCD$ is a rectangle; P is a point such that the sum of PA and PC is equal to the sum of PB and PD: shew that the locus of P consists of the two straight lines through the centre of the rectangle parallel to its sides.

III. 1 to 37.

507. Describe a circle which shall pass through a given point and touch a given straight line at a given point.

508. Describe a circle which shall pass through a given point and touch a given circle at a given point.

509. Describe a circle which shall touch a given circle at a given point and touch a given straight line.

510. AD, BE are perpendiculars from the angles A and B of a triangle on the opposite sides; BF is perpendicular to ED or ED produced: shew that the angle FBD is equal to the angle EBA.

511. If ABC be a triangle, and BE, CF the perpendiculars from the angles on the opposite sides, and K the middle point of the third side, shew that the angles FEK, EFK are each equal to A.

512. AB is a diameter of a circle; AC and AD are two chords meeting the tangent at B at E and F respectively: shew that the angles FCE and FDE are equal.

513. Shew that the four straight lines bisecting the angles of any quadrilateral form a quadrilateral which can be inscribed in a circle.

514. Find the shortest distance between two circles which do not meet.

515. Two circles cut one another at a point A: it is required to draw through A a straight line so that the extreme length of it intercepted by the two circles may be equal to that of a given straight line.

516. If a polygon of an even number of sides be inscribed in a circle, the sum of the alternate angles together with two right angles is equal to as many right angles as the figure has sides.

517. Draw from a given point in the circumference of a circle, a chord which shall be bisected by its point of intersection with a given chord of the circle.

518. When an equilateral polygon is described about a circle it must necessarily be equiangular if the number of sides be odd, but not otherwise.

519. AB is the diameter of a circle whose centre is C, and DCE is a sector having the arc DE constant; join AE, BD intersecting at P; shew that the angle APB is constant.

520. If any number of triangles on the same base BC, and on the same side of it have their vertical angles equal, and perpendiculars, intersecting at D, be drawn from B and C on the opposite sides, find the locus of D; and shew that all the straight lines which bisect the angle BDC pass through the same point.

521. Let O and C be any fixed points on the circumference of a circle, and OA any chord; then if AC be joined and produced to B, so that OB is equal to OA, the locus of B is an equal circle.

522. From any point P in the diagonal BD of a parallelogram $ABCD$, straight lines PE, PF, PG, PH are drawn perpendicular to the sides AB, BC, CD, DA: shew that EF is parallel to GH.

523. Through any fixed point of a chord of a circle other chords are drawn; shew that the straight lines from the middle point of the first chord to the middle points of the others will meet them all at the same angle.

524. ABC is a straight line, divided at any point B

into two parts; ADB and CDB are similar segments of circles, having the common chord BD; CD and AD are produced to meet the circumferences at F and E respectively, and AF, CE, BF, BE are joined: shew that ABF and CBE are isosceles triangles, equiangular to one another.

525. If the centres of two circles which touch each other externally be fixed, the common tangent of those circles will touch another circle of which the straight line joining the fixed centres is the diameter.

526. A is a given point: it is required to draw from A two straight lines which shall contain a given angle and intercept on a given straight line a part of given length.

527. A straight line and two circles are given: find the point in the straight line from which the tangents drawn to the circles are of equal length.

528. In a circle two chords of given length are drawn so as not to intersect, and one of them is fixed in position; the opposite extremities of the chords are joined by straight lines intersecting within the circle: shew that the locus of the point of intersection will be a portion of the circumference of a circle, passing through the extremities of the fixed chord.

529. A and B are the centres of two circles which touch internally at C, and also touch a third circle, whose centre is D, externally and internally respectively at E and F: shew that the angle ADB is double of the angle ECF.

530. C is the centre of a circle, and CP is a perpendicular on a chord APB: shew that the sum of CP and AP is greatest when CP is equal to AP.

531. AB, BC, CD are three adjacent sides of any polygon inscribed in a circle; the arcs AB, BC, CD are bisected at L, M, N; and LM cuts BA, BC respectively at P and Q: shew that BPQ is an isosceles triangle; and that the angles ABC, BCD are together double of the angle LMN.

532. In the circumference of a given circle determine a point so situated that if chords be drawn to it from the extremities of a given chord of the circle their difference shall be equal to a given straight line less than the given chord.

533. Construct a triangle, having given the sum of the

sides, the difference of the segments of the base made by the perpendicular from the vertex, and the difference of the base angles.

534. On a straight line AB as base, and on the same side of it are described two segments of circles; P is any point in the circumference of one of the segments, and the straight line BP cuts the circumference of the other segment at Q: shew that the angle PAQ is equal to the angle between the tangents at A.

535. AKL is a fixed straight line cutting a given circle at K and L; APQ, ARS are two other straight lines making equal angles with AKL, and cutting the circle at P, Q and R, S: shew that whatever be the position of APQ and ARS, the straight line joining the middle points of PQ and RS always remains parallel to itself.

536. If about a quadrilateral another quadrilateral can be described such that every two of its adjacent sides are equally inclined to that side of the former quadrilateral which meets them both, then a circle may be described about the former quadrilateral.

537. Two circles touch one another internally at the point A: it is required to draw from A a straight line such that the part of it between the circles may be equal to a given straight line, which is not greater than the difference between the diameters of the circles.

538. $ABCD$ is a parallelogram; AE is at right angles to AB, and CE is at right angles to CB: shew that ED, if produced, will cut AC at right angles.

539. From each angular point of a triangle a perpendicular is let fall on the opposite side: shew that the rectangles contained by the segments into which each perpendicular is divided by the point of intersection of the three are equal to each other.

540. The two angles at the base of a triangle are bisected by two straight lines on which perpendiculars are drawn from the vertex: shew that the straight line which passes through the feet of these perpendiculars will be parallel to the base and will bisect the sides.

541. In a given circle inscribe a rectangle equal to a given rectilineal figure.

542. In an acute-angled triangle ABC perpendiculars AD, BE are let fall on BC, CA respectively; circles

described on AC, BC as diameters meet BE, AD respectively at F, G and H, K: shew that F, G, H, K lie on the circumference of a circle.

543. Two diameters in a circle are at right angles; from their extremities four parallel straight lines are drawn; shew that they divide the circumference into four equal parts.

544. E is the middle point of a semicircular arc AEB, and CDE is any chord cutting the diameter at D, and the circle at C: shew that the square on CE is twice the quadrilateral $AEBC$.

545. AB is a fixed chord of a circle, AC is a moveable chord of the same circle; a parallelogram is described of which AB and AC are adjacent sides: find the locus of the middle points of the diagonals of the parallelogram.

546. AB is a fixed chord of a circle, AC is a moveable chord of the same circle; a parallelogram is described of which AB and AC are adjacent sides: determine the greatest possible length of the diagonal drawn through A.

547. If two equal circles be placed at such a distance apart that the tangent drawn to either of them from the centre of the other is equal to a diameter, shew that they will have a common tangent equal to the radius.

548. Find a point in a given circle from which if two tangents be drawn to an equal circle, given in position, the chord joining the points of contact is equal to the chord of the first circle formed by joining the points of intersection of the two tangents produced; and determine the limit to the possibility of the problem.

549. AB is a diameter of a circle, and AF is any chord; C is any point in AB, and through C a straight line is drawn at right angles to AB, meeting AF, produced if necessary at G, and meeting the circumference at D: shew that the rectangle FA, AG, and the rectangle BA, AC, and the square on AD are all equal.

550. Construct a triangle, having given the base, the vertical angle, and the length of the straight line drawn from the vertex to the base bisecting the vertical angle.

551. A, B, C are three given points in the circumference of a given circle: find a point P such that if AP, BP, CP meet the circumference at D, E, F respectively, the arcs DE, EF may be equal to given arcs.

552. Find the point in the circumference of a given circle, the sum of whose distances from two given straight lines at right angles to each other, which do not cut the circle, is the greatest or least possible.

553. On the sides of a triangle segments of a circle are described *internally*, each containing an angle equal to the excess of two right angles above the opposite angle of the triangle: shew that the radii of the circles are equal, that the circles all pass through one point, and that their chords of intersection are respectively perpendicular to the opposite sides of the triangle.

IV. 1 to 16.

554. From the angles of a triangle ABC perpendiculars are drawn to the opposite sides meeting them at D, E, F respectively: shew that DE and DF are equally inclined to AD.

555. The points of contact of the inscribed circle of a triangle are joined; and from the angular points of the triangle so formed perpendiculars are drawn to the opposite sides: shew that the triangle of which the feet of these perpendiculars are the angular points has its sides parallel to the sides of the original triangle.

556. Construct a triangle having given an angle and the radii of the inscribed and circumscribed circles.

557. Triangles are constructed on the same base with equal vertical angles; shew that the locus of the centres of the escribed circles, each of which touches one of the sides externally and the other side and base produced, is an arc of a circle, the centre of which is on the circumference of the circle circumscribing the triangles.

558. From the angular points A, B, C of a triangle perpendiculars are drawn on the opposite sides, and terminated at the points D, E, F on the circumference of the circumscribing circle: if L be the point of intersection of the perpendiculars, shew that LD, LE, LF are bisected by the sides of the triangle.

559. $ABCDE$ is a regular pentagon; join AC and BD intersecting at O: shew that AO is equal to DO, and that the rectangle AC, CO is equal to the square on BC.

560. A straight line PQ of given length moves so that its ends are always on two fixed straight lines CP, CQ; straight lines from P and Q at right angles to CP and CQ respectively intersect at R; perpendiculars from P and Q on CQ and CP respectively intersect at S: shew that the loci of R and S are circles having their common centre at C.

561. Right-angled triangles are described on the same hypotenuse: shew that the locus of the centres of the inscribed circles is a quarter of the circumference of a circle of which the common hypotenuse is a chord.

562. On a given straight line AB any triangle ACB is described; the sides AC, BC are bisected and straight lines drawn at right angles to them through the points of bisection to intersect at a point D; find the locus of D.

563. Construct a triangle, having given its base, one of the angles at the base, and the distance between the centre of the inscribed circle and the centre of the circle touching the base and the sides produced.

564. Describe a circle which shall touch a given straight line at a given point, and bisect the circumference of a given circle.

565. Describe a circle which shall pass through a given point and bisect the circumferences of two given circles.

566. Within a given circle inscribe three equal circles, touching one another and the given circle.

567. If the radius of a circle be cut as in II. 11, the greater segment will be the side of a regular decagon inscribed in the circle.

568. If the radius of a circle be cut as in II. 11, the square on its greater segment, together with the square on the radius, is equal to the square on the side of a regular pentagon inscribed in the circle.

569. From the vertex of a triangle draw a straight line to the base so that the square on the straight line may be equal to the rectangle contained by the segments of the base.

570. Four straight lines are drawn in a plane forming four triangles; shew that the circumscribing circles of these triangles all pass through a common point.

571. The perpendiculars from the angles A and B of a triangle on the opposite sides meet at D; the circles described round ADC and DBC cut AB or AB produced at the points E and F: shew that AE is equal to BF.

572. The four circles each of which passes through the centres of three of the four circles touching the sides of a triangle are equal to one another.

573. Four circles are described so that each may touch internally three of the sides of a quadrilateral: shew that a circle may be described so as to pass through the centres of the four circles.

574. A circle is described round the triangle ABC, and from any point P of its circumference perpendiculars are drawn to BC, CA, AB, which meet the circle again at D, E, F: shew that the triangles ABC and DEF are equal in all respects, and that the straight lines AD, BE, CF are parallel.

575. With any point in the circumference of a given circle as centre, describe another circle, cutting the former at A and B; from B draw in the described circle a chord BD equal to its radius, and join AD, cutting the given circle at Q: shew that QD is equal to the radius of the given circle.

576. A point is taken without a square, such that straight lines being drawn to the angular points of the square, the angle contained by the two extreme straight lines is divided into three equal parts by the other two straight lines: shew that the locus of the point is the circumference of the circle circumscribing the square.

577. Circles are inscribed in the two triangles formed by drawing a perpendicular from an angle of a triangle on the opposite side; and analogous circles are described in relation to the two other like perpendiculars: shew that the sum of the diameters of the six circles together with the sum of the sides of the original triangle is equal to twice the sum of these perpendiculars.

578. Three concentric circles are drawn in the same plane: draw a straight line, such that one of its segments between the inner and outer circumference may be bisected at one of the points at which the straight line meets the middle circumference.

VI. 1 to D.

579. AB is a diameter, and P any point in the circumference of a circle; AP and BP are joined and produced if necessary; from any point C in AB a straight line is drawn at right angles to AB meeting AP at D and BP at E, and the circumference of the circle at F: shew that CD is a third proportional to CE and CF.

580. A, B, C are three points in a straight line, and D a point at which AB and BC subtend equal angles: shew that the locus of D is the circumference of a circle.

581. If a straight line be drawn from one corner of a square cutting off one-fourth from the diagonal it will cut off one-third from a side. Also if straight lines be drawn similarly from the other corners so as to form a square, this square will be two-fifths of the original square.

582. The sides AB, AC of a given triangle ABC are produced to any points D, E, so that DE is parallel to BC. The straight line DE is divided at F so that DF is to FE as BD is to CE: shew that the locus of F is a straight line.

583. A, B, C are three points in order in a straight line: find a point P in the straight line so that PB may be a mean proportional between PA and PC.

584. A, B are two fixed points on the circumference of a given circle, and P is a moveable point on the circumference; on PB is taken a point D such that PD is to PA in a constant ratio, and on PA is taken a point E such that PE is to PB in the same ratio: shew that DE always touches a fixed circle.

585. ABC is an isosceles triangle, the angle at A being four times either of the others: shew that if BC be bisected at D and E, the triangle ADE is equilateral.

586. Perpendiculars are let fall from two opposite angles of a rectangle on a diagonal: shew that they will divide the diagonal into equal parts, if the square on one side of the rectangle be double that on the other.

587. A straight line AB is divided into any two parts at C, and on the whole straight line and on the two parts of it equilateral triangles ADB, ACE, BCF are described, the two latter being on the same side of the straight

line, and the former on the opposite side; G, H, K are the centres of the circles inscribed in these triangles: shew that the angles AGH, BGK are respectively equal to the angles ADC, BDC, and that GH is equal to GK.

588. On the two sides of a right-angled triangle squares are described: shew that the straight lines joining the acute angles of the triangle and the opposite angles of the squares cut off equal segments from the sides, and that each of these equal segments is a mean proportional between the remaining segments.

589. Two straight lines and a point between them are given in position: draw two straight lines from the given point to terminate in the given straight lines, so that they shall contain a given angle and have a given ratio.

590. With a point A in the circumference of a circle ABC as centre, a circle PBC is described cutting the former circle at the points B and C; any chord AD of the former meets the common chord BC at E, and the circumference of the other circle at O: shew that the angles EPO and DPO are equal for all positions of P.

591. ABC, ABF are triangles on the same base in the ratio of two to one; AF and BF produced meet the sides at D and E; in FB a part FG is cut off equal to FE, and BG is bisected at O: shew that BO is to BE as DF is to DA.

592. A is the centre of a circle, and another circle passes through A and cuts the former at B and C; AD is a chord of the latter circle meeting BC at E, and from D are drawn DF and DG tangents to the former circle: shew that G, E, F lie on one straight line.

593. In AB, AC, two sides of a triangle, are taken points D, E; AB, AC are produced to F, G such that BF is equal to AD, and CG equal to AE; BG, CF are joined meeting at H: shew that the triangle FHG is equal to the triangles BHC, ADE together.

594. In any triangle ABC if BD be taken equal to one-fourth of BC, and CE one-fourth of AC, the straight line drawn from C through the intersection of BE and AD will divide AB into two parts, which are in the ratio of nine to one.

595. Any rectilineal figure is inscribed in a circle: shew that by bisecting the arcs and drawing tangents to the points of bisection parallel to the sides of the recti-

lineal figure, we can form a similar rectilineal figure circumscribing the circle.

596. Find a mean proportional between two similar right-angled triangles which have one of the sides containing the right angle common.

597. In the sides AC, BC of a triangle ABC points D and E are taken, such that CD and CE are respectively the third parts of AC and BC; BD and AE are drawn intersecting at O: shew that EO and DO are respectively the fourth parts of AE and BD.

598. CA, CB are diameters of two circles which touch each other externally at C; a chord AD of the former circle, when produced, touches the latter at E, while a chord BF of the latter, when produced, touches the former at G: shew that the rectangle contained by AD and BF is four times that contained by DE and FG.

599. Two circles intersect at A, and BAC is drawn meeting them at B and C; with B, C as centres are described two circles each of which intersects one of the former at right angles: shew that these circles and the circle whose diameter is BC meet at a point.

600. $ABCDEF$ is a regular hexagon: shew that BF divides AD in the ratio of one to three.

601. ABC, DEF are triangles, having the angle A equal to the angle D; and AB is equal to DF: shew that the areas of the triangles are as AC to DE.

602. If M, N be the points at which the inscribed and an escribed circle touch the side AC of a triangle ABC; shew that if BM be produced to cut the escribed circle again at P, then NP is a diameter.

603. The angle A of a triangle ABC is a right angle, and D is the foot of the perpendicular from A on BC; DM, DN are perpendiculars on AB, AC: shew that the angles BMC, BNC are equal.

604. If from the point of bisection of any given arc of a circle two straight lines be drawn, cutting the chord of the arc and the circumference, the four points of intersection shall also lie in the circumference of a circle.

605. The side AB of a triangle ABC is touched by the inscribed circle at D, and by the escribed circle at E: shew that the rectangle contained by the radii is equal to the rectangle AD, DB and to the rectangle AE, EB.

606. Shew that the locus of the middle points of straight lines parallel to the base of a triangle and terminated by its sides is a straight line.

607. A parallelogram is inscribed in a triangle, having one side on the base of the triangle, and the adjacent sides parallel to a fixed direction: shew that the locus of the intersection of the diagonals of the parallelogram is a straight line bisecting the base of the triangle.

608. On a given straight line AB as hypotenuse a right-angled triangle is described; and from A and B straight lines are drawn to bisect the opposite sides: shew that the locus of their intersection is a circle.

609. From a given point outside two given circles which do not meet, draw a straight line such that the portions of it intercepted by each circle shall be respectively proportional to their radii.

610. In a given triangle inscribe a rhombus which shall have one of its angular points coincident with a point in the base, and a side on that base.

611. ABC is a triangle having a right angle at C; $ABDE$ is the square described on the hypotenuse; F, G, H are the points of intersection of the diagonals of the squares on the hypotenuse and sides: shew that the angles DCE, GFH are together equal to a right angle.

MISCELLANEOUS.

612. O is a fixed point from which any straight line is drawn meeting a fixed straight line at P; in OP a point Q is taken such that the rectangle OP, OQ is constant: shew that the locus of Q is the circumference of a circle.

613. O is a fixed point on the circumference of a circle, from which any straight line is drawn meeting the circumference at P; in OP a point Q is taken such that the rectangle OP, OQ is constant: shew that the locus of Q is a straight line.

614. The opposite sides of a quadrilateral inscribed in a circle when produced meet at P and Q: shew that the square on PQ is equal to the sum of the squares on the tangents from P and Q to the circle.

615. $ABCD$ is a quadrilateral inscribed in a circle; the opposite sides AB and DC are produced to meet at F; and the opposite sides BC and AD at E: shew that the circle described on EF as diameter cuts the circle $ABCD$ at right angles.

616. From the vertex of a right-angled triangle a perpendicular is drawn on the hypotenuse, and from the foot of this perpendicular another is drawn on each side of the triangle: shew that the area of the triangle of which these two latter perpendiculars are two of the sides cannot be greater than one-fourth of the area of the original triangle.

617. If the extremities of two intersecting straight lines be joined so as to form two vertically opposite triangles, the figure made by connecting the points of bisection of the given straight lines, will be a parallelogram equal in area to half the difference of the triangles.

618. AB, AC are two tangents to a circle, touching it at B and C; R is any point in the straight line which joins the middle points of AB and AC; shew that AR is equal to the tangent drawn from R to the circle.

619. AB, AC are two tangents to a circle; PQ is a chord of the circle which, produced if necessary, meets the straight line joining the middle points of AB, AC at R; shew that the angles RAP, AQR are equal to one another.

620. Shew that the four circles each of which passes through the middle points of the sides of one of the four triangles formed by two adjacent sides and a diagonal of any quadrilateral all intersect at a point.

621. Perpendiculars are drawn from any point on the three straight lines which bisect the angles of an equilateral triangle: shew that one of them is equal to the sum of the other two.

622. Two circles intersect at A and B, and CBD is drawn through B perpendicular to AB to meet the circles; through A a straight line is drawn bisecting either the interior or exterior angle between AC and AD, and meeting the circumferences at E and F: shew that the tangents to the circumferences at E and F will intersect in AB produced.

623. Divide a triangle by two straight lines into three

parts, which, when properly arranged, shall form a parallelogram whose angles are of given magnitude.

624. $ABCD$ is a parallelogram, and P is any point: shew that the triangle PAC is equal to the difference of the triangles PAB and PAD, if P is within the angle BAD or that which is vertically opposite to it; and that the triangle PAC is equal to the sum of the triangles PAB and PAD, if P has any other position.

625. Two circles cut each other, and a straight line $ABCDE$ is drawn, which meets one circle at A and D, the other at B and E, and their common chord at C: shew that the square on BD is to the square on AE as the rectangle BC, CD is to the rectangle AC, CE.

THE END.

CAMBRIDGE: PRINTED AT THE UNIVERSITY PRESS.

Made in the USA